Spatiotemporal Data Analytics and Modeling of Land Systems: Shaping Sustainable Landscape

Spatiotemporal Data Analytics and Modeling of Land Systems: Shaping Sustainable Landscape

Guest Editors

Wenwu Tang
Jianxin Yang
Minrui Zheng
Jingye Li

MDPI

Basel • Beijing • Wuhan • Barcelona • Belgrade • Novi Sad • Cluj • Manchester

Guest Editors
Wenwu Tang
Department of Earth, Environmental, and Geographical Sciences
University of North Carolina at Charlotte
Charlotte
USA

Jianxin Yang
Department of Land Resource Management
China University of Geosciences
Wuhan
China

Minrui Zheng
School of Public Administration and Policy
Renmin University of China
Beijing
China

Jingye Li
School of Public Administration
Hohai University
Nanjing
China

Editorial Office
MDPI AG
Grosspeteranlage 5
4052 Basel, Switzerland

This is a reprint of the Special Issue, published open access by the journal *Land* (ISSN 2073-445X), freely accessible at: https://www.mdpi.com/journal/land/special_issues/57P2R7ZM7N.

For citation purposes, cite each article independently as indicated on the article page online and as indicated below:

Lastname, A.A.; Lastname, B.B. Article Title. *Journal Name* **Year**, *Volume Number*, Page Range.

ISBN 978-3-7258-4829-4 (Hbk)
ISBN 978-3-7258-4830-0 (PDF)
https://doi.org/10.3390/books978-3-7258-4830-0

Cover image courtesy of Jianxin Yang

Contents

About the Editors

Wenwu Tang

Wenwu Tang is a Professor at the Department of Earth, Environmental and Geographical Sciences and the Executive Director of the Center for Applied Geographic Information Science (gis.charlotte.edu) at the University of North Carolina at Charlotte. Tang holds a Ph.D. degree from the University of Iowa and was a Post-Doc and Research Scientist at the U.S. National Center for Supercomputing Applications and the University of Illinois at Urbana-Champaign. His research interests include cyberinfrastructure and high-performance geocomputation, agent-based modeling, land change modeling, web GIS, artificial intelligence, and their geospatial applications. Tang has over 100 peer-reviewed publications and three edited books. He has taught various relevant courses, including CyberGIS and Big Data, Spatial Statistics, and Web GIS.

Jianxin Yang

Jianxin Yang, Ph.D., is an Associate Professor at the Department of Land Resources Management at the School of Public Administration, the China University of Geosciences (Wuhan). He also received his Ph.D. there as part of his doctoral training as a visiting scholar at the University of North Carolina at Charlotte. His research pertains to land change modeling, spatiotemporal analytics, national spatial planning, and the ecological impacts of land systems. He specializes in advanced methodologies including spatiotemporal statistics, agent-based modeling, and complex network analysis to address pressing issues in sustainable land management. Dr. Yang has authored numerous high-impact papers in journals such as *Landscape and Urban Planning*, *Land*, and *Computers, Environment and Urban Systems*. He is an active reviewer for over 50 academic journals and serves on the editorial boards of *China Land Science*, *Geographical Science*, *Land*, and *Social Sciences & Humanities Open*. His research has been supported by multiple grants, including projects funded by the National Natural Science Foundation of China and the Ministry of Natural Resources.

Minrui Zheng

Dr. Minrui Zheng is an Associate Professor at the Department of Land Resources Management at the School of Public Administration and Policy, the Renmin University of China. She serves as a Young Editorial Board Member of China Land Science, a Member of the Professional Committee of Natural Resources Information System under the Chinese Society of Natural Resources, and a Member of the Resource Big Data Branch of the Chinese Society of Natural Resources. Her research focuses on the spatial governance of big data, spatiotemporal analysis and modeling, territorial spatial planning and governance, and the application of emerging technologies. She has published over 40 papers in SSCI, EI, and CSSCI journals, such as *Landscape and Urban Planning*, *International Journal of Geographic Information Science*, *Acta Geographica Sinica*, *Acta Ecologica Sinica*, and *China Land Science*. Multiple grants have supported her research, including funding from the National Natural Science Foundation of China, the Beijing Social Science Foundation, and the Fundamental Research Funds for the Central Universities.

Jingye Li

Dr. Jingye Li is a Lecturer at the Department of Land Resource Management at the School of Public Administration, Hohai University (HHU). He holds a Postdoctoral Fellowship in Geography from the Nanjing Institute of Geography and Limnology, the Chinese Academy of Sciences (CAS), and earned his Ph.D. from the China University of Geosciences (Wuhan) as part of his doctoral training as a visiting scholar at the Department of Urban and Regional Planning at The Ohio State University. Dr. Li serves as a Member of the Professional Committee of Resource Rights Research within the Chinese Society of Natural Resources. His research explores urban expansion and spatial optimization, the spatiotemporal simulation and prediction of land-use change, and land-use planning. He has authored over 20 peer-reviewed publications in journals including *Journal of Environmental Management, Habitat International, Remote Sensing,* and *Land Science of China.* Dr. Li has led multiple research projects, notably a Youth Project funded by the Ministry of Education's Humanities and Social Sciences Foundation, an Open Fund project from the Hubei Provincial Key Laboratory of Regional Ecological Processes and Environmental Evolution, and a Fundamental Research Funds for the Central Universities project.

Preface

The evolution of land systems under the influence of human activities and natural processes has emerged as a critical area of research in recent decades. With growing pressures from urbanization, climate change, and socio-economic transformation, the need for innovative tools and methodologies to analyze, model, and manage these systems has never been more urgent. This reprint, "Spatiotemporal Data Analytics and Modeling of Land Systems: Shaping Sustainable Landscapes", seeks to address this gap by presenting a curated collection of cutting-edge studies that advance our understanding of spatiotemporal dynamics in land-use systems. This reprint spans a broad spectrum of topics, from the delineation of urban growth boundaries and forecasting transportation land demand to assessing ecological environmental quality and applying machine learning in land-use suitability analysis. It incorporates diverse methodologies, including GIS-based multi-criteria evaluation, remote sensing ecological indices, and spatiotemporal modeling. The case studies range from the Qinghai–Tibet Plateau and the Yangtze River Economic Belt in China to urban contexts like Wuhan and Kelowna, Canada, offering insights into both global and localized land-use challenges.

The primary aim of this reprint is to provide insights into spatiotemporal analysis and modeling in the face of the number of challenges that emerge when applying this method to the study of land systems. It intends to provide researchers, practitioners, and policymakers with a comprehensive understanding of the tools and frameworks available to address contemporary land system challenges. By integrating theoretical advancements with practical applications, the works compiled here emphasize the value of spatiotemporal data analytics in shaping sustainable landscapes. Readers will find detailed explorations of the drivers of land-use changes, innovative modeling approaches for urban development, and methodologies for ecological resilience and sustainability planning. The motivation for this compilation arises from the pressing need to balance human development with nature conservation. As urbanization continues to transform natural landscapes, the ability to predict, evaluate, and manage these changes becomes paramount. This reprint aims to equip its audience with the knowledge and methodologies necessary to navigate these complexities, promoting informed decision-making and sustainable land management practices. This reprint is intended for an audience that includes academic researchers, graduate students, urban planners, environmental managers, and policymakers. The diversity of topics and methods ensures relevance for those working across disciplines such as geography, urban studies, environmental science, and data science.

We are grateful to the authors whose contributions form the backbone of this reprint. Their innovative approaches and rigorous analyses provide the foundation for this collection. The studies featured here represent the forefront of research in land system modeling and spatiotemporal data analytics, reflecting the collaborative efforts of experts from diverse academic and professional backgrounds. We hope this reprint will serve as a valuable resource for those engaged in understanding and managing the intricate dynamics of land systems, and that it may inspire further research, innovation, and action towards achieving sustainable landscapes and resilient communities.

Wenwu Tang, Jianxin Yang, Minrui Zheng, and Jingye Li
Guest Editors

land

MDPI

Editorial

Spatiotemporal Data Analytics and the Modeling of Land Systems: Shaping Sustainable Landscape

Wenwu Tang [1], Jianxin Yang [2,*], Minrui Zheng [3] and Jingye Li [4]

1 Center for Applied GIScience, Department of Earth, Environmental and Geographical Sciences, School of Data Science, University of North Carolina at Charlotte, Charlotte, NC 28223, USA; wenwutang@charlotte.edu
2 School of Public Administration, China University of Geosciences, Wuhan 430074, China
3 School of Public Administration and Policy, Renmin University of China, Beijing 100872, China; minruizheng@ruc.edu.cn
4 School of Public Administration, Hohai University, Nanjing 211100, China; jingye.li@hhu.edu.cn
* Correspondence: yangjianxin@cug.edu.cn

1. Introduction

Dynamics in land systems are pivotal in driving socioeconomic development, biodiversity protection, and the provision of ecosystem services. However, land use activities such as urban sprawl, deforestation, and agricultural practices may lead to a series of challenges across ecological, social, or economic dimensions [1–4]. The dynamics of land systems are often influenced by an interplay of biophysical and socioeconomic factors [5,6]. Biophysical factors relating to the climate, topography, or even soil provide foundational conditions for land use activities, while anthropogenic factors—including, but not limited to, population variability, economic development, and policy—play a driving role in the land uses and land cover changes that shape our landscape across varying spatiotemporal scales. Exploring the interactions among these factors and thereby gaining deeper insight into the complexity of land systems often requires the support of spatiotemporal data analysis and modeling capabilities. These capabilities are typically based on the integration of Geographic Information Systems (GIS), remote sensing, and computational models [7,8]. Computational models include generic approaches (e.g., statistics, optimization, simulation, or Artificial Intelligence (AI)) and domain-specific models. With this integration, we can obtain the high-resolution data that are becoming increasingly available to investigate spatiotemporal patterns and mechanisms related to land systems at various scales [9,10]. The use of spatiotemporal data analysis and modeling capabilities can enhance our understanding of how land systems respond to their internal drivers or external events (e.g., disasters), which is key to providing informed decision-making support for stakeholders such as policy makers [11,12].

The modeling of land systems allows for the projection of future land development dynamics in response to different scenarios, such as the impact of alternative policy interventions or external events. These land system models are often dynamic [13–15], empowering the study of the short- or long-term impact of land use activities driven by constraints from the population, economy, and environment. This aids in identifying potential challenges and opportunities associated with the development of land systems, which require sustainable land management and ecosystem resilience.

The aim of this Special Issue is to evaluate the role of spatiotemporal data analytics and modeling in the study of land systems, and thus to contribute to the resolution of current land development challenges. This Special Issue may offer insights into the development of sustainable landscapes in terms of scientific advances and practical implications. Our

https://doi.org/10.3390/land14071428

Special Issue includes 13 research papers, covering three thematic topics: (1) ecological and environmental functioning, (2) urban development, and (3) land change dynamics. Table 1 summarizes these papers in terms of their study region, model type, and factor type. A variety of models are noted, including statistical models, simulation models, and optimization models. The spatial statistics model is also highlighted [16], which is a special type of statistical model. Machine learning was also included in the table, as machine learning algorithms [17–19] can be used to support both statistical analysis (e.g., regression) and optimization (the search for optimal solutions). The factor type is determined based on a typology that includes biophysical factors, regulation and policy, infrastructure and accessibility, and socio-economic factors. Biophysical factors cover those drivers related to, for example, topographic, environmental, and ecological dimensions. Anthropogenic factors are more complicated, and are thus separated into regulation and policy, infrastructure and accessibility, and socio-economic categories.

Table 1. Summary of the types of models and factors used in the articles in this Special Issue.

Author (Year)	Study Region	Type of Models	Type of Factors
Shen and Gong (2024)	Qinghai–Tibet Plateau, China	Statistical Model, Machine Learning	Biophysical Factors
Zou, Fan et al. (2024)	Wanning City, China	Statistical Model	Regulation and Policy, Infrastructure and Accessibility, Biophysical Factors
Li, Yang et al. (2024)	Guiyang, China	Statistical Model, Spatial Statistics	Biophysical, Socioeconomic Factors
Lin, Li et al. (2024)	Central Yunnan Urban Agglomeration, China	Statistical Model	Regulation and Policy, Socioeconomic Factors
Cai, Song et al. (2024)	Shanxi–Shaanxi–Inner Mongolia Energy Zone, China	Simulation, Spatial Statistics	Regulation and Policy, Biophysical, Infrastructure and Accessibility
Wang, Zeng et al. (2024)	Urban Agglomerations of the Yellow River Basin, China	Spatial Statistics	Socioeconomic, Biophysical Factors
Zhang, Lin et al. (2024)	Guangdong-Hong Kong-Macao Greater Bay Area, China	Statistical Model, Spatial Statistics	Biophysical Factors
Wang, Wang et al. (2024)	Yangtze River Economic Belt, China	Machine Learning, Statistical Model	Infrastructure and Accessibility, Socioeconomic Factors
Zhao, Ni et al. (2024)	Mianning County, Eastern Edge of the Qinghai–Tibet Plateau, China	Statistical Model	Infrastructure and Accessibility, Biophysical Factors
Cai, Li et al. (2024)	Hangzhou City, China	Statistical Model, Machine Learning	Biophysical Factors
Zhang, Xia et al. (2024)	Wuhan City, China	Statistical Model, Simulation	Biophysical Factors, Regulation and Policy, Socioeconomic, Infrastructure and Accessibility Factors
Bilintoh, Pontius et al. (2024)	The Plum Island Ecosystems of northeastern Massachusetts, USA	Statistical Model	Biophysical Factors
Zhao, van Duynhoven et al. (2024)	Kelowna, BC, Canada	Statistical Model, Machine Learning	Biophysical, Infrastructure and Accessibility Factors

2. Topics

2.1. Ecological and Environmental Functioning

This theme encompasses studies of ecological or environmental functioning that investigate the interactions between biophysical mechanisms and anthropogenic activities at a landscape scale. Spatiotemporal analysis and modeling allow us to explore ecosystem dynamics and how these ecosystems adapt to ecological or environmental stressors [6,20,21]. For example, Shen and Gong (List of Contributions, 1) presented a space–time analytics framework to study how ecological quality on the Qinghai–Tibet Plateau varies over time due to the impact of climate change and anthropogenic activities, including policies. This space–time analytics framework drives the study of the ways in which this ecologically vulnerable region are modified by various dynamic processes. In Cai, Song et al.'s (List of Contributions, 2) study of ecosystems' adaptive capacity, multi-scenario simulations were used to assess ecosystems' response to different disturbances in the Shanxi–Shaanxi Inner Mongolia Energy Zone. Cai, Song et al. stressed the important role of resilience in the mitigation of environmental degradation in their study of ecosystems' adaptive capacity. Li, Yang et al. (List of Contributions, 3) explored how ecosystem services are influenced by land change activities in Guiyang, China. With support from their spatiotemporal analysis, the impact of land cover change on the value of ecosystem services was quantified, further highlighting the importance of land management practices in ecosystem services' valuation. In their ecological connectivity study, Zou, Fan et al. (List of Contributions, 4) investigated the ecological networks in Wanning city, China. A framework that identifies ecological corridors was developed by Zou et al. to evaluate ecological resilience in their study region. Landscape metrics such as the largest patch index and degree of landscape division were used to quantify landscape patterns in their study region. Cai, Li et al. (List of Contributions, 5) conducted an assessment of urban ecological health in Hangzhou, China. A suite of environmental indicators was combined in their assessment to guide the development of landscape management practices. These studies demonstrate that spatiotemporal data analytics are methodologically essential to the integration of multi-data sources and modeling capabilities, which can increase our understanding of ecological quality and resilience [21,22].

2.2. Urban Development

Urban development is another theme that receives considerable benefits from spatiotemporal analytics and modeling within the context of land systems [23–25]. Zhang, Lin et al. (List of Contributions, 6) conducted a study of urban development in the Guangdong–Hong Kong–Macao Greater Bay Area, focusing on the impact of biophysical factors, including topography, climate, soil, water bodies, and fault. Multicriteria evaluation and a spatial statistical model were used to evaluate the land suitability and carrying-capacity potential in their study region. Zhang, Xia et al. (List of Contributions, 7) delineated urban growth boundaries over time under different development scenarios. A cellular automata-driven urban simulation model was used in Zhang et al.'s study, which takes into account factors from the biophysical and socioeconomic dimensions (e.g., population, GDP, topography, proximity to transportation infrastructure, and land cover). Wang, Zeng et al. (List of Contributions, 8) assessed the impact of urban expansion on carbon emissions in urban agglomerations of the Yellow River Basin, China. The spatiotemporal patterns of urban development and carbon emissions from 2000 to 2020 were evaluated using kernel density estimation, Gini coefficient, landscape metrics (e.g., aggregation index, patch density, and landscape shape index), and a geographically temporally weighted regression model. Lin, Li et al. (List of Contributions, 9) investigated the spatially interacting dynamics of Central Yunan Urban Agglomeration in China from 2000 to 2020 within the theoretical framework of urban symbiosis. The functioning and interactions of urban development at the county level were evaluated in terms of production, living, and ecological functions. Each of these functions was characterized by an indexing system of relevant factors (similar to multicriteria evaluation) [26].

2.3. Land Change Dynamics

While urban development is a form of land change with a focus on urban dimensions, land change is a broader and more inclusive theme. Zhao, Ni et al. (List of Contributions, 10) analyzed spatiotemporal changes in construction land in Mianning county, located on the eastern side of the Qinghai–Tibet Plateau, China. The landscape expansion index and geographically weighted regression were used to investigate the changes in construction land from 1990 to 2020 by considering influential factors from five dimensions: geomorphology, geology, climate, river and vegetation environment, and socioeconomy.

Bilintoh, Pontius et al. (List of Contributions, 11) applied a Total Operating Characteristics (TOC) [27] approach to quantify temporal changes in one land cover type (marsh) in an ecological research site in Massachusetts, USA. Gains and losses of marsh, with reference to the distance-to-marsh boundary or elevation, were evaluated over three time periods (1938, 1972, and 2013). Bilintoh et al. demonstrated the importance of applying TOC to assess the spatiotemporal characteristics of land gain and loss.

Zhao, van Duynhoven et al. (List of Contributions, 12) discussed the use of three machine learning approaches (random forest, extreme gradient boosting, and support vector machine) to generate land suitability maps for the city of Kelowna, British Columbia, Canada. Land cover data in 2015 and land change data from 2015 to 2020 were used to train these machine learning models to estimate the weights of alternative criteria. These machine learning-derived land suitability maps were compared against traditional approaches, relying on expert knowledge such as the Analytical Hierarchy Process [28].

The study of land change dynamics can be conducted at the regional level. Wang, Wang et al. (List of Contributions, 13) proposed a framework that integrates meta-analysis, statistical analysis, and neural network modeling to estimate land demand for the transportation needs of the Yangtze River Economic Belt, China. Transportation land demands for 127 cities in the study region were predicted based on the use of a suite of influential factors related to socio-economic development (e.g., GDP, population). This analysis framework facilitates the exploration of spatiotemporal patterns of land demand for transportation and their driving mechanisms at the regional level.

3. Summary and Perspectives

The studies collected in this Special Issue highlight the use of spatiotemporal data analysis and modeling in the investigation of the dynamics in complex land systems. These studies concentrate on three themes: ecological and environmental functioning, urban development, and land change dynamics. Geospatial data at various spatial resolutions (e.g., 30 m, 1000 m, and 3.5 km) were used to study dynamics in land systems at various scales (from local to regional). A suite of nonspatial and spatial metrics (including landscape metrics) were extracted from these geographic data and used in their corresponding applications. The spatiotemporal analytics and modeling capabilities used in these studies include spatial statistics (e.g., spatial autocorrelation analysis, geographically weighted regression [29]), multicriteria evaluation, machine learning (e.g., neural networks, random forests), and spatial simulation [30,31].

Spatiotemporal data analytics and modeling serve as a data-intensive scientific approach to the exploration of land system dynamics across various scales [32]. Spatiotemporal data analytics and modeling allow us to document and analyze what happened in the past and in various regions. Additionally, these capabilities (such as spatial simulation) provide spatially explicit modeling support for exploring what may happen in the future or in alternative scenarios [33–35]. Further, spatiotemporal data analytics and modeling capabilities hold great promise for representing and investigating the complexity of land systems, such as feedback loops, scale effect, emergence, and adaptation [36–39].

The use of AI [40,41] techniques such as neural networks or random forests is reported in this Special Issue. These AI techniques are applied in a traditional way and most AI techniques used in these studies remain conventional. As AI, exemplified by generative AI and agentic AI [42–44], continues to advance, modern AI techniques will highly likely catalyze a new wave of applications of spatiotemporal data analytics and modeling in the study of land systems in the near future. Modern AI techniques hold great potential in boosting the efficiency and effectiveness of spatiotemporal data analytics and modeling. Benefiting from the autonomy of emerging AI techniques, such as generative AI and agentic AI, the steps of spatiotemporal data analytics and modeling (e.g., preprocessing, model development and integration, post-processing, and evaluation) can be substantially automated. This will lead to a significant reduction in the time and cost required for modeling cycles cost, i.e., efficiency will increase. Furthermore, emerging AI techniques provide increasingly extensive support for new or novel modeling algorithms (e.g., foundation models; see [45]). These emerging AI algorithms may be of assistance when using spatiotemporal data analytics and modeling to obtain a better representation of the complex properties of land systems, such as nonlinearity, self-organization, scaling effects, and adaptation.

Conflicts of Interest: The authors declare no conflict of interest.

List of Contributions:

1. Shen, Z.; Gong, J. Spatial–Temporal Changes and Driving Mechanisms of Ecological Environmental Quality in the Qinghai–Tibet Plateau, China. *Land* **2024**, *13*, 2203. https://doi.org/10.3390/land13122203.
2. Cai, X.; Song, Y.; Xue, D.; Ma, B.; Liu, X.; Zhang, L. Spatial and Temporal Changes in Ecological Resilience in the Shanxi–Shaanxi–Inner Mongolia Energy Zone with Multi-Scenario Simulation. *Land* **2024**, *13*, 425. https://doi.org/10.3390/land13040425.
3. Li, Q.; Yang, L.; Jiao, H.; He, Q. Spatiotemporal Analysis of the Impacts of Land Use Change on Ecosystem Service Value: A Case from Guiyang, China. *Land* **2024**, *13*, 211. https://doi.org/10.3390/land13020211.
4. Zou, S.; Fan, R.; Gong, J. Spatial Optimization and Temporal Changes in the Ecological Network: A Case Study of Wanning City, China. *Land* **2024**, *13*, 122. https://doi.org/10.3390/land13010122.
5. Cai, C.; Li, J.; Wang, Z. Long-Term Ecological and Environmental Quality Assessment Using an Improved Remote-Sensing Ecological Index (IRSEI): A Case Study of Hangzhou City, China. *Land* **2024**, *13*, 1152. https://doi.org/10.3390/land13081152.
6. Zhang, Y.; Lin, T.; Zhang, J.; Lin, M.; Chen, Y.; Zheng, Y.; Wang, X.; Liu, Y.; Ye, H.; Zhang, G. Potential and Influencing Factors of Urban Spatial Development under Natural Constraints: A Case Study of the Guangdong-Hong Kong-Macao Greater Bay Area. *Land* **2024**, *13*, 783. https://doi.org/10.3390/land13060783.
7. Zhang, Y.; Xia, X.; Li, J.; Xing, L.; Yang, C.; Wang, H.; Dai, X.; Wang, J. Simulation of Urban Growth Boundary under the Guidance of Stock Development: A Case Study of Wuhan City. *Land* **2024**, *13*, 1174. https://doi.org/10.3390/land13081174.
8. Wang, Z.; Zeng, Y.; Wang, X.; Gu, T.; Chen, W. Impact of Urban Expansion on Carbon Emissions in the Urban Agglomerations of Yellow River Basin, China. *Land* **2024**, *13*, 651. https://doi.org/10.3390/land13050651.
9. Lin, S.; Li, C.; Li, Y.; Chen, L. Exploring Integrative Development of Urban Agglomeration from the Perspective of Urban Symbiosis and Production–Living–Ecological Function. *Land* **2024**, *13*, 258. https://doi.org/10.3390/land13020258.

10. Zhao, Y.; Ni, Z.; Zhang, Y.; Wan, P.; Geng, C.; Yu, W.; Li, Y.; Long, Z. Exploring the Spatiotemporal Evolution Patterns and Determinants of Construction Land in Mianning County on the Eastern Edge of the Qinghai–Tibet Plateau. *Land* **2024**, *13*, 993. https://doi.org/10.3390/land13070993.
11. Bilintoh, T.M.; Pontius, R.G., Jr.; Liu, Z. Analyzing the Losses and Gains of a Land Category: Insights from the Total Operating Characteristic. *Land* **2024**, *13*, 1177. https://doi.org/10.3390/land13081177.
12. Zhao, L.Q.; van Duynhoven, A.; Dragićević, S. Machine Learning for Criteria Weighting in GIS-Based Multi-Criteria Evaluation: A Case Study of Urban Suitability Analysis. *Land* **2024**, *13*, 1288. https://doi.org/10.3390/land13081288.
13. Wang, K.; Wang, L.; Zhang, J. Towards a Comprehensive Framework for Regional Transportation Land Demand Forecasting: Empirical Study from Yangtze River Economic Belt, China. *Land* **2024**, *13*, 847. https://doi.org/10.3390/land13060847.

References

1. Rindfuss, R.R.; Walsh, S.J.; Turner, B.L., 2nd; Fox, J.; Mishra, V. Developing a Science of Land Change: Challenges and Methodological Issues. *Proc. Natl. Acad. Sci. USA* **2004**, *101*, 13976–13981. [CrossRef] [PubMed]
2. Lambin, E.F.; Turner, B.L.; Geist, H.J.; Agbola, S.B.; Angelsen, A.; Bruce, J.W.; Coomes, O.T.; Dirzo, R.; Fischer, G.; Folke, C. The Causes of Land-Use and Land-Cover Change: Moving Beyond the Myths. *Glob. Environ. Change* **2001**, *11*, 261–269. [CrossRef]
3. Verburg, P.H.; Erb, K.-H.; Mertz, O.; Espindola, G. Land System Science: Between Global Challenges and Local Realities. *Curr. Opin. Environ. Sustain.* **2013**, *5*, 433–437. [CrossRef] [PubMed]
4. Council; National Research; Division of Behavioral; Social Sciences; Board on Environmental Change; Committee on the Human Dimensions of Global Change. *People and Pixels: Linking Remote Sensing and Social Science*; National Academies Press: Washington, DC, USA, 1998.
5. Turner, B.L., 2nd; Lambin, E.F.; Reenberg, A. The Emergence of Land Change Science for Global Environmental Change and Sustainability. *Proc. Natl. Acad. Sci. USA* **2007**, *104*, 20666–20671. [CrossRef]
6. Liu, J.; Dietz, T.; Carpenter, S.R.; Alberti, M.; Folke, C.; Moran, E.; Pell, A.N.; Deadman, P.; Kratz, T.; Lubchenco, J.; et al. Complexity of Coupled Human and Natural Systems. *Science* **2007**, *317*, 1513–1516. [CrossRef]
7. Brown, D.G.; Riolo, R.; Robinson, D.T.; North, M.; Rand, W. Spatial Process and Data Models: Toward Integration of Agent-Based Models and Gis. *J. Geogr. Syst.* **2005**, *7*, 25–47. [CrossRef]
8. Parker, D.C.; Manson, S.M.; Janssen, M.A.; Hoffmann, M.J.; Deadman, P. Multi-Agent Systems for the Simulation of Land-Use and Land-Cover Change: A Review. *Ann. Assoc. Am. Geogr.* **2003**, *93*, 314–337. [CrossRef]
9. Irwin, E.G.; Geoghegan, J. Theory, Data, Methods: Developing Spatially Explicit Economic Models of Land Use Change. *Agric. Ecosyst. Environ.* **2001**, *85*, 7–24. [CrossRef]
10. Verburg, P.H.; Neumann, K.; Nol, L. Challenges in Using Land Use and Land Cover Data for Global Change Studies. *Glob. Change Bio* **2011**, *17*, 974–989. [CrossRef]
11. An, L. Modeling Human Decisions in Coupled Human and Natural Systems: Review of Agent-Based Models. *Ecol. Model.* **2012**, *229*, 25–36. [CrossRef]
12. *Council;* National Research Advancing Land Change Modeling: Opportunities and Research Requirements; The National Academies Press: Washington, DC, USA, 2014.
13. Costanza, R.; Voinov, A. *Landscape Simulation Modeling: A Spatially Explicit, Dynamic Approach*; Springer: New York, NY, USA, 2004.
14. Pontius, R.G.; Boersma, W.; Castella, J.-C.; Clarke, K.; de Nijs, T.; Dietzel, C.; Duan, Z.; Fotsing, E.; Goldstein, N.; Kok, K.; et al. Comparing the Input, Output, and Validation Maps for Several Models of Land Change. *Ann. Reg. Sci.* **2008**, *42*, 11–37. [CrossRef]
15. Tang, W.; Yang, J. Agent-Based Land Change Modeling of a Large Watershed: Space-Time Locations of Critical Threshold. *J. Artif. Soc. Soc. Simul.* **2020**, *23*, 15. [CrossRef]
16. Ripley, B.D. *Spatial Statistics*; John Wiley & Sons: Hoboken, NJ, USA, 2005; Volume 272.
17. Mitchell, T.; Hill, M.G. *Machine Learning: A Guide to Current Research*; Springer Science & Business Media: Berlin/Heidelberg, Germany, 1997; Volume 12.
18. LeCun, Y.; Bengio, Y.; Hinton, G. Deep Learning. *Nature* **2015**, *521*, 436–444. [CrossRef]
19. Heaton, J. Ian Goodfellow, Yoshua Bengio, and Aaron Courville: Deep Learning. *Genet. Program. Evolvable Mach.* **2018**, *19*, 305–307. [CrossRef]
20. Turner, M.G.; Gardner, R.H. *Landscape Ecology in Theory and Practice*; Springer: New York, NY, USA, 2015.
21. Gunderson, L.H. Ecological Resilience—In Theory and Application. *Annu. Rev. Ecol. Evol. Syst.* **2000**, *31*, 425–439. [CrossRef]

22. Cumming, G.S. ; *Spatial Resilience in Social-Ecological Systems*; Springer Nature: Dordrecht, The Netherlands, 2011.
23. Benenson, I.; Torrens, P.M. Geosimulation: Object-Based Modeling of Urban Phenomena. *Comput. Environ. Urban Syst.* **2004**, *28*, 1–8. [CrossRef]
24. Batten, D.F. *Cities and Complexity: Understanding Cities with Cellular Automata, Agent-Based Models, and Fractals, by Michael Batty*; The MIT Press: Cambridge, MA, USA, 2007; Volume 47, pp. 624–627.
25. Batty, M. ; *The New Science of Cities*; MIT press: Cambridge, UK, 2013.
26. Eastman, J.R. Multi-Criteria Evaluation and Gis. *Geogr. Inf. Syst.* **1999**, *1*, 493–502.
27. Pontius, R.G., Jr.; Si, K. The Total Operating Characteristic to Measure Diagnostic Ability for Multiple Thresholds. *Int. J. Geogr. Inf. Sci.* **2014**, *28*, 570–583. [CrossRef]
28. Vaidya, O.S.; Kumar, S. Analytic Hierarchy Process: An Overview of Applications. *Eur. J. Oper. Res.* **2006**, *169*, 1–29. [CrossRef]
29. Coburn, T.C. Statistical Methods for Spatial Data Analysis. *Math. Geol.* **2006**, *38*, 511–513. [CrossRef]
30. White, R.; Engelen, G.; Uljee, I. The Use of Constrained Cellular Automata for High-Resolution Modelling of Urban Land-Use Dynamics. *Environ. Plan. B* **1997**, *24*, 323–343. [CrossRef]
31. Clarke KC, Gazulis N, Dietzel CK, Goldstein NC. A decade of SLEUTHing: Lessons learned from applications of a cellular automaton land use change model. In *Classics in IJGIS: Twenty Years of the International Journal of Geographical Information Science and Systems*; Taylor and Francis: Boca Raton, FL, USA, 2007; pp. 413–425.
32. Veldkamp, A.; Lambin, E.F. Predicting Land-Use Change. *Agric. Ecosyst. Environ.* **2001**, *85*, 1–6. [CrossRef]
33. Meentemeyer, R.K.; Tang, W.; Dorning, M.A.; Vogler, J.B.; Cunniffe, N.J.; Shoemaker, D.A. Futures: Multilevel Simulations of Emerging Urban–Rural Landscape Structure Using a Stochastic Patch-Growing Algorithm. *Ann. Assoc. Am. Geogr.* **2013**, *103*, 785–807. [CrossRef]
34. Batty, M.; Milton, R. A New Framework for Very Large-Scale Urban Modelling. *Urban Stud.* **2021**, *58*, 3071–3094. [CrossRef]
35. Manson, S.M. Simplifying Complexity: A Review of Complexity Theory. *Geoforum* **2001**, *32*, 405–414. [CrossRef]
36. Lansing, J.S. Complex Adaptive Systems. *Annu. Rev. Anthropol.* **2003**, *32*, 183–204. [CrossRef]
37. Verburg, P.H. Simulating Feedbacks in Land Use and Land Cover Change Models. *Landsc. Ecol.* **2006**, *21*, 1171–1183. [CrossRef]
38. Verburg, P.H.; Alexander, P.; Evans, T.; Magliocca, N.R.; Malek, Z.; Rounsevell, M.D.A.; van Vliet, J. Beyond Land Cover Change: Towards a New Generation of Land Use Models. *Curr. Opin. Environ. Sustain.* **2019**, *38*, 77–85. [CrossRef]
39. Russell, S.; Norvig, P. *Artificial Intelligence: A Modern Approach, Global Edition*; Pearson Education: London, UK, 2021.
40. Niu, J.; Tang, W.; Xu, F.; Zhou, X.; Song, Y. Global Research on Artificial Intelligence from 1990–2014: Spatially-Explicit Bibliometric Analysis. *ISPRS Int. J. Geo-Inf.* **2016**, *5*, 66. [CrossRef]
41. Bail, C.A. Can Generative Ai Improve Social Science? *Proc. Natl. Acad. Sci. USA* **2024**, *121*, e2314021121. [CrossRef]
42. Acharya, D.B.; Kuppan, K.; Divya, B. Agentic Ai: Autonomous Intelligence for Complex Goals—A Comprehensive Survey. *IEEE Access* **2025**, *13*, 18912–18936. [CrossRef]
43. Epstein, Z.; Hertzmann, A. Art and the science of generative AI. *Science* **2023**, *380*, 1110–1111. [CrossRef]
44. Bommasani, R.; Hudson, D.A.; Adeli, E.; Altman, R.; Arora, S.; von Arx, S.; Bernstein, M.S.; Bohg, J.; Bosselut, A.; Brunskill, E.; et al. On the Opportunities and Risks of Foundation Models. *arXiv* **2021**, arXiv:2108.07258.
45. Moor, M.; Banerjee, O.; Abad, Z.S.H.; Krumholz, H.M.; Leskovec, J.; Topol, E.J.; Rajpurkar, P. Foundation Models for Generalist Medical Artificial Intelligence. *Nature* **2023**, *616*, 259–265. [CrossRef]

land

MDPI

Article

Analyzing the Losses and Gains of a Land Category: Insights from the Total Operating Characteristic

Thomas Mumuni Bilintoh [1,*], Robert Gilmore Pontius, Jr. [1] and Zhen Liu [2]

[1] Graduate School of Geography, Clark University, 950 Main Street, Worcester, MA 01610, USA; rpontius@clarku.edu

[2] Department of Geographical Sciences, University of Maryland, College Park, MD 20742, USA; zliu1997@terpmail.umd.edu

* Correspondence: tbilintoh@clarku.edu; Tel.: +1-517-303-7904

Abstract: This manuscript provides guidance concerning how to use the Total Operating Characteristic (TOC) when (1) analyzing change through time, (2) ranking a categorical independent variable, and (3) constraining the extent for a gaining category. The illustrative variable is the marsh land-cover category in the Plum Island Ecosystems of northeastern Massachusetts, USA. The data are an elevation map and maps showing the land categories of water, marsh, and upland in 1938, 1971, and 2013. There were losses and gains near the edge of the marsh between 1938 and 1972 and between 1972 and 2013. The TOC curves show that marsh gained most intensively at intermediate elevations during the first time interval and then had a weaker association with elevation during the second time interval. Marsh gains more intensively from water than from upland during both time intervals. The TOC curves also demonstrate that the marsh gains occurred where marsh was previously lost, a phenomenon called Alternation. Furthermore, eliminating far distances and extreme elevations from the spatial extent decreased the area under the curve (AUC) for distance and increased the AUC for elevation. We invite scientists to use the TOC because the TOC is easier to interpret and shows more information than the Relative Operative Characteristic.

Keywords: alternation; land change; marsh; TOC; AUC

1. Introduction

The Total Operating Characteristic (TOC) can help researchers understand land change because the TOC analyzes the relationships between a ranked independent variable, such as distance, and a binary variable, such as the presence or absence of a land category. The TOC has gained increasing attention in the field of spatiotemporal analysis. This is because the TOC shows the total information in an error or change matrix, which other popular methods like the Receiver Operating Characteristic (ROC) fail to show. Pontius Jr and Si [1] first proposed the TOC as a modification of the ROC, which has been popular in diverse fields such as genetics [2,3], radiology [4], psychology [5,6], machine learning [7,8], and remote sensing [9,10].

Pontius and Si [1] described the use of the TOC to analyze the change in a land-cover category during a time interval. Subsequently, Bilintoh et al. [11] described how to use the TOC to analyze the losses and gains of land-cover categories during two intervals. Analyzing land change during more than one time interval provides an opportunity to compare the patterns of losses and gains during consecutive time intervals, which can reveal a change pattern called Alternation [12]. Alternation derives from pairing losses and gains through time at the same location.

The TOC compares a binary variable to a ranked index variable. A threshold of the ranked variable determines the diagnosis of an observation's presence or absence based on whether an observation's index value exceeds a threshold [1]. For our application, we diagnosed the presence or absence of gains and losses of marsh based on multiple indexes. The TOC can analyze ranked continuous variables such as elevation and distance. Furthermore, the TOC can analyze categorical variables [13]. For instance, scientists might want to analyze how a category such as urban gains from other categories such as forest or agriculture. Using the TOC to analyze the relationship between land-cover categories can reveal information about the categories that a gaining category targets or avoids.

Figure 1 shows the matrix for threshold t. The reference and the diagnosis can agree in two ways: Hits (H_t) and Correct Rejections (C_t) at threshold t. Similarly, the reference and the diagnosis can disagree in two ways: Misses (M_t) and False Alarms (F_t) at threshold t. Figure 1 shows the two types of agreements and disagreements and the total number of observations, which is $P + Q$ (Pontius Jr and Si 2014 [1]). Scientists require four bits of information to complete the matrix for threshold t. These pieces of information could be H_t; M_t; F_t; and C_t. Alternatively, the bits could be $P + Q$; P; $H_t + F_t$; and H_t. Other combinations of the four bits of information are possible. Table 1 defines the mathematical notation in Figure 1.

		Reference		Diagnosis Total
		Presence	Absence	
Diagnosis	Presence	H_t	F_t	$H_t + F_t$
	Absence	M_t	C_t	$M_t + C_t$
Reference Total		$H_t + M_t = P$	$F_t + C_t = Q$	$P + Q$

Figure 1. Contingency table showing the number of observations for a threshold (modified after [1]).

Table 1. Mathematical notation (modified after Pontius Jr and Si 2014 [1]).

Symbol	Meaning
T	Index for a threshold
H_t	Hits, which is the number of observations that are reference presence and diagnosed presence at threshold t
M_t	Misses, which is the number of observations that are reference presence and diagnosed absence at threshold t
F_t	False Alarms, which is the number of observations that are reference absence and diagnosed presence at threshold t
C_t	Correct Rejections, which is the number of observations that are reference absence and diagnosed absence at threshold t
P	Number of observations that are reference presence, also known as Abundance
Q	Number of observations that are reference absence

The area under the curve (AUC) is a popular metric among scientists using the ROC and the TOC. Some scientists consider particular AUC values to designate the results as good, which is problematic because an arbitrary spatial extent of absence influences the AUC. For example, Naghibi et al. [14] used the TOC to evaluate the accuracy of an urban gain model in a region of Iran. However, we suspect that they failed to mask urban areas at the initial time point, given the shape of the curve. Failing to mask pixels that are not candidates for change can result in inflated AUC values, which may lead to a flawed interpretation of the TOC curves. Similarly, Chakraborti et al. [15] used the TOC to analyze LULCC in the Siliguri region of India, but we suspect they also failed to mask urban areas at the first time point. Another situation exists where scientists include vast regions that have zero probability of change, which inflates the AUC. This causes some scientists to question the usefulness of the AUC, because the AUC increases when the scientists arbitrarily include places where change is not plausible. The arbitrariness of the spatial extent causes confusion when comparing the AUC values across case studies [16].

The TOC's ability to provide detailed information about the relationship between a ranked variable and a reference binary variable makes the TOC a valuable method for evaluating land changes and the accuracy of models [13,17]. However, authors must apply the TOC appropriately. This manuscript illustrates universal mathematical concepts by using the TOC to analyze how the losses and gains of land-cover categories during two time intervals relate to three types of variables: distance; elevation; and two land-cover categories. In addition, this manuscript addresses two crucial concepts in TOC analysis: (1) ranking a categorical independent variable and (2) constraining the extent of a gaining category.

2. Materials and Methods

2.1. Study Region

Figure 2 shows the Plum Island Ecosystem (PIE) Long-Term Ecological Research site in northeastern Massachusetts. The PIE's marshes provide several ecosystem services, including storm protection, biodiversity habitats, nutrient cycling, and carbon storage. Rising sea levels threaten these ecosystem services. For example, sea-level rise could cause the cordgrass *Spartina alterniflora* to become flooded, thus causing *Spartina alterniflora* to shift to higher elevations. Measuring and visualizing these changes is crucial for understanding the relationship between changes and the ecosystem function.

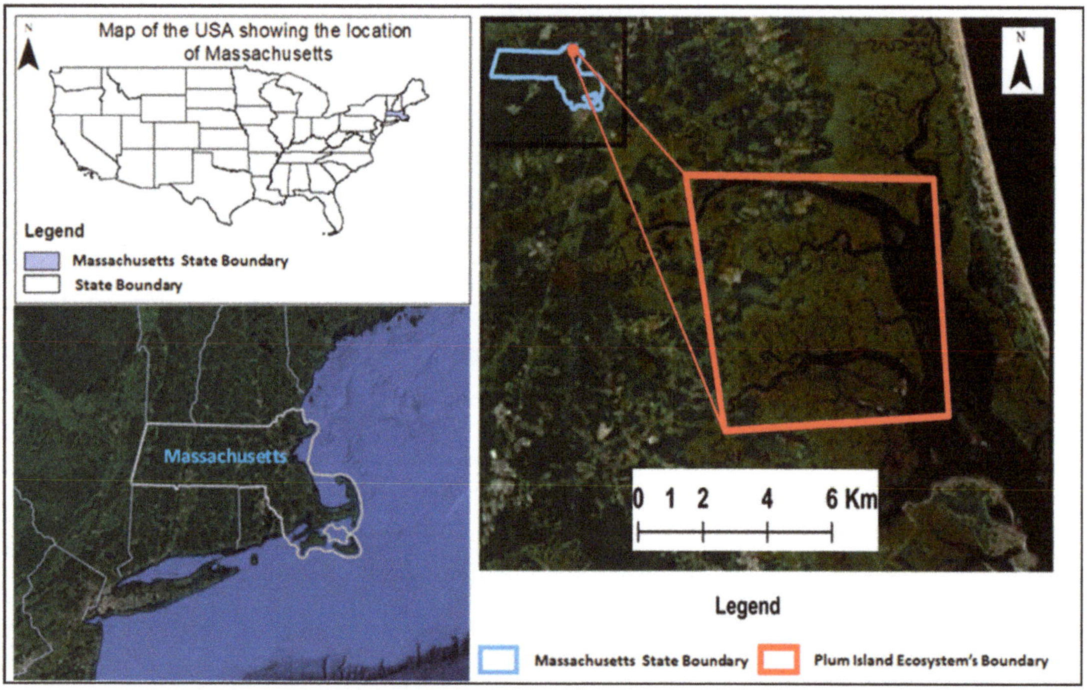

Figure 2. Maps showing the location of the Plum Island Ecosystem site.

2.2. Data

Figures 3 and 4 show the data, which are land-cover maps in 1938, 1972, and 2013 and an elevation map. Each land-cover map has a spatial resolution of 10 m by 10 m, where each pixel shows one of three land-cover categories: water; marsh; and upland. The land-cover maps were derived from the Georgia Coastal Ecosystems website [18]. The distance to the edge of the marsh in the 1938 map in Figure 3 ranges between 0 and 865 m. A distance of 0 m indicates marsh, while a distance of 865 m indicates the furthest distance away from the edge of the marsh.

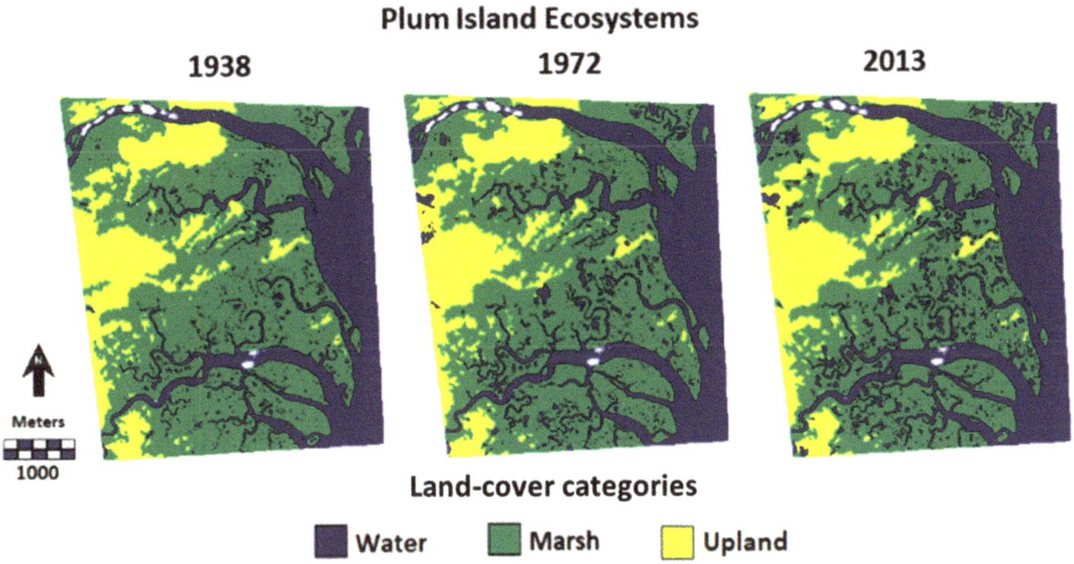

Figure 3. Plum Island Ecosystems' land-cover maps at three points. White indicates no data.

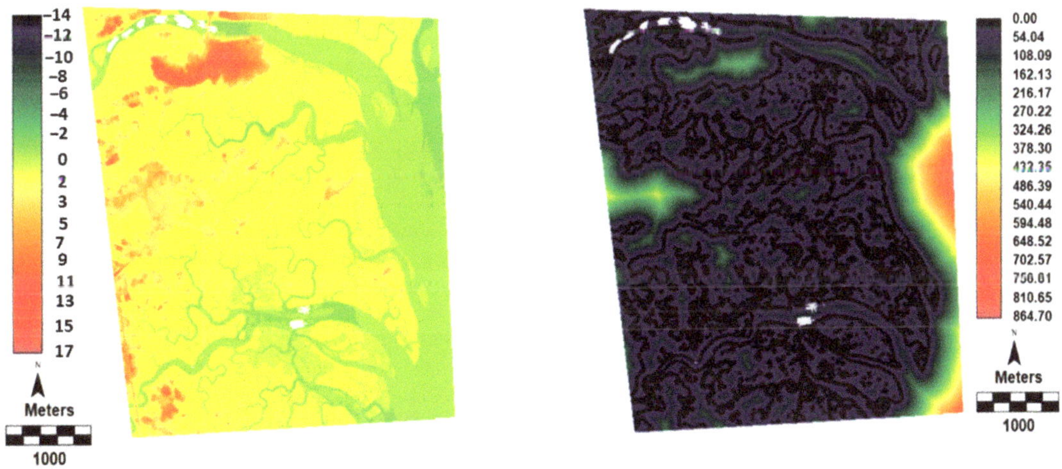

Figure 4. Maps of elevation and distance to the edge of the marsh in 1938. White indicates no data.

2.3. Methods

We used the TOC Curve Generator version 1.2.7 [19] to generate the TOC curves. The software requires the following inputs to generate a TOC curve: a binary variable; an index variable; and an optional mask. The binary variable contains information about a category's presence or absence, where 1 shows the presence and 0 shows the absence of the category. A mask restricts the spatial extent of the analysis. Figure 5 shows the binary variables in the form of maps of marsh change, including gains, losses, and persistence during each time interval. We then segmented each time interval's maps into four binary maps: gain and loss during the first time interval and gain and loss during the second time interval. Thus, each pixel uses 1 to denote gain or loss while using 0 to denote other. The second step is to create index variables. We used maps of the distance from the edges of the marsh at the start of each time interval, a map of elevation, and a map showing the two non-marsh land categories at the beginning of each time interval.

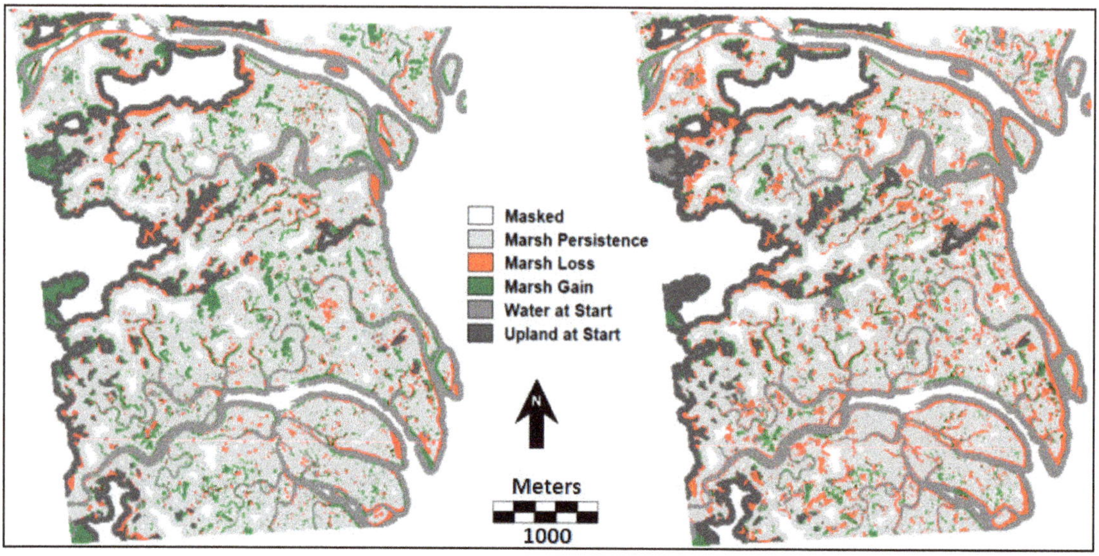

Figure 5. Maps of change between 1938 and 1972 on the left and between 1972 and 2013 on the right.

Pontius [13] described the procedure for a TOC analysis involving a categorical index variable. We adopted this procedure to create the TOC curves that show the relationship between marsh gain and the other two categories during each time interval. We excluded marsh from the categorical TOC curve because an observation must be non-marsh at the initial time point to experience marsh gain during the time interval. We had two land categories to rank after eliminating marsh. The land categories were water and upland. We computed each category's intensity of marsh gain to rank the water and upland. The greater intensities received earlier ranks. An intensity is a ratio where the numerator is the size of the marsh's gain from the losing category while the denominator is the size of the losing category [13].

The next step was to create masks to eliminate locations that were not plausible candidates for the gain of marsh. Our approach examines the thresholds for the index variable where the TOC curve touches the upper or lower bounds of the TOC parallelogram. Thresholds to the right of the point that touches the upper bound must be on the upper bound, so we considered those locations to be not plausible. Thresholds to the left of the point that leaves the lower bound must be on the lower bound, so we also considered those locations to be not plausible. This step is relevant for the TOC curves of a gaining land-cover class. We therefore conducted this step for gains of marsh during the first and second time intervals where the index variables were elevation and distance to the edge of the marsh.

Figure 6a shows the TOC curve for distance to the edge of the marsh and marsh gain during the second time interval, without applying a mask. We use the red rectangles to zoom in on segments of the distances that did not experience marsh change. The zoomed upper-right corner shows that the curve touches the upper bound at a distance of 310 m and continues to hug the upper bound until the farthest distance. Any distance beyond 310 m did not experience change, and so we excluded those large distances from the analysis. Figure 6c,d show the analysis for the masking of elevation. Figure 6c shows the entire TOC before the masking of elevation, while Figure 6d shows the zoomed lower-left and upper-right corners of Figure 6c.

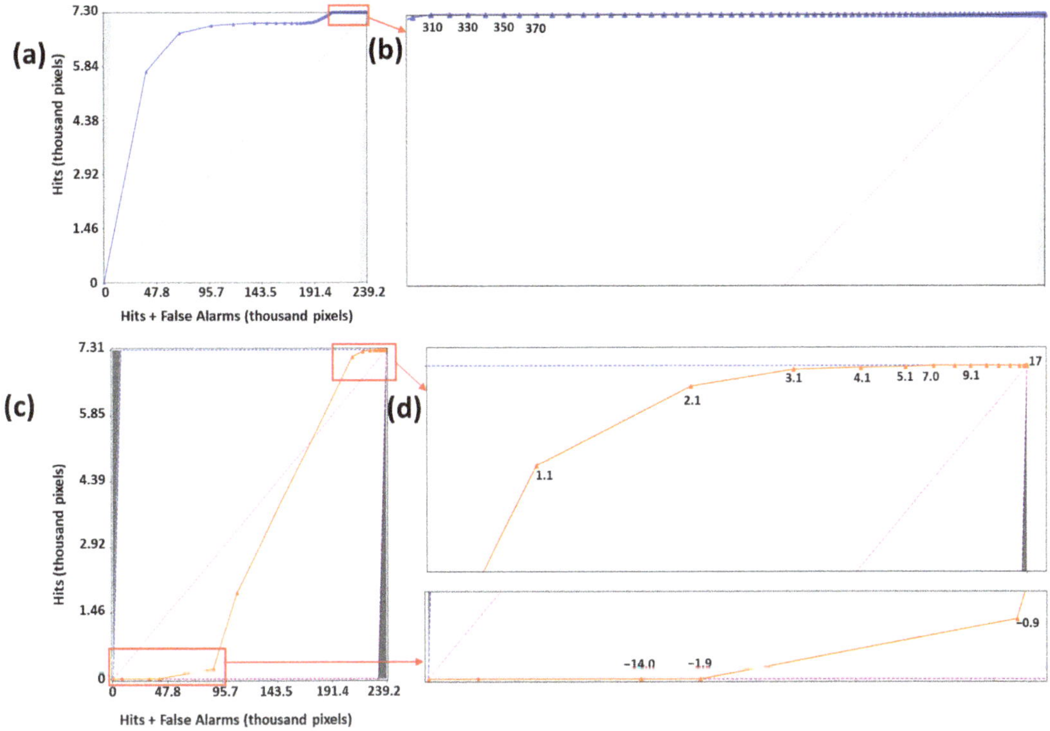

Figure 6. (**a**) The TOC curve for marsh gain between 1972 and 2013 and the distance to the marsh edge in 1972; (**b**) zoomed segments showing the threshold at which the TOC curve reaches the upper bound of the parallelogram; (**c**) the TOC curve for marsh gain between 1972 and 2013 and elevation; and (**d**) zoomed segments showing the threshold at which the TOC curve leaves the left bound and reaches the upper bounds of the parallelogram.

The zoomed version shows that the curve starts from the origin and hugs the horizontal axis until after an elevation of −1.9. Any elevation below −1.9 m did not experience change; thus, we excluded elevations below −1.9 m from the analysis. Similarly, the zoomed upper-right corner shows that the curve touches the upper bound at an elevation of 7 m and continues to hug the upper bound until the highest elevation. Thus, any elevation above 7 m did not experience change. We therefore excluded elevations above 7 m from the analysis.

3. Results

Figure 7 provides the results for the marsh's loss, while Figure 8 provides the TOC curves for the marsh's gain. The TOC curves start at the lower-left corner of the parallelogram with coordinates (0,0), where the number of Hits and False Alarms is zero. Each TOC curve consists of segments, where two thresholds bound each segment. Labels on the segments give numerical thresholds when the index variable is numerical. Labels are words when the index variable is categorical. The gray regions of Figures 7 and 8 show regions where it is impossible for a TOC curve to reside. The left gray triangle is an impossible region because Hits cannot be greater than Hits plus False Alarms. The right gray triangle is an impossible region because Hits cannot be less than Hits plus False Alarms. The slope of each TOC curve's segment is the intensity with which change occurs. Scientists must compare the steepness of a TOC curve's segment to the uniform line. If a curve's segment is steeper than the uniform line, then the change occurs more intensively than across the spatial extent in that segment. Conversely, if the segment is flatter than the uniform line, then the change occurs less intensively than across the spatial extent in that segment.

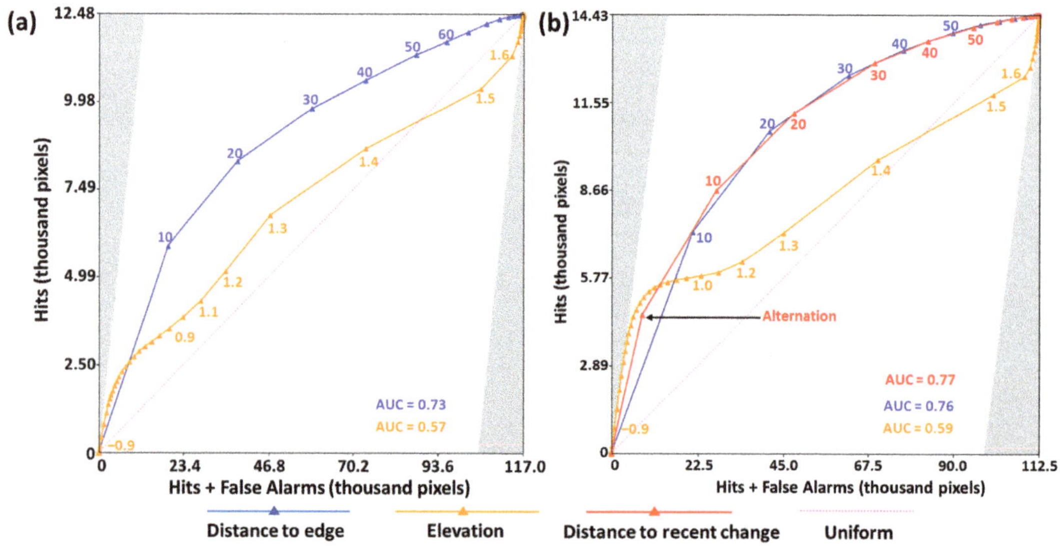

Figure 7. The TOC curves for marsh loss between (**a**) 1938 and 1972 and (**b**) 1972 and 2013.

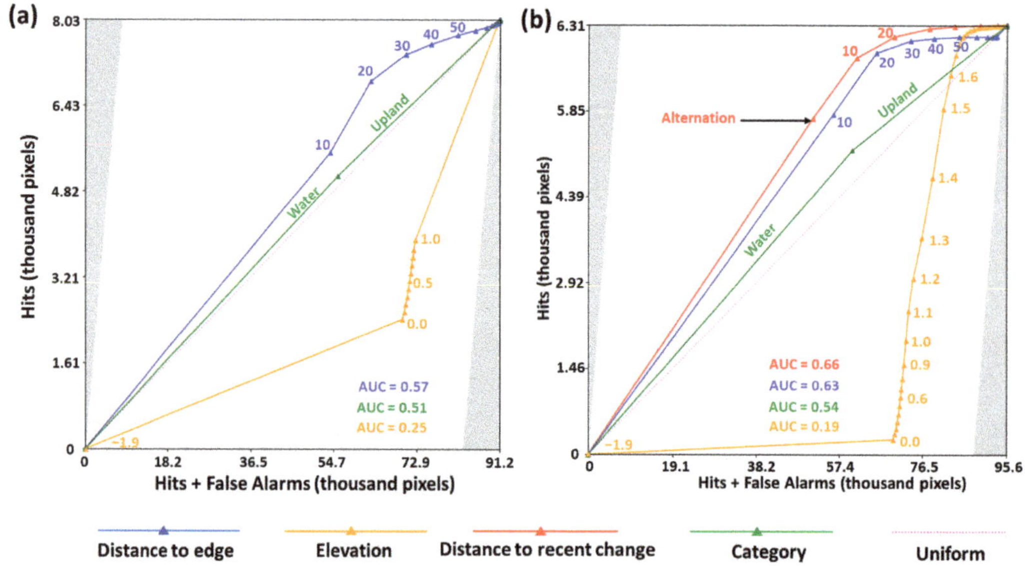

Figure 8. The TOC curves for marsh gain between (**a**) 1938 and 1972 and (**b**) 1972 and 2013.

Figure 7a shows the TOC curves for marsh loss between 1938 and 1972, while Figure 8b shows the TOC curves for marsh loss between 1972 and 2013. The steeper parts of the blue curves show that marsh losses were most intensive closer to the marsh edge during both time intervals. The steep parts of the orange curves show that marsh losses were most intensive at the lowest and highest elevations during both time intervals. The steepest part of the red curve shows that marsh losses were most intensive during the second time interval, where marsh gained during the first time interval, which is at the threshold labeled Alternation.

Figure 8a shows the TOC curves for marsh gain between 1938 and 1972, while Figure 8b shows the TOC curves for marsh gain between 1972 and 2013. The steepest parts of the blue curves show that marsh gains were most intensive between 10 and 20 m from the marsh's edge. The steep part of the orange curves shows intensive marsh gains at intermediate elevations. The green curves show that marsh gains slightly more intensively from water than from upland during both time intervals. Finally, the first segment from the origin for the red curve in Figure 8b shows that marsh gains were most intensive during the second time interval, where marsh was lost during the first time interval. The TOC curve for elevation shows the largest absolute deviation from 0.5.

4. Discussion

Examining changes during at least two time intervals presents an opportunity to observe patterns that would be impossible within a single time interval [20,21]. Figures 7b and 8b display the relationship between the distance to change during the first time interval and marsh change during the second time interval. Figure 7b demonstrates that marsh losses were most intensive where marsh was previously gained, which demonstrates Alternation. Alternation is a pattern where a category alternates between presence and absence through time. Figure 8b shows that marsh gains were most intensive where marsh was previously lost, which is also Alternation. Figures 7b and 8b show the thresholds at which Alternation occurs. Alternation has several implications for interpreting change patterns. For instance, Alternation may indicate a map error. For example, if marsh persists on the ground during both time intervals, while the map erroneously shows an absence of marsh at the middle time point, then the time series of maps will erroneously show loss followed by gain. Alternatively, Alternation on the ground occurs for some categories. Erosion followed by accretion along the edges of the marsh will cause Alternation of the marsh. Cropland alternates when farming practices include sequential cultivation and fallow years [22,23]. Increasing the number of time points increases the possibility of observing Alternation, regardless of whether Alternation is a true change or a result of map errors.

The AUC is a metric that measures the strength of a monotonic relationship between the binary variable and the ranked variable [13]. The maximum AUC is 1, while the uniform line has an AUC of 0.5. A common practice in the profession is to compare 0.5 to the AUC of a ranked variable. However, the AUC does not provide detailed information about the shape of the TOC curve; thus, scientists have cautioned against using the AUC in assessing the performance of models [24,25]. Imagine a situation where a TOC curve crosses the uniform line. This scenario indicates a non-monotonic relationship between the dependent and ranked index variables, potentially resulting in an AUC of 0.5. This scenario highlights the risk of exclusively using the AUC to assess the overall relationship. Scientists may overlook a substantial non-monotonic relationship between the independent and ranked index variables, evident from the shape of the TOC curve but not necessarily conveyed by the AUC alone.

Some authors may be inclined to establish universal rules and anoint the AUC values as poor, acceptable, good, excellent, etc. However, such universal rules do not address particular research questions or applications precisely because they are universally applied. Instead, these universal rules tend to cater more to the psychological desires of scientists rather than serving scientific purposes [13]. Scientists should, therefore, focus on aligning the interpretation of the AUC to the intended purpose of their research. Researchers must decide whether the AUC is a relevant metric. For our example, the AUC is not appropriate to compare the strength of the relationship with distance to the strength of the relationship with elevation because elevation has a non-monotonic relationship with the change intensity of the marsh. Therefore, we take a different approach to interpreting the relationship between the uniform line and the other TOC curves. Specifically, we compare the slope of each curve's segments to the slope of the uniform line to determine the change intensities along the TOC curve.

Figure 7a,b show a monotonic decreasing trend in the relationship between marsh loss intensity and distance to the edge during both time intervals. Figure 8b shows a non-monotonic decreasing trend for the relationship between marsh gain intensity and distance to the edge. Several factors could account for the observed patterns, ranging from geomorphic processes to the misregistration of images at different time points. For instance, this relationship could be because marsh changes along its edges, which a process of erosion and accretion would cause. Elevation plays a critical role in a plethora of ecological studies [26,27]. Figure 7a,b show that marsh loss occurs most intensively at lower and higher elevations, avoiding intermediate elevations of marsh. The study region has three distinct elevation patterns: (1) lower elevation dominated by water; (2) intermediate elevation occupied by marsh; and (3) high elevation occupied by upland. Figure 8a,b show that marsh gains occur most intensively at intermediate elevations during the first and second time intervals. The green lines in Figure 8a,b show that marsh gains slightly more intensively from water than from upland. The horizontal distance of a segment for a categorical variable shows the size of the category. The TOC curves show that the size of water is greater than that of upland during both time intervals.

Masking the spatial extent to show data only for locations where the phenomenon is plausible helps interpret the AUC. The masked parts of the maps in Figure 5 show where the landscape has not experienced marsh change during the two time intervals. These locations are primarily at the lowest and the highest elevations. Figure 5 shows that marsh does not gain at the lowest elevations or the highest elevations. Failure to mask these areas from the TOC analysis may result in flat segments in the TOC curve near the origin or the upper-right corner of the parallelogram, thus impacting the AUC values. Several studies exhibit this pattern [14,28]. Interpreting the AUC values resulting from the TOC analysis that failed to mask the implausible region of interest leads to misleading conclusions. The next logical question is how to define implausible regions for masking. A straightforward approach is to examine the thresholds for the index variable at locations where the curve leaves the lower bound and touches the upper bound of the parallelogram. Hits equal zero at all thresholds to the left point where the TOC curve leaves the lower bound of the parallelogram. Hits equal Prevalence at all thresholds to the right of where the curve first arrives at the upper bound of the parallelogram. We gain no insight by the inclusion of thresholds to the left of the threshold where the TOC curve leaves the lower bound or to the right of where the curve arrives at the upper bound.

Section 2.3 described the procedure to eliminate the distances to the edge and elevations that do not experience the gain of marsh. Comparatively, Figure 6a has an AUC of 0.9, while the TOC curve for distance to the edge of the marsh in Figure 8b has an AUC value of 0.6.

The green curves in Figure 8 analyze the relationship between marsh gains and land cover in 1938 and 1972. The land-cover maps in 1938 and 1972 have the same categories: water; marsh; and upland. The TOC requires ranking to establish a hierarchy among the categories that lose. The hierarchy derives from the intensities with which the category experiences gain of marsh; thus, the category with the greatest intensity is ranked first, while the category with the least intensity is ranked last. The intensity of each category is calculated as a ratio, with the numerator representing the size of the marsh's gain from the category and the denominator representing the size of the category. Water and upland have, respectively, 9% and 8% intensities during the first time interval. Similarly, water and upland have 8% and 6% intensities during the second time interval. Water has the greater intensity during both time intervals; thus, the green TOC curves show the segment for water first and upland second.

5. Conclusions

Our manuscript provides scientists with a blueprint for using the TOC to analyze the spatial and temporal patterns of losses and gains of a land category. We show the association between change and four variables to illustrate the TOC's applicability to varying factors that may influence change in a landscape. Our manuscript shows a change pattern called Alternation, which pairs the losses and gains of a category during sequential time intervals at the same location. We show how to rank a categorical variable for use in the TOC. In addition, we provide a methodology to constrain the TOC analysis to relevant parts of the spatial extent when analyzing a category's gain. The constraint influences the AUC values and the interpretation of the relationship between the binary and index variables.

Author Contributions: Conceptualization, T.M.B. and R.G.P.J.; methodology, T.M.B.; software, Z.L.; validation, T.M.B.; formal analysis, T.M.B.; investigation, T.M.B.; resources, R.G.P.J.; data curation, T.M.B.; writing—original draft preparation, T.M.B.; writing—review and editing, T.M.B., R.G.P.J. and Z.L.; visualization, T.M.B. and R.G.P.J.; supervision, R.G.P.J.; project administration, R.G.P.J.; funding acquisition, T.M.B. and R.G.P.J. All authors have read and agreed to the published version of the manuscript.

Funding: The United States National Science Foundation's Division of Environmental Biology supported this work via grants OCE-1637630 and OCE-2224608 for the Plum Island Ecosystem Long-Term Ecological Research site. The Edna Bailey Sussman Trust supplied additional funding.

Data Availability Statement: The data that support the findings of this study are available at https://gce-lter.marsci.uga.edu/public/app/dataset_details.asp?accession=GIS-GCET-1810 (accessed on 11 September 2021). The code that supports the findings of this study is available at https://lazygis.github.io/projects/TOCCurveGenerator (accessed on 22 July 2024).

Acknowledgments: The authors thank the anonymous reviewers for their comments that helped to improve this manuscript. This manuscript contributes to the Global Land Programme (https://glp.earth (accessed on 22 July 2024)).

Conflicts of Interest: The authors have declared no conflicts of interest.

References

1. Pontius, R.G.; Si, K. The Total Operating Characteristic to Measure Diagnostic Ability for Multiple Thresholds. *Int. J. Geogr. Inf. Sci.* **2014**, *28*, 570–583. [CrossRef]
2. Dybowski, R.; Weller, P.; Chang, R.; Gant, V. Prediction of Outcome in Critically Ill Patients Using Artificial Neural Network Synthesised by Genetic Algorithm. *Lancet* **1996**, *347*, 1146–1150. [CrossRef]
3. Gu, Y.; Li, J.; Guo, D.; Chen, B.; Liu, P.; Xiao, Y.; Yang, K. Identification of 13 Key Genes Correlated With Progression and Prognosis in Hepatocellular Carcinoma by Weighted Gene Co-Expression Network Analysis. *Front. Genet.* **2020**, *11*, 153. [CrossRef]
4. Lindahl, D.; Lanke, J.; Lundin, A.; Palmer, J.; Edenbrandt, L. Improved Classifications of Myocardial Bull's-Eye Scintigrams with Computer-Based Decision Support System. *J. Nucl. Med.* **1999**, *40*, 96–101.
5. Sun, X.; Li, H.; Song, W.; Jiang, S.; Shen, C.; Wang, X. ROC Analysis of Three-dimensional Psychological Pain in Suicide Ideation and Suicide Attempt among Patients with Major Depressive Disorder. *J. Clin. Psychol.* **2020**, *76*, 210–227. [CrossRef]
6. Ögren, J.Å.; Ögren, J.Å.; Sjöblom, T. Exact Probability Distribution for the ROC Area under Curve. *Cancers* **2023**, *15*, 1788. [CrossRef]
7. Hung, M.; Voss, M.W.; Rosales, M.N.; Li, W.; Su, W.; Xu, J.; Bounsanga, J.; Ruiz-Negrón, B.; Lauren, E.; Licari, F.W. Application of Machine Learning for Diagnostic Prediction of Root Caries. *Gerodontology* **2019**, *36*, 395–404. [CrossRef]
8. Chiu, P.; Tang, H.; Wei, C.; Zhang, C.; Hung, G. NMD-12: A New Machine-Learning Derived Screening Instrument to Detect Mild Cognitive Impairment and Dementia. *PLoS ONE* **2019**, *14*, e0213430. [CrossRef]
9. Alatorre, L.C.; Sánchez-Andrés, R.; Cirujano, S.; Beguería, S.; Sánchez-Carrillo, S. Identification of Mangrove Areas by Remote Sensing: The ROC Curve Technique Applied to the Northwestern Mexico Coastal Zone Using Landsat Imagery. *Remote Sens.* **2011**, *3*, 1568–1583. [CrossRef]
10. Liu, Z.; Pontius, R.G. The Total Operating Characteristic from Stratified Random Sampling with an Application to Flood Mapping. *Remote Sens.* **2021**, *13*, 3922. [CrossRef]
11. Bilintoh, T.M.; Ishola, J.I.; Akansobe, A. Deploying the Total Operating Characteristic to Assess the Relationship between Land Cover Change and Land Surface Temperature in Abeokuta South, Nigeria. *Land* **2022**, *11*, 1830. [CrossRef]
12. Bilintoh, T.M.; Korah, A.; Opuni, A.; Akansobe, A. Comparing the Trajectory of Urban Impervious Surface in Two Cities: The Case of Accra and Kumasi, Ghana. *Land* **2023**, *12*, 927. [CrossRef]

13. Pontius, R.G.J. *Metrics That Make a Difference*; Shivanand, B., Dragicevic, S., Eds.; Springer Nature: Cham, Switzerland, 2022; ISBN 978-3-030-70764-4.
14. Naghibi, F.; Delavar, M.R.; Pijanowski, B. Urban Growth Modeling Using Cellular Automata with Multi-Temporal Remote Sensing Images Calibrated by the Artificial Bee Colony Optimization Algorithm. *Sensors* **2016**, *16*, 2122. [CrossRef]
15. Chakraborti, S.; Das, D.N.; Mondal, B.; Shafizadeh-Moghadam, H.; Feng, Y. A Neural Network and Landscape Metrics to Propose a Flexible Urban Growth Boundary: A Case Study. *Ecol. Indic.* **2018**, *93*, 952–965. [CrossRef]
16. Lobo, J.M.; Jiménez-valverde, A.; Real, R. AUC: A Misleading Measure of the Performance of Predictive Distribution Models. *Glob. Ecol. Biogeogr.* **2008**, *17*, 145–151. [CrossRef]
17. e Silva, L.P.; Xavier, A.P.C.; da Silva, R.M.; Santos, C.A.G. Modeling Land Cover Change Based on an Artificial Neural Network for a Semiarid River Basin in Northeastern Brazil. *Glob. Ecol. Conserv.* **2020**, *21*, e00811. [CrossRef]
18. Burns, C.; Alber, M.; Alexander, C. GCE-LTER Data Set Summary. Available online: https://gce-lter.marsci.uga.edu/public/app/dataset_details.asp?accession=GIS-GCET-1810 (accessed on 22 July 2024).
19. Liu, Z. TOC Curve Generator 2020. Available online: https://lazygis.github.io/projects/TOCCurveGenerator (accessed on 22 July 2024).
20. Gómez, C.; White, J.C.; Wulder, M.A. Optical Remotely Sensed Time Series Data for Land Cover Classification: A Review. *ISPRS J. Photogramm. Remote Sens.* **2016**, *116*, 55–72. [CrossRef]
21. Pontius, R.G.; Krithivasan, R.; Sauls, L.; Yan, Y.; Zhang, Y. Methods to Summarize Change among Land Categories across Time Intervals. *J. Land Use Sci.* **2017**, *12*, 218–230. [CrossRef]
22. Houet, T.; Loveland, Æ.T.R.; Napton, Æ.D.; Barnes, Æ.C.A.; Sayler, K. Exploring Subtle Land Use and Land Cover Changes: A Framework for Future Landscape Studies. *Landsc. Ecol.* **2010**, *25*, 249–266. [CrossRef]
23. Pierre, C.; Hiernaux, P.; Rajot, J.L.; Kergoat, L.; Webb, N.P.; Touré, A.A.; Marticorena, B.; Bouet, C. Wind Erosion Response to Past and Future Agro-Pastoral Trajectories in the Sahel (Niger). *Landsc. Ecol.* **2022**, *37*, 529–550. [CrossRef]
24. Dodd, L.E.; Pepe, M.S. Partial AUC Estimation and Regression. *Biometrics* **2003**, *59*, 614–623. [CrossRef] [PubMed]
25. Peterson, A.T.; Papeş, M.; Soberón, J. Rethinking Receiver Operating Characteristic Analysis Applications in Ecological Niche Modeling. *Ecol. Model.* **2008**, *213*, 63–72. [CrossRef]
26. Jackson, J.M.; Pimsler, M.L.; Jeannet, K.; James, J.B.K.; James, D.H.; Dillon, M.E.; Lozier, J.D. Distance, Elevation and Environment as Drivers of Diversity and Divergence in Bumble Bees across Latitude and Altitude. *Mol. Ecol.* **2018**, *27*, 2926–2942. [CrossRef] [PubMed]
27. Muñoz, M.; Klanderud, K.; Finegan, B.; Veintimilla, D.; Bermeo, D.; Murrieta, E.; Delgado, D.; Sheil, D. Forest Ecology and Management How Forest Structure Varies with Elevation in Old Growth and Secondary Forest in Costa Rica. *For. Ecol. Manag.* **2020**, *469*, 118191. [CrossRef]
28. Shafizadeh-Moghadam, H.; Tayyebi, A.; Ahmadlou, M.; Delavar, M.R.; Hasanlou, M. Integration of Genetic Algorithm and Multiple Kernel Support Vector Regression for Modeling Urban Growth. *Comput. Environ. Urban Syst.* **2017**, *65*, 28–40. [CrossRef]

land MDPI

Article

Towards a Comprehensive Framework for Regional Transportation Land Demand Forecasting: Empirical Study from Yangtze River Economic Belt, China

Ke Wang [1], Li Wang [1,*] and Jianjun Zhang [2,3,*]

1 Key Laboratory of Remote Sensing and Digital Earth, Aerospace Information Research Institute, Chinese Academy of Sciences, Beijing 100101, China; wangke11260@163.com or wangke@aircas.ac.cn
2 School of Land Science and Technology, China University of Geosciences, 29, Xueyuan Road, Haidian District, Beijing 100083, China
3 Key Laboratory of Land Consolidation and Rehabilitation, Ministry of Natural Resources, Beijing 100083, China
* Correspondence: wangli@aircas.ac.cn (L.W.); zhangjianjun_bj@126.com or zhangjianjun@cugb.edu.cn (J.Z.)

Abstract: China is currently experiencing rapid expansion in its transportation land. To promote sustainable land use, accurately estimating transportation land demand is crucial. This study aims to develop a comprehensive framework for urban transportation land forecasting within the Yangtze River Economic Belt (YREB), providing support for optimizing regional land allocation. Employing methods such as meta-analysis, statistical analysis, and BP neural network analysis, this study forecasts the transportation land demand of 127 cities in the YREB. The study findings indicate that cities with high transportation land demand are mainly distributed in the middle and upper reaches of the Yangtze River. Moreover, the growth rate of transportation land in the upper reaches significantly outstrips that in the middle and lower reaches, suggesting a focus shift in transportation infrastructure construction toward the upper regions. Additionally, some cities within the YREB face a mismatch between the supply and demand of transportation land, necessitating proactive adjustments to their land supply plans to achieve a balance between supply and demand. The main contribution of this study is the development of a comprehensive and adaptable framework that guides the development of future strategies for optimal land allocation by forecasting transportation land demand at a regional level.

Keywords: transportation land; influencing factors; land demand forecasting; land supply

1. Introduction

With the deepening of industrialization and urbanization, land scarcity has become increasingly pronounced [1]. Transportation land, which is used for transport activities, is essential infrastructure for daily life and economic production. The rapid concentration of populations and urban expansion have intensified travel and logistics demands, leading to a significant increase in the need for transportation infrastructure and, consequently, an expansion of transportation land [2]. The demand for global land transportation infrastructure is rapidly increasing [3]. According to statistics from the International Energy Agency (IEA), by 2050, it is projected that there will be at least 25 million kilometers of new roads and 335,000 km of railways globally, representing a 60% increase in total land transportation network length compared to 2010. In terms of land area requirements for transportation infrastructure, roads, railways, and parking facilities are estimated to encompass an area of 250,000 to 350,000 square kilometers by 2050, which is approximately equivalent to the size of the UK and Germany, respectively [4]. Over the past decade, China has experienced rapid development in its transportation sector, accompanied by a substantial expansion in the land area allocated for transportation [5]. Since 2015, the area allocated to transportation land in China has surpassed all other types of construction land. By 2020, urban transportation land comprised 12.8% of the total urban construction land area. In comparison, developed countries such as Germany (36.9% in 2015) and Japan (41.5% in 2020) allocate a larger percentage of their urban land to transportation, indicating that China has a greater need for transportation land supply. Therefore, accurately forecasting transportation land demand in China is crucial. It enables policymakers and planners to understand future trends and the scale of transportation land needs, assisting them in developing more effective land use plans.

Previous research has shown that the actual developed and utilized area of land can significantly deviate from the planned area by government [6]. This deviation sometimes reflects the insufficient consideration given during the planning process, highlighting the need for timely adjustments to land use plans to better align with local development needs [7]. Therefore, as the conflict between land supply and demand intensifies, accurately estimating transportation land demand becomes crucial for devising targeted land supply strategies. Investigating and modeling land use change patterns are essential for improving land demand forecasting. Typically, land demand forecasting is mathematically modeled by systematically considering the trends, correlations, similarities, and probabilities of various factors [8–10]. Given the complex decision-making process involving socio-economic systems in land use studies, identifying factors that influence the transportation system is essential for the accurate forecasting of transportation land demand. The relationship between transportation infrastructure development and socio-economic factors has been widely discussed [11,12]. The positive interaction between transportation infrastructure construction and economic development is widely recognized [13–16]. Moreover, factors such as population size, income level [17], consumption level [18], infrastructure investment [19], industrial development [20], employment [21], urbanization [22], car ownership [23], ecological protection [24,25], and policy directives [26] have an important impact on the transportation development. Research on land demand forecasting has predominantly focused on urban construction land [27], with fewer studies dedicated to transportation land demand forecasting, which often emphasizes forecasting traffic volumes and transport infrastructure demand [28–31]. As the availability of urban land continues to decrease, there is growing attention to the demand for specific types of land to support the development of detailed supply plans. These include demand forecasts for residential, commercial, industrial, public service, and transportation land [32–34].

Research in land use forecasting typically falls into two main categories: One involves constructing mathematical models using statistical methods to forecast the scale and structure of land use [35]. The other simulates the spatial and temporal expansion of land uses incorporating spatial data from remote sensing and geographic information technology [36,37]. Statistical methods such as trend extrapolation, indicator-based methods, and multiple regression analysis are widely employed for forecasting the scale and type of land demand [38]. Trend extrapolation forecasts future urban built-up land area based on historical usage trends [39], focusing solely on land area changes and often neglecting socio-economic drivers, which limits its use to short-term forecasting with considerable uncertainty. Indicator-based methods establish quantitative models linking land use demand to key influencers, setting fixed indicators for each land type based on historical data analysis [40], such as residential land per capita, industrial land per unit of GDP, and transportation land density [41]. Regression analysis is an effective method for forecasting future land demand by modeling the relationships between land use and various economic, demographic, and environmental factors [42–45]. Moreover, as research into land use change intensifies, advanced algorithmic models such as system dynamics, decision trees, neural networks, support vector machines, and random forests are increasingly being applied to forecast land demand [46–48]. With advances in remote sensing and GIS technology, modeling land use changes using satellite data has become a significant research focus [49]. Current studies primarily focus on forecasting the expansion of urban impervious surfaces and simulating the structural evolution of land use [50,51]. However, forecasting the expansion of transportation land use through remote sensing is challenging due to the complex factors that influence it [52]. Therefore, statistical methods are still better suited for macro-level land supply decision-making and development planning.

Unlike land demand driven primarily by economic objectives, such as residential, commercial, and industrial purposes, transportation land demand focuses on supporting socio-economic benefits by facilitating services essential for human production and lifestyle needs [53–55]. Consequently, accurately forecasting transportation land demand and developing an effective land supply strategy are essential for the sustainable development of regional transportation. The literature reviews reveal a predominant focus on urban construction land demand forecasting, with limited empirical investigation into transportation land demand at a regional level. Moreover, despite extensive discourse on the relationship between transportation demand and its influencing factors, a systematic review and comparison of these influencing factors are notably absent. This deficiency has resulted in insufficient research on regional transportation land supply and demand, thereby complicating the support for integrated regional development strategies and the formulation of regional territorial spatial plans. The growth of transportation land is a nonlinear and complex process influenced by various factors, necessitating the use of specific rules, indicators, or models in the forecasting process. Therefore, this study aims to develop a framework to forecast regional urban transportation land demand, thereby assisting governmental bodies in strategic planning. The specific objectives of this study are as follows: (1) to construct a framework using meta-analysis and data modeling for forecasting regional urban transportation land demand and (2) to analyze the supply and demand characteristics of urban transportation land at the regional level, thus guiding future governmental land supply decisions. The main contribution of this study is the development of a comprehensive framework for forecasting transportation land demand at the regional level, offering a holistic perspective for decision-makers to effectively develop and update land supply plans. The findings provide valuable insights for optimizing the allocation of regional urban land resources.

The remainder of this paper is organized as follows: Section 2 outlines the research framework, data sources, and methodology employed. Section 3 details the main factors influencing transportation land demand, forecasts the area of urban transportation land in the study area, and analyzes the supply and demand characteristics of transportation land. Sections 4 and 5 discuss the study's findings and provide conclusions, respectively.

2. Materials and Methods

2.1. Overview of the Study Area

The Yangtze River Economic Belt (YREB) includes nine provinces and two centrally administered municipalities—Shanghai, Jiangsu, Zhejiang, Anhui, Jiangxi, Hubei, Hunan, Chongqing, Sichuan, Yunnan, and Guizhou—encompassing one hundred and twenty-seven prefecture-level units across approximately 2.05 million square kilometers (Figure 1). This region is a key strategic development area in China, focusing on constructing a comprehensive three-dimensional transportation network that integrates land, water, and air, anchored by the Yangtze River, which cuts across eastern and western China. The population and economic output of the YREB collectively account for over 45 percent of China's totals, yet the region's land area represents just 21.4 percent of the national territory, highlighting significant land resource scarcity that constrains regional development. As of 2020, the YREB boasted 44,620 km of operational railway, accounting for 30.49% of China's total. The highway network spanned 2.35 million kilometers, making up 45.14% of the national total, while inland waterways reached 90,833 km, constituting 71.14% of the nation's total. Notably, the YREB handles nearly half of China's passenger and freight traffic, with both cargo volumes and passenger counts surpassing 40% of national levels. As China's golden waterway, the YREB needs to further rationalize the allocation of transportation land to achieve the construction of a coordinated and sustainable integrated three-dimensional transportation network.

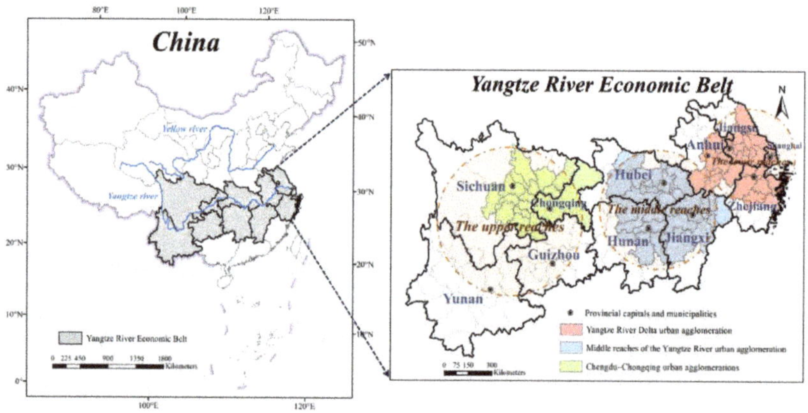

Figure 1. Study area.

2.2. Policy Timeline for the Transportation Development of the YREB

Since China's reform and opening up, the transportation system in the Yangtze River Economic Belt has undergone rapid development, with the length of transportation routes expanding significantly and consistently accounting for over 40% of the nation's total. The evolution of transportation policy in the YREB can be categorized into three stages (Figure 2). The first stage is the early conception stage (1980–1992). In the 1980s, China's central government put forward the strategic concept of "one line, one axis". One line refers to the coastal line, and one axis refers to the Yangtze River. The second stage, the mid-term exploration stage (1992–2012), marked the YREB's inclusion in a major national development strategy for the first time. During the 14th National Congress of the Communist Party of China (CPC) in 1992, it was proposed that Pudong's development in Shanghai should spearhead the further opening of Yangtze River cities, boosting the economic growth of the Yangtze River Delta and the entire basin. In 2005, seven provinces and two cities along the river signed the "YREB Cooperation Agreement" to enhance transportation development, though administrative barriers often led to fragmented progress. The "Rise of Central China" strategy, introduced by the State Council in 2006, emphasized that accelerating transportation development was crucial for the region's growth and its coordinated development. In 2010, Hubei Province released the "Master Plan for the Opening and Development of the YREB", aiming to establish an integrated transportation system centered around the Yangtze River waterway. The third stage, the comprehensive development stage (2013 to present), began following the 18th National Congress of the CPC, marking a period of accelerated transportation development within the YREB. In 2013, General Secretary Xi Jinping highlighted the vision of transforming the entire Yangtze River basin into a golden waterway during an inspection in Wuhan. Subsequently, in 2014, the State Council unveiled the "YREB Comprehensive Three-dimensional Transportation Corridor Plan", aiming to establish a complete three-dimensional transportation system by 2020. Following this, the Chinese government introduced several policies and plans that significantly boosted the transportation development of the YREB.

An analysis of the YREB's policy timeline reveals that transportation construction has been a crucial component of the region's development. For this study, the YREB has been chosen as the study area, aligning with China's major strategic development initiatives. Moreover, studying the demand for transportation land serves as a fundamental element in supporting the coordinated and integrated development of regional transportation [56].

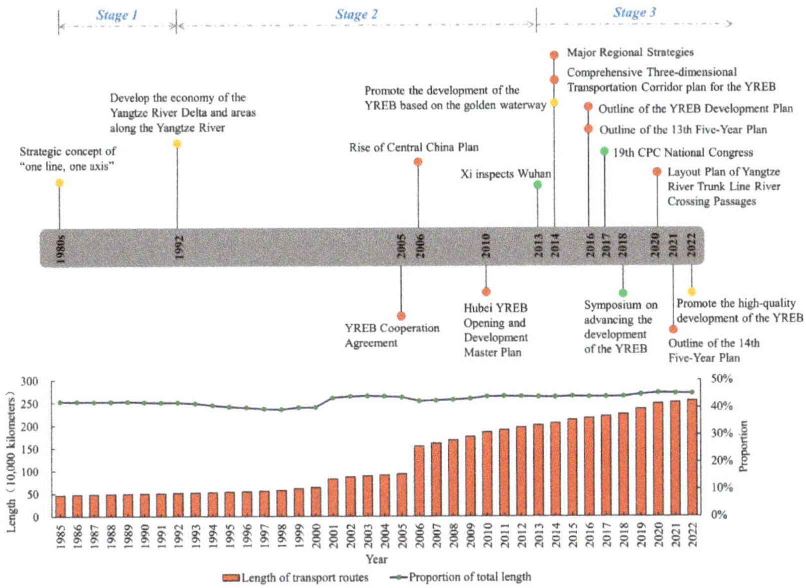

Figure 2. Policy development timeline of the YREB.

2.3. Considerations for the Study Framework

Under the new economic normal, the contradiction between the supply and demand of land resources in China remains a significant challenge. The scale of transportation land supply directly affects the potential and direction of urban transportation development. Over the past decade, transportation infrastructure in China has expanded rapidly, making the area of land allocated for transportation the largest among all types of urban construction land. Quantitative analysis of urban transportation land demand is essential for optimizing the allocation of regional land, providing decision-makers with guidance for future land supply plans to improve regional land use efficiency. This study aims to analyze the demand for urban transportation land in the Yangtze River Economic Belt (YREB) across four dimensions: identifying influencing factors, constructing models, verifying results, and conducting empirical analysis. The main contribution of this research framework is the development of an effective transportation land demand forecasting method system that assists decision-makers in formulating or adjusting future land use plans more effectively.

Identifying the influencing factors is a crucial initial step in estimating urban transportation land demand. To mitigate subjectivity in selecting factors, this study adopts a meta-analysis and develops a structured framework for screening main influencing factors. Initially, it involves reviewing the relevant literature and extracting essential information to pinpoint factors affecting transportation land demand. Preliminary selection of these factors is conducted through frequency analysis, with their validity further confirmed using statistical methods, ultimately determining the main influencing factors (Figure 3). Building on this, this study constructs a gray–BP neural network model under planning constraints and forecasts the transportation land demand of 127 cities in the YREB for the target year through influencing factor input, model training, and forecasting result output. In addition, a spatial overlay analysis between the forecasted transportation land demand and the planned transportation infrastructure network is conducted to further assess the model's reliability. Lastly, a matching analysis between the forecasted transportation land demand and historical transportation land supply is performed, revealing supply–demand characteristics and informing future transportation land supply strategies. Forecasts of transportation land demand in regional cities offer valuable insights into the dynamics of transportation land use and identify cities with a greater need for transportation land supply. For cities experiencing a mismatch between transportation land supply and demand, these forecasts can guide the optimization of land allocation by adjusting land supply strategies accordingly.

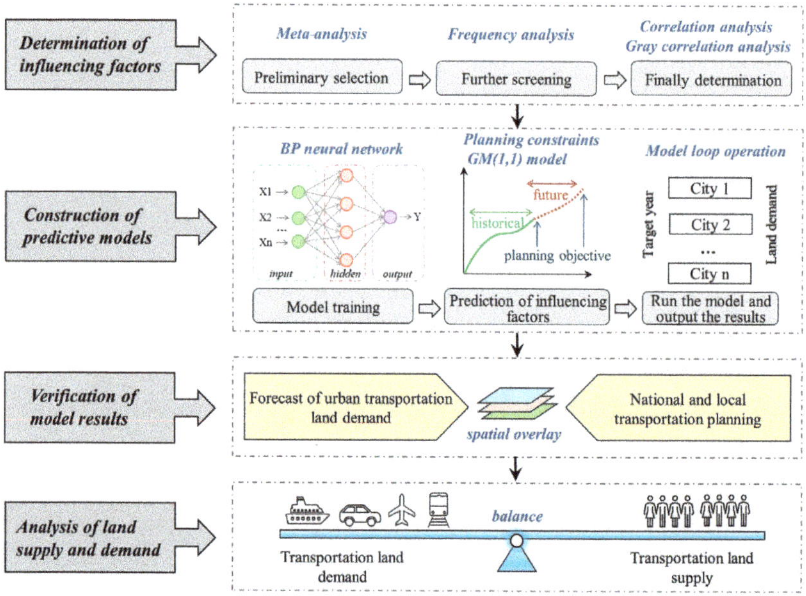

Figure 3. Research framework.

2.4. Data Sources

The data used in this article are categorized into three types: official statistical data, spatial information data, and text data. The primary data for this study include transportation land area and land supply area. In this context, "transportation land" encompasses roads, railways, airports, ports, terminals, and transportation stations, excluding rural roads (Figure 4). The transportation land area data (2009–2020) were sourced from the land survey results shared via the application service platform of the Ministry of Natural Resources of China (https://gtdc.mnr.gov.cn/Share#/, accessed on 30 December 2023). Data on transportation land supply (2011–2020) were obtained from the China Land Market (https://landchina.com/#/, accessed on 30 December 2023). Socio-economic-related statistical data (2009–2020) were collected from the statistical yearbooks and statistical communiqué of the provinces and cities within the study area. Spatial information data include vector and raster data: vector data primarily define the administrative boundaries of the study area (national, provincial, and city boundaries) and are sourced from the standard map service of the Surveying and Mapping Department of the Ministry of Natural Resources of China (http://bzdt.ch.mnr.gov.cn/, accessed on 30 December 2023). Raster data mainly comprise transportation and land use planning maps that include spatial information about planned road networks, railway networks, waterways, and airports. Text data encompass provincial and municipal land spatial planning, land use master plans, and state-owned construction land supply plans, all of which are publicly released by the government.

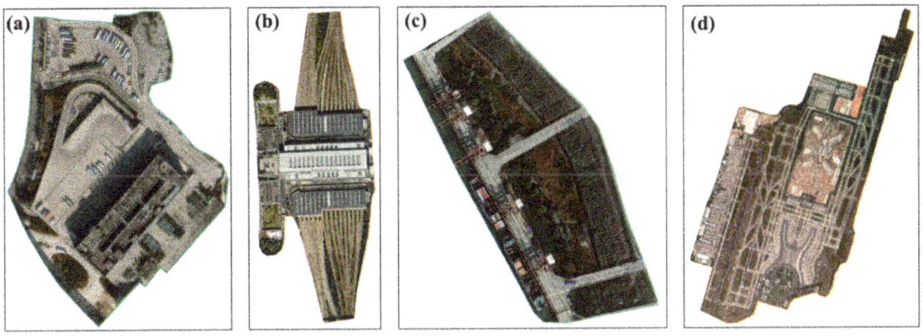

Figure 4. Sample map of transportation land. (**a**) Highway land (Guiyang Longdongbao Bus Terminal); (**b**) railway land (Shanghai Hongqiao Railway Station); (**c**) port terminal land (Wuhan Yangluo New Port); (**d**) airport land (Chongqing Jiangbei Airport).

2.5. Method

2.5.1. Determination of Main Factors Affecting Transportation Land Demand

We developed an empirical analysis methodology that integrates a systematic review of the literature to identify and synthesize the factors influencing urban transportation land demand. Initially, we adhered to the Preferred Reporting Items for Systematic Reviews and Meta-Analyses (PRISMA) protocol to identify potentially relevant studies from the literature databases [57]. Subsequently, factors impacting urban transportation land demand were derived through detailed literature review and critical information extraction. These influencing factors were then screened through frequency analysis and expert judgment. Finally, grey relation analysis and correlation analysis were employed to validate the relevance and applicability of these impact factors.

(1) Methodology of the systematic literature review

The PRISMA protocol is extensively applied across diverse research fields, including socio-economic, environmental, land management, and sustainable development studies. It employs a structured approach to analyze data and evidence from existing research, significantly reducing author bias and thus enhancing the scientific integrity of the findings [58]. The PRISMA protocol outlines a four-stage systematic screening process for including and excluding publications in a review. First, using a defined search strategy, a total of 4055 papers were selected from the Web of Science (WoS) and China National Knowledge Infrastructure (CNKI) databases, with search terms applied exclusively to titles. The considered studies were limited to articles published in English or Chinese, including those in other languages with English translations, across peer-reviewed journals, conferences, and dissertations. The specific search terms used and the corresponding number of articles identified are detailed in Table 1. Secondly, duplicates were removed, and titles and abstracts were screened for relevance. Publications not directly related to our study objectives were then excluded. Finally, we read the full text of the remaining studies, and included 69 studies in our systematic review.

Table 1. Search terms, sources, and the corresponding number of papers identified.

Search Term	Number of Publications from WoS	Number of Publications from CNKI	Search Date
TITLE ((Transportation OR Transport OR Traffic OR Highway OR Road OR Railway OR Airport OR Port) AND land)	3170	885	6 November 2023

(2) Grey correlation analysis

Grey correlation analysis, a multivariate statistical method, assesses the presence of connections and the degree of correlation among factors by comparing their change trends and differentiation magnitudes. It introduces the grey correlation degree as a quantitative index to gauge the correlation level between factors. Grey correlation analysis involves three main steps: First, dimensionless processing on the original data of factors necessary for this research is conducted. Second, the grey correlation coefficients between factors are calculated. Third, the grey correlation degree among factors is determined. The gray correlation degree is then utilized to compare and analyze the degree of connection between factors. The specific calculation process is as follows:

$$x_i'(k) = \frac{x_i(k)}{x_i(1)} \tag{1}$$

$$\varepsilon_i(k) = \frac{\text{minimink}\left|y'(k) - x_i'(k)\right| + \rho\text{maximaxk}\left|y'(k) - x_i'(k)\right|}{\left|y'(k) - x_i'(k)\right| + \rho\text{maximaxk}\left|y'(k) - x_i'(k)\right|} \tag{2}$$

$$r_i = \frac{1}{n}\sum_{k=1}^{n} \varepsilon_i(k), k = 1, 2, \ldots n \tag{3}$$

where x_i' is the normalized value of variable i. x_i is the raw value of variable i. $x_i(1)$ is the value of the first sample of variable i. y(k) is an ideal data set, and x(k) are the alternative data sets of the same length. n is the number of samples. k is the k-th sample. $\varepsilon_i(k)$ denotes the grey relational coefficients. ρ denotes the dynamic distinguishing coefficient, in the interval of 0–1; the value in this study is 0.5. r_i denotes the grey relational degree.

2.5.2. Construction of Land Demand Forecasting Model of Gray–BP Neural Network under the Constraints of Planning Objectives

The backpropagation (BP) neural network algorithm is a multilayer feedforward network trained using the error backpropagation algorithm and stands as one of the most commonly used neural network models [59]. Compared to traditional statistical methods commonly used for land demand forecasting, such as trend extrapolation, regression analysis, and indicator-based methods, the BP neural network model excels in iterative learning through sample training and an error-limiting output mechanism, making it exceptionally suitable for nonlinear prediction tasks. Moreover, it can handle multiple input variables simultaneously, providing more comprehensive and accurate predictions. Existing research has also confirmed that the BP neural network model achieves high accuracy in complex land area demand forecasting [60]. However, forecasting land demand presents a complex socio-economic challenge, as nonlinear forecasting methods often struggle to accurately reflect the primary driving factors and their mechanisms. To address this, this study integrates grey system theory with the BP neural network model to improve the accuracy of land demand forecasting. Additionally, the GM(1,1) grey model is well adapted to constraints such as short data series, small sample sizes, and limited information, facilitating more precise predictions of short-term changes in sample data [61]. The specific forecasting process of this study involves training the historical data (2009–2020) on transportation land area and its influencing factors using a BP neural network to verify the model's accuracy. Based on this, the influencing factors for the target year (2025) are predicted using GM(1,1) grey prediction and planning constraints, and these factors are then input into the trained neural network model to forecast the transportation land area for the target year. The computational procedures and formulas for both the BP neural network model and the GM(1,1) prediction model are detailed below:

1. BP neural network model

(1) Network initialization: Assign random numbers in the range of (−1, 1) to each connection weight, set the error function e, and give the calculation accuracy ε and the maximum number of learning times M. The formula is as in (4).

$$f(x) = \frac{1}{2}\sum_{j=1}^{n}\left(y_i - y_j^{'}\right)^2 \tag{4}$$

where n represents the output node, y_i represents the actual output value, and $y_j^{'}$ represents the target output value.

(2) Weight correction: Calculate the output for each unit in the hidden and output layers based on the input sample values and expected output values; then, adjust the weights of each neuron's input nodes according to the gradient direction. The formula is as follows:

$$\Delta\omega_{ij} = -\eta \times \frac{\partial_\varepsilon}{\partial_{\omega_{ij}}} = -\eta \times \frac{\partial_\varepsilon}{\partial_{I_j}} \times \frac{\partial_{I_j}}{\partial_{\omega_{ij}}} \tag{5}$$

where ω_{ij} is the connected weights from node *i* to node *j* in the output layer; η is the learning efficiency value, and I_j is the transfer function of the jth hidden layer. The output layer to the hidden layer are *tansig* functions, while the hidden layer to the output layer are *trainlm* functions.

(3) Iteration: Select the next input mode and return to the second step. If the output layer does not produce the desired result, it sends an error signal back along the original connection pathway. Throughout the iteration process, the weights of each neuron are adjusted to minimize the error signal, and the process continues until the output error meets the specified accuracy requirement.

2. GM(1,1) prediction model

Based on grey system theory, the GM(1,1) prediction model can utilize limited information to construct a model that closely approximates complete information by transforming time series data into differential equations using the differential fitting method. This model is extensively used in parameter prediction across socio-economic and ecological fields, demonstrating high accuracy for short- to medium-term forecasts, particularly for parameters with brief historical time series. The formula is as follows:

$$\hat{x}^{(1)}(k) = \left(\hat{x}^{(0)}(1) - \frac{b}{a}\right)e^{-a(k-1)} + \frac{b}{a}, k = 1,2,\ldots,n \tag{6}$$

where $\hat{x}^{(0)}(1)$ is the original data sequence in the urban transportation land demand system. $\hat{x}^{(1)}(k)$ is the accumulated value of the original data sequence. k is the time series. a is the development coefficient, which mainly controls the development trend of the system. b is the gray action quantity, which reflects the relationship between data changes.

To ensure the accuracy of the grey model, an error test is essential. The GM(1,1) model typically employs a post-test method for error testing. The formula used is as follows:

$$C = S_1/S_0 \tag{7}$$

$$P = P\left(\left|\varepsilon^{(0)}(k) - \overline{\varepsilon}^{(0)}\right| < 0.6745S_0\right) \tag{8}$$

where C is a posterior difference ratio, P is the small error probability, S_0 is the variance in the original sequence, S_1 is the variance in the residuals, and $\varepsilon^{(0)}$ is the residual. When $C < 0.35$ and $P > 0.95$, the model is generally considered to be reliable, and the model can be used for prediction.

3. Results

3.1. Main Factors Affecting Transportation Land Demand

Through a literature meta-analysis, we extracted factors that influence the demand for transportation land and grouped similar factors together. For instance, resident income and per capita disposable income were categorized under income levels. Using frequency analysis, we then filtered out infrequently mentioned factors, ultimately identifying 10 main influencing factors: economic development level, population size, investment level, urbanization level, urban construction level, industrial development level, income level, industrial structure, employment, and consumption level. These factors cover economic, demographic, social, and industrial aspects and commonly affect transportation land demand. Specifically, areas with a high level of economic development typically exhibit more frequent internal and external connections, thereby generally requiring more transportation land. Additionally, areas with high population densities and significant levels of urbanization see increased demand for travel. Social development levels, reflecting human well-being, directly impact the development of transportation infrastructure. Variations in industry types also influence inputs to the transportation sector.

Based on the screening of commonly influencing factors, a correlation matrix between multiple variables is constructed using statistical data from 127 prefecture-level administrative units in the YREB (Figure 5). By analyzing the correlation coefficients and significance levels, we can verify the applicability of these main influencing factors and explore the presence of multicollinearity among them. The results showed that, except for the correlation between investment and urban construction level, which was significant at the 0.05 level ($p < 0.05$), all other correlations were significant at the 0.01 level ($p < 0.01$), indicating a significant linear relationship between these influencing factors and transportation land demand. Comparing correlation coefficients, the strongest linear relationship was with urban construction level, with a coefficient exceeding 0.95. This was followed by the urbanization rate, total industrial output value, and the number of urban employees, all with coefficients above 0.6. Additionally, GDP, investment, and industrial structure had correlation coefficients above 0.5. Correlation analysis reveals that all the 10 main influencing factors obtained in this study have significant correlation with the transportation land area. In addition, the correlation analysis revealed a degree of multicollinearity among the influencing factors, necessitating the elimination of influencing factors with strong covariance to enhance the model's accuracy. Consequently, we further refined the influencing factors through grey correlation analysis.

The results of the grey correlation analysis indicate that fixed asset investment (X3) has the most substantial impact, followed by population size (X2), urban construction level (X4), industrial development (X6), consumption level (X8), GDP (X1), employment (X10), income level (X7), urbanization rate (X5), and, finally, industrial structure (X9). Generally, a grey relational degree above 0.8 signifies a high correlation between two factors, a degree between 0.5 and 0.8 suggests a moderate correlation, and a degree below 0.5 implies a negligible or non-existent correlation. The grey relational degrees of all influencing factors exceed 0.5, confirming the suitability of the factors identified through the literature meta-analysis. All factors, except for X9, demonstrated a grey relational degree above 0.8 with transportation land demand. Due to the relatively low grey relational degree of X9 and its correlation coefficient with multiple other independent variables exceeding 0.8, this factor was excluded from further analysis. Consequently, this study identifies nine factors (X1, X2, X3, X4, X5, X6, X7, X8, X10) as determinants in developing a demand forecast model for urban transportation land (Table 2).

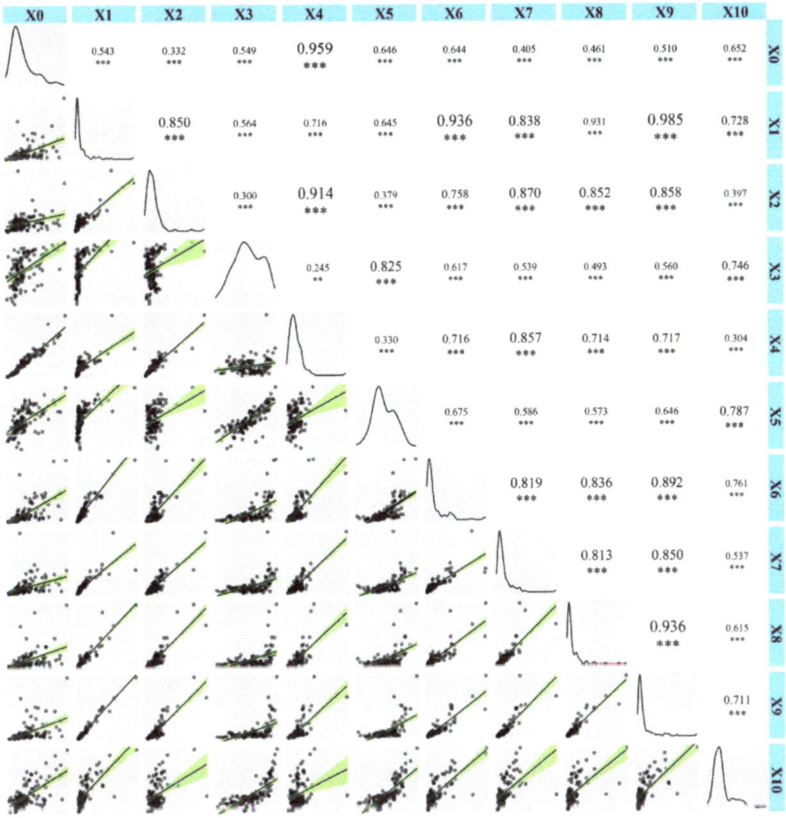

Figure 5. Correlation matrix of main factors influencing urban transportation land demand. (X0, X1, X2, X3, X4, X5, X6, X7, X8, X9, and X10 represent transportation land area, GDP, population, investment, urban construction level, urbanization, industrial output, income level, consumption level, industrial structure, and employment, respectively.) Arabic numerals represent the Pearson correlation coefficient between the influencing factors. ***, ** are significance levels of 1%, 5%, respectively. The curves in the figure are kernel density estimates for different influencing factors.

Table 2. Grey relational degree of factors affecting transportation land demand.

Influencing Factors	Grey Relational Degree	Description
Economic development level (X1)	0.85529	GDP
Population size (X2)	0.89608	Number of individuals present
Investment (X3)	0.90119	Fixed-asset investment
Urban construction level (X4)	0.88750	Urban construction land area
Urbanization rate (X5)	0.80576	Ratio of urban population to total population
Industrial development (X6)	0.87040	Gross industrial output value
Income level (X7)	0.83050	Per capita disposable income
Consumption level (X8)	0.85628	Total retail sales of consumer goods
Industrial structure (X9)	0.67864	Ratio of secondary and tertiary industries to GDP
Employment (X10)	0.85312	Employed population

3.2. Forecasting Urban Transportation Land Demand in the YREB

3.2.1. An Empirical Result of Urban Transportation Land Demand Forecasting Models—Shanghai Example

To elucidate the process of urban transportation land demand forecasting, this study utilizes Shanghai, a leading city in the YREB, as a case study for empirical analysis. The nine main influencing factors outlined in Section 3.1 serve as input neurons for the neural network. Socio-economic data from Shanghai, covering the years 2009 to 2020, are employed as training samples. The BP neural network's simulation and prediction code, developed using MATLAB R2020b, was executed with a maximum of 5000 training iterations and a minimum learning rate of 0.05. Following extensive debugging, the model achieved a minimal fitting residual of 0.000077 with an optimal hidden layer size of 12. The results from the trained BP neural network model demonstrate a low error margin, with the relative error of the simulated values for each year being less than 0.01, an average error of 0.0035, and a standard deviation of errors at 0.0031. These results indicate the high accuracy of the prediction model trained using the BP neural network, demonstrating its suitability for forecasting urban transportation land demand (Figure 6).

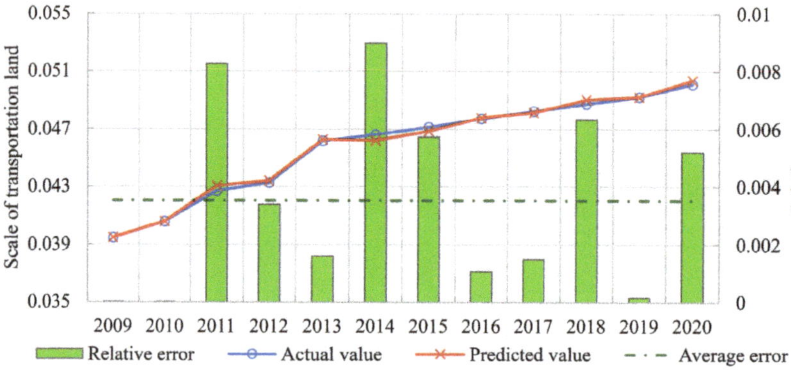

Figure 6. BP neural network sample simulation training results.

Leveraging the existing planning objectives of Shanghai and a grey prediction model, this study establishes the future values for each influencing factor for the year 2025 (Table 3). Specifically, the constrained target values of GDP, total retail sales of consumer goods, and per capita disposable income for 2025 are sourced from the Outline of Shanghai's 14th Five-Year Plan. Meanwhile, the constrained target values of population size and urban construction land size are based on the Shanghai City Master Plan (2017–2035) and the Shanghai Territory Development Plan (2021–2035), respectively. For influencing factors without explicit planning targets, this study employs the grey prediction model to estimate their predicted values. On this basis, the planned or forecasted values of the influencing factors are substituted into the developed neural network simulation model to forecast the theoretical demand area of transportation land in Shanghai. According to the forecasted outcomes of this study, the urban transportation land area in Shanghai is predicted to reach 346.16 square kilometers by 2025, indicating that about 30 square kilometers of urban transportation land will need to be supplied during the 14th Five-Year Plan period (2021–2025).

Table 3. Forecasting results of transportation land area and related influencing factors in Shanghai.

Factors [1]	Planned or Forecasted Value	Methods and Constraints
X1	49,392.84	14th Five-Year Plan
X2	2500	City's master plan
X3	3200	Territorial spatial plan
X4	0.9	GM (1,1) model
X5	39,136	GM (1,1) model
X6	10,265.15	GM (1,1) model
X7	1400	GM (1,1) model
X8	20,000	14th Five-Year Plan
X9	98,648.39	14th Five-Year Plan
Transportation land area in 2025	346.16	Gray–BP neural network model under planning constraints

[1] The influencing factors X1, X2, X3, X4, X5, X6, X7, X8, and X9 represent GDP, population size, urban construction land area, urbanization rate, gross industrial output value, fixed-asset investment, employed population, total retail sales of consumer goods, and per capita disposable income, respectively.

3.2.2. Comparative Analysis of the Differential Characteristics of Urban Transportation Land Demand in the YREB

Utilizing the grey-BP neural network land demand forecasting model, developed under the constraints of the planning objectives, this study forecasts the transportation land area for 127 cities in the YREB for the year 2025. By integrating these data with the transportation land area data from 2020, this study derives the projected transportation land area demand for these cities over the next five years. To eliminate potential biases in assessing urban transportation land use due to administrative division size, this study quantifies the scale of urban transportation land use through the ratio of urban transportation land area to total urban area.

Examining the spatial distribution of transportation land within the YREB reveals that the scale of urban transportation land in the lower reaches (Shanghai, Jiangsu, Zhejiang, and Anhui) significantly exceeds that in the middle (Jiangxi, Hubei, and Hunan) and upper reaches (Chongqing, Guizhou, Sichuan, and Yunnan). Furthermore, urban transportation land exhibits marked spatial agglomeration, with high-scale transportation land primarily located in three national urban agglomerations: the Yangtze River Delta urban agglomeration, the Yangtze River middle reaches urban agglomeration, and the Chengdu–Chongqing urban agglomeration. The scale of urban transportation land within each province of the YREB also presents a pattern of spatial reduction from central and sub-central cities to their surrounding cities. Furthermore, the scale of transportation land in cities within the YREB exhibits a marked left-skewed distribution, suggesting that urban transportation land scale across the YREB is predominantly low. This highlights significant polarization among cities and an uneven distribution of transportation land resources across the region.

Cities exhibiting higher demand for urban transportation land are predominantly located in the lower reaches of the Yangtze River. However, the analysis of the transportation land area growth rate between 2020 and 2025 indicates a faster expansion of urban transportation land in the middle and upper reaches of the Yangtze River. This suggests a strategic shift in the focus of transportation infrastructure development within the YREB towards these regions. The frequency curve of transportation land area growth rates within the YREB exhibits a bimodal distribution, with most cities distributed in the two ranges of 10–15% and 20–25%. Cities within the 10–15% range are mainly located in the middle and lower reaches of the Yangtze River, whereas those in the 20–25% range tend to be in the upper reaches. Notably, 15 cities exhibited a transportation land area growth rate exceeding 25%, with over 70% of these cities situated in the upper reaches. The coordinated integration strategy for the YREB has catalyzed a pronounced regional radiating effect in transportation land demand, leading to significant growth in the transportation land area of cities surrounding these national urban agglomerations (Figure 7).

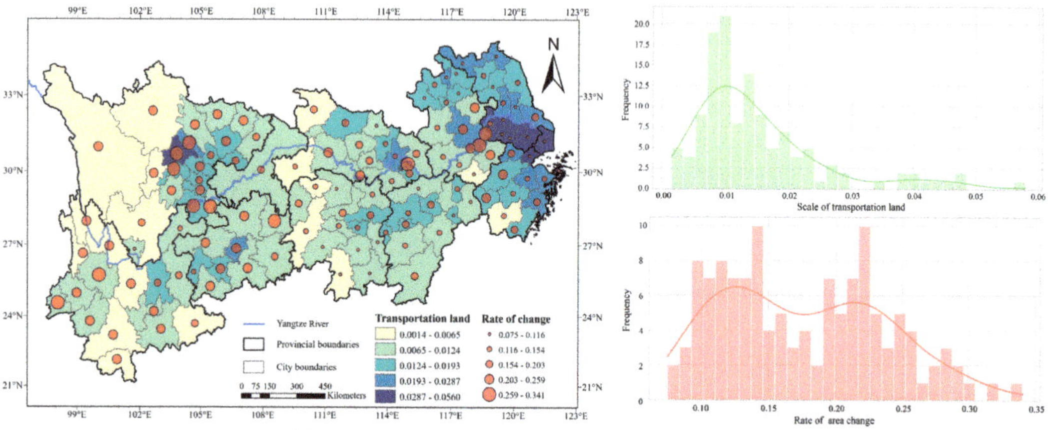

(a) (b)

Figure 7. Forecast of urban transportation land demand in the YREB in 2025. (**a**) Spatial distribution characteristics of transportation land demand in the YREB; (**b**) frequency density curve of transportation land scale and its change rate.

3.3. Validation of the Reasonableness of the Forecasted Results

To validate the rationality of the forecasted results from this study, the forecasted transportation land areas were spatially overlaid with the planned transportation infrastructure for the 14th Five-Year Plan period, including the planned highway network, railway network, high-grade waterways, and airports. A subsequent correlation analysis was conducted between the forecasted transportation land areas and the planned transportation infrastructure. The findings indicate that this study's forecasted results largely align with the planning objectives set by government departments (Figure 8). Specifically, Chongqing emerges as the city with the highest demand for transportation land and the most extensive land transportation network planned from 2020 to 2025. Additionally, Chongqing is expected to construct two new civil airports within this period. In contrast, cities with lower demand for transportation land tend to have a more limited transportation infrastructure layout, and most of these cities are third- or fourth-tier cities, such as Maanshan, Huaibei, Xinyu, and Suining. Notably, underdeveloped marginal cities such as Liangshan Yi Autonomous Prefecture, Aba Tibetan and Qiang Autonomous Prefecture, and Garze Tibetan Autonomous Prefecture have smaller forecasted transportation land areas but larger planned transportation infrastructure networks. On the contrary, regional core cities such as Chengdu and Guiyang have larger forecasted transportation land areas but smaller planned transportation infrastructure networks.

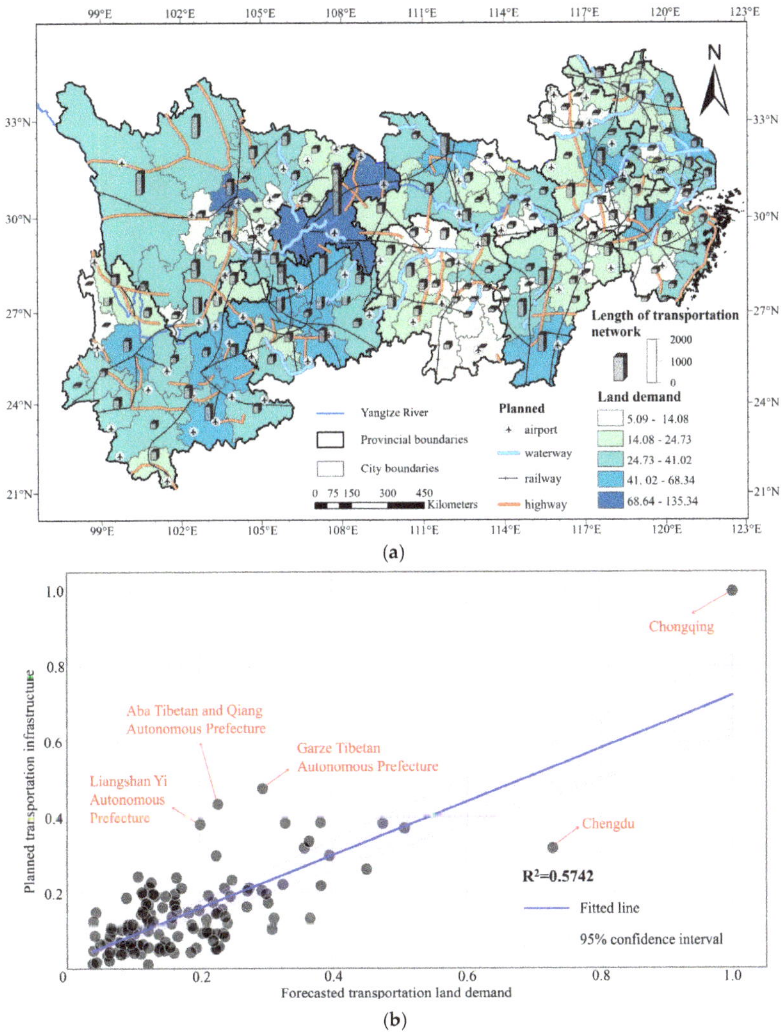

Figure 8. Validation of forecasted results. (**a**) Spatial characteristics of forecasted transportation land demand and planned transportation network; (**b**) correlation analysis between forecasted transportation land demand and the planned transportation network.

3.4. Analysis of Supply and Demand of Urban Transportation Land in the YREB

This study aligns the average annual demand for urban transportation land in the YREB from 2020 to 2025 with the average annual land supply from the 2011–2020 period (12th and 13th Five-Year Plans), providing a reference for future transportation land supply decisions in cities within the YREB. The results indicate that the demand for transportation land generally matches the historical supply in the YREB, with over half of the cities showing demand levels consistent with past supply. This suggests that these cities can draw from their historical land supply experiences when formulating new land supply plans. Specifically, cities with higher transportation land demand typically have higher historical land supplies and are often economically more developed or function as major regional transportation hubs, such as Chongqing, Hangzhou, Hefei, Changsha, Wuhan, Chengdu, and Kunming. Conversely, cities with lower transportation land demand, which have historically smaller land supplies, tend to be distant from the region's core development areas, such as Yaan, Panzhihua, Nujiang Lisu Autonomous Prefecture, and Diqing Tibetan Autonomous Prefecture (Figure 9).

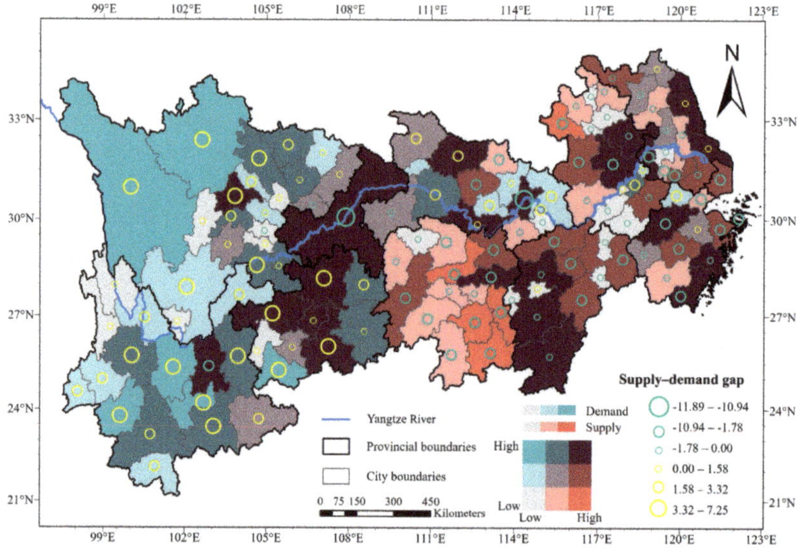

Figure 9. Characteristics of transportation land supply and demand in the YREB.

Despite efforts to match the demand and supply of transportation land in the YREB, significant mismatches still exist in some cities, with notable spatial disparities. Cities facing a high demand but low supply of transportation land are predominantly located in the upper reaches of the Yangtze River. These cities are often remote from core development areas and face challenges in expanding transportation infrastructure due to geographical constraints. Conversely, cities with low demand and high supply are mainly located in the middle and lower reaches of the Yangtze River, often near regional core cities. The influence of these core cities has intensified transportation infrastructure development in the surrounding areas over the past decade. There is a significant spatial aggregation in the mismatch between the demand for transportation land and the historical supply across the YREB. Cities with a demand significantly exceeding supply are in the upper reaches, whereas those with supply exceeding demand are primarily in the middle and lower reaches. Most provincial capitals have a historical average supply of transportation land that exceeds future demand. Thus, these cities should consider controlling their transportation land supply when formulating land supply plans. In less developed cities, it is common for the supply of transportation land to fall short of demand. The future land supply plans in these areas could consider increasing the supply of transportation land.

4. Discussion

4.1. Implications and Applications

China is a country of public ownership of land, where land is centrally managed and controlled by the State, and the government is responsible for the allocation, planning, and approval of land use [62]. As the conflict between urban growth and land availability in China intensifies, the rational allocation of land resources has become a crucial challenge for government departments. The question of how to utilize limited land resources to achieve sustainable development is a subject of ongoing debate [63,64]. Transportation land serves as vital infrastructure supporting both people's livelihoods and economic activities, playing a significant role in a city's capacity for population mobility, its economic growth potential, and the direction of its industrial development. Therefore, to promote coordinated and integrated regional transportation development, it is essential to optimize land use by addressing the variations in regional land demand. Developing the YREB as a strategic priority for China, the sustainable use of land resources plays a crucial role in supporting high-quality regional development. Furthermore, the YREB is actively engaged in constructing a comprehensive three-dimensional transport system to enhance both internal and external connections [65]. Consequently, accurately forecasting urban transportation land demand in the region is essential for crafting effective land supply strategies. However, studies focusing on regional urban transportation land demand are scarce, posing challenges to decision-making for optimal land resource allocation at the regional level. This study establishes a comprehensive framework for forecasting urban transportation land at the regional level, serving as a reference for the YREB to devise future transportation land supply plans and providing a foundation for systematic analyses of urban transportation land demand in other regions.

The foundation for accurately estimating regional transportation land demand lies in identifying the main factors that influence it. Compared to other types of built-up land, transportation land serves a broader array of functions, leading to a more complex set of factors influencing its demand. Therefore, decision-makers should thoroughly consider future regional socio-economic development when formulating land use plans. The core findings of this study reveal that socio-economic factors such as economic development, population size, industrial development, urbanization level, scale of urban construction, infrastructure investment levels, and the consumption and income of residents are the primary factors driving future demand for transportation land. Development plans formulated by central and local governments in China typically outline specific values for these key socio-economic factors. Almost every city presents expected economic and population sizes for the medium to long term within their national economic and social development plans, providing crucial data for forecasting future demand for transportation land. This study considers the factors affecting transportation land demand identified in previous research and refines and synthesizes these factors through meta-analysis. Compared to existing studies, the framework of influencing factors developed in this study is more applicable and comprehensive [5]. Furthermore, the model constructed in this study incorporates the constraints of future socio-economic development objectives, ensuring that the forecasted area of land for transportation aligns with future socio-economic goals and is deemed reasonable.

The results of forecasting regional urban transportation land demand can elucidate the spatial and temporal patterns of transportation development changes and provide essential guidelines for developing targeted land supply plans for the region. Although China is actively constructing a comprehensive three-dimensional transportation system in the YREB, this study reveals significant spatial disparities in the demand for transportation land across the region's cities, aligning with the phased patterns of transportation development in China [66]. Initially, when transportation infrastructure is underdeveloped, the regional demand for transportation land is substantial, aiming primarily to expand the land allocation area. As transportation land expands, regional transportation development gradually matures, shifting the focus from increasing land supply to enhancing land use efficiency. For instance, core cities or major transportation hubs in the YREB have largely completed their comprehensive transportation networks, and their demand for transportation land is gradually decreasing as they shift towards improving land use efficiency. However, most cities in the YREB, characterized by limited transportation facilities and a single mode of transport, are experiencing a significant surge in transportation land demand driven by regional integration initiatives. Therefore, the allocation of regional transportation land should be tailored to the specific stages of urban transportation development, implementing differentiated land supply strategies. However, promoting integrated regional transportation development does not imply a large-scale supply of transportation land to areas with underdeveloped transportation infrastructure; instead, it requires ensuring that the transportation land supply plans can accommodate future development needs. The excessive supply of transportation land can lead to large tracts of unused land, thereby diminishing land use efficiency.

The alignment of supply and demand for regional transportation land constitutes the foundation for decision-making in developing future land supply strategies. While historical land supply data can serve as a reference for transportation land supply planning, it is crucial to thoroughly assess whether past supply levels align with future land demand. Therefore, analyzing the balance between supply and demand and identifying gaps in regional transportation land availability are critical steps for policymakers in developing future land supply strategies. By comparing the matching levels of transportation land supply and demand across the region, it becomes clearer which cities need to adjust their transportation land supply strategies. In cities with a high historical land supply but low future demand, this suggests that the city's transportation infrastructure is gradually completing, and the area allocated for transportation land should be carefully controlled in the future to prevent excess supply. Conversely, in cities with a low historical land supply but high future demand, it indicates that transportation development is accelerating, and the area dedicated to transportation land should be increased to ensure that the supply meets the upcoming needs for transportation infrastructure construction.

4.2. Limitations and Prospects

Transportation land demand forecasting is a multifaceted study that necessitates a comprehensive consideration of socio-economic development and policy orientation, which leads to inevitable uncertainties in this study. To reduce the uncertainty associated with forecasting, this study presents a robust forecasting method for urban transportation land demand at the regional level, which enables decision-makers to identify areas with urgent transportation land demand and those where supply should be controlled. This study conducts transportation land demand forecasting for cities in the YREB using a unified framework that comprehensively considers the impact of socio-economic factors on land demand. However, the impact of policy factors on urban transportation land demand was not considered, which could lead to discrepancies between forecasted and actual demands. Typically, shifts in policy orientation can cause transportation land demand to diverge from normal socio-economic development patterns. For instance, increased government support for the development of regions with underdeveloped transportation infrastructure can lead to a short-term surge in transportation land demand. Such forecasts are subject to potential uncertainties due to policy changes. Guided by the principles of coordinated and integrated development, the YREB is actively constructing a comprehensive transportation system. We assume that local policy orientation within the YREB will remain relatively stable during the study period, with policy factors having a pervasive and significant influence across the region. This leads us to conclude that the primary determinant of urban transportation land demand is the region's own development needs. Future research should further investigate the impact of policy changes on transportation land demand.

It is crucial to note that the forecasting framework in this study ensures high accuracy for short- and medium-term transportation land demand forecasts. However, the model's uncertainty increases when applied to long-term land use forecasting (beyond ten years). Despite this, short- and medium-term forecasts provide practical guidance for development planning, while long-term planning focuses more on envisioning the future. Since this study accounts for the common influencing factors affecting transportation land demand and demonstrates robustness in short- and medium-term forecasts, the forecasting framework developed here can be applied to other regions in China, such as the Beijing–Tianjin–Hebei region, the Yellow River High-Quality Development Area, and the Pearl River Delta urban agglomeration. However, for practical applications, it is advisable for decision-makers to thoroughly consider the local development context and policy orientations to enhance the accuracy of the forecasting framework. Moreover, this study's forecast of urban transportation land was limited to considering each city's own future development, without including spatial influences from the model. Future research should explore the spatial interactions between cities and their neighboring or central counterparts, integrating these spatial impacts into the forecasting model to enhance its interpretability.

5. Conclusions

Clarifying land demand is crucial for the optimal allocation of regional land resources. In China, transportation land has been the largest category of construction land supplied over the last decade. Consequently, the accurate forecasting of transportation land demand is essential for developing effective future land supply strategies. This study presents a comprehensive framework for forecasting regional urban transportation land by employing a literature meta-analysis and statistical forecasting models and aligning with regional development goals, supporting the sustainable development of transportation infrastructure in the region.

The main achievement of this paper is the development of a comprehensive framework for forecasting transportation land demand at the regional level, which significantly guides the optimal allocation of regional transportation resources. This study finds that socio-economic factors, including economic development level, population size, urbanization level, urban construction scale, industrial structure, industrial development level, investment level, and resident income and consumption levels, predominantly influence transportation land demand. By 2025, cities with large-scale transportation land in the YREB will still be mainly concentrated in the lower reaches of the Yangtze River national urban agglomeration, and most of these cities are municipalities directly under the central government, provincial capitals, or core cities of regional development. However, transportation land demand in these cities is decelerating, and optimizing land use structure is becoming critical for enhancing land use efficiency. A high demand for transportation land is increasingly found in the middle and upper reaches of the Yangtze River, indicating a shift in the focus of transportation infrastructure development from the lower to the middle and upper reaches. The forecasting results of this study align with the goals of comprehensive transportation development planning set by government bodies, demonstrating the feasibility and applicability of the proposed forecasting framework. Cities with significant transportation land demand over the next five years are identified as major centers for transportation development in the YREB.

Despite general stability in the supply and demand levels for transportation land across most YREB cities, mismatches remain in some areas. Cities with historically high land supply but low current demand are predominantly distributed in the middle reaches of the Yangtze River. These cities have significantly increased their investment in transportation infrastructure over the past decade. However, the decreasing demand for land reminds these cities to control transportation land supply and prevent resource wastage by avoiding idle land. Conversely, cities with historically low land supply but high current demand are found in the upper reaches of the Yangtze River, typically distant from regional development cores and with slower transportation growth. As regional integration and development accelerate, enhancing transportation infrastructure in these cities is becoming crucial, and the area allocated for transportation land should be appropriately increased in the future.

Author Contributions: Conceptualization, L.W. and K.W.; methodology, J.Z. and K.W.; software, K.W.; validation, K.W. and L.W.; formal analysis, K.W.; data curation, K.W.; writing—original draft preparation, K.W.; writing—review and editing, K.W., J.Z. and L.W.; visualization, K.W. and L.W; supervision, L.W. and J.Z.; project administration, K.W.; funding acquisition, K.W. All authors have read and agreed to the published version of the manuscript.

Funding: This research was funded by the China Postdoctoral Science Foundation (Grant Number: 2023M733579) and the National Natural Science Foundation of China (Grant Number: 42301340).

Data Availability Statement: The data presented in this study are available upon request from the corresponding author. The data are not publicly available due to privacy.

Acknowledgments: The authors would like to thank the anonymous reviewers for their comments and suggestions.

Conflicts of Interest: The authors declare no conflicts of interest.

References

1. Chen, G.; Li, X.; Liu, X.; Chen, Y.; Liang, X.; Leng, J.; Xu, X.; Liao, W.; Qiu, Y.a.; Wu, Q.; et al. Global projections of future urban land expansion under shared socioeconomic pathways. *Nat. Commun.* **2020**, *11*, 537. [CrossRef] [PubMed]
2. Schuckmann, S.W.; Gnatzy, T.; Darkow, I.-L.; von der Gracht, H.A. Analysis of factors influencing the development of transport infrastructure until the year 2030—A Delphi based scenario study. *Technol. Forecast. Soc. Chang.* **2012**, *79*, 1373–1387. [CrossRef]
3. Laurance, W.F.; Clements, G.R.; Sloan, S.; O'Connell, C.S.; Mueller, N.D.; Goosem, M.; Venter, O.; Edwards, D.P.; Phalan, B.; Balmford, A.; et al. A global strategy for road building. *Nature* **2014**, *513*, 229–232. [CrossRef] [PubMed]
4. Dulac, J. *Global Land Transport Infrastructure Requirements: Estimating Road and Railway Infrastructure Capacity and Costs to 2050*; International Energy Agency: Paris, France, 2013.
5. Zeng, L.; Li, H.; Wang, X.; Yu, Z.; Hu, H.; Yuan, X.; Zhao, X.; Li, C.; Yuan, D.; Gao, Y. China's transport land: Spatiotemporal expansion characteristics and driving mechanism. *Land* **2022**, *11*, 1147. [CrossRef]
6. Li, C.; Wu, K.; Gao, X. Manufacturing industry agglomeration and spatial clustering: Evidence from Hebei Province, China. *Environ. Dev. Sustain.* **2020**, *22*, 2941–2965. [CrossRef]
7. Huang, Z.; Du, X. Government intervention and land misallocation: Evidence from China. *Cities* **2017**, *60*, 323–332. [CrossRef]
8. Ewing, R.; Bartholomew, K. Comparing Land Use Forecasting Methods: Expert Panel versus Spatial Interaction Model. *J. Am. Plan. Assoc.* **2009**, *75*, 343–357. [CrossRef]
9. Han, H.; Yang, C.; Song, J. Scenario Simulation and the Prediction of Land Use and Land Cover Change in Beijing, China. *Sustainability* **2015**, *7*, 4260–4279. [CrossRef]
10. Li, B.; Cao, X.; Xu, J.; Wang, W.; Ouyang, S.; Liu, D. Spatial–Temporal Pattern and Influence Factors of Land Used for Transportation at the County Level since the Implementation of the Reform and Opening-Up Policy in China. *Land* **2021**, *10*, 833. [CrossRef]
11. Aljoufie, M. Toward an effective analysis of public transportation demand factors in car-dependent cities: Case of Makkah City, Saudi Arabia. *Urban Plan. Transp. Res.* **2021**, *9*, 257–275. [CrossRef]
12. Ma, L.; Liang, J.; Gao, D.; Sun, J.; Li, Z. The Future Demand of Transportation in China: 2030 Scenario based on a Hybrid Model. *Procedia—Soc. Behav. Sci.* **2012**, *54*, 428–437. [CrossRef]
13. Magazzino, C.; Mele, M. On the relationship between transportation infrastructure and economic development in China. *Res. Transp. Econ.* **2020**, *88*, 100947. [CrossRef]
14. Pradhan, R.P.; Bagchi, T.P. Effect of transportation infrastructure on economic growth in India: The VECM approach. *Res. Transp. Econ.* **2013**, *38*, 139–148. [CrossRef]
15. Banerjee, A.; Duflo, E.; Qian, N. On the road: Access to transportation infrastructure and economic growth in China. *J. Dev. Econ.* **2020**, *145*, 102442. [CrossRef]
16. Du, Q.; Wang, X.; Li, Y.; Zou, P.X.W.; Han, X.; Meng, M. An analysis of coupling coordination relationship between regional economy and transportation: Empirical evidence from China. *Environ. Sci. Pollut. Res.* **2022**, *29*, 34360–34378. [CrossRef]
17. Liddle, B. Long-run relationship among transport demand, income, and gasoline price for the US. *Transp. Res. Part D Transp. Environ.* **2009**, *14*, 73–82. [CrossRef]
18. Kim, E. Determinants of Optimal Level of Transportation Infrastructure. *J. Urban Plan. Dev.* **2002**, *128*, 150–163. [CrossRef]
19. Fisch-Romito, V.; Guivarch, C. Transportation infrastructures in a low carbon world: An evaluation of investment needs and their determinants. *Transp. Res. Part D Transp. Environ.* **2019**, *72*, 203–219. [CrossRef]
20. Feng, X.; Sun, Q.; Qian, K.; Liu, M. Cointegration Relationship of Regional Integrated Transport Demand and Industrial Structure. *J. Transp. Syst. Eng. Inf. Technol.* **2012**, *12*, 10–16. [CrossRef]
21. Albalate, D.; Fageda, X. High-Technology Employment and Transportation: Evidence from the European Regions. *Reg. Stud.* **2016**, *50*, 1564–1578. [CrossRef]
22. Maparu, T.S.; Mazumder, T.N. Transport infrastructure, economic development and urbanization in India (1990–2011): Is there any causal relationship? *Transp. Res. Part A Policy Pract.* **2017**, *100*, 319–336. [CrossRef]
23. Yang, Z.; Jia, P.; Liu, W.; Yin, H. Car ownership and urban development in Chinese cities: A panel data analysis. *J. Transp. Geogr.* **2017**, *58*, 127–134. [CrossRef]
24. Asher, S.; Garg, T.; Novosad, P. The Ecological Impact of Transportation Infrastructure. *Econ. J.* **2020**, *130*, 1173–1199. [CrossRef]
25. Xi, F.; He, H.S.; Hu, Y.; Bu, R.; Chang, Y.; Wu, X.; Liu, M.; Shi, T. Simulating the impacts of ecological protection policies on urban land use sustainability in Shenyang-Fushun, China. *Int. J. Urban Sustain. Dev.* **2009**, *1*, 111–127. [CrossRef]
26. Mishalani, R.G.; Goel, P.K.; Westra, A.M.; Landgraf, A.J. Modeling the relationships among urban passenger travel carbon dioxide emissions, transportation demand and supply, population density, and proxy policy variables. *Transp. Res. Part D Transp. Environ.* **2014**, *33*, 146–154. [CrossRef]
27. Xiong, Y.; Chen, Y.; Peng, F.; Li, J.; Yan, X. Analog simulation of urban construction land supply and demand in Chang-Zhu-Tan Urban Agglomeration based on land intensive use. *J. Geogr. Sci.* **2019**, *29*, 1346–1362. [CrossRef]

28. Plakandaras, V.; Papadimitriou, T.; Gogas, P. Forecasting transportation demand for the U.S. market. *Transp. Res. Part A Policy Pract.* **2019**, *126*, 195–214. [CrossRef]
29. Benjamin, J. A time-series forecast of average daily traffic volume. *Transp. Res. Part A Gen.* **1986**, *20*, 51–60. [CrossRef]
30. Mrówczyńska, B.; Łachacz, K.; Haniszewski, T.; Sładkowski, A. A comparison of forecasting the results of road transportation needs. *Transport* **2012**, *27*, 73–78. [CrossRef]
31. Li, Z.; Hensher, D.A. Toll Roads in Australia: An Overview of Characteristics and Accuracy of Demand Forecasts. *Transp. Rev.* **2010**, *30*, 541–569. [CrossRef]
32. Li, C.; Gao, X.; Wu, J.; Wu, K. Demand prediction and regulation zoning of urban-industrial land: Evidence from Beijing-Tianjin-Hebei Urban Agglomeration, China. *Environ. Monit. Assess.* **2019**, *191*, 412. [CrossRef] [PubMed]
33. Batista e Silva, F.; Koomen, E.; Diogo, V.; Lavalle, C. Estimating Demand for Industrial and Commercial Land Use Given Economic Forecasts. *PLoS ONE* **2014**, *9*, e91991. [CrossRef] [PubMed]
34. Shi, Y.; Li, M.; Fu, Y.; Wang, L.; Sun, M.; Hao, J. Multi-scenario traffic land demand forecasting based on grey system-BP neural network model: A case study of urban agglomeration in the middle reaches of the Yangtze River. *J. China Agric. Univ* **2020**, *25*, 142–153. [CrossRef]
35. Jun, M.-J. Forecasting Urban Land-Use Demand Using a Metropolitan Input-Output Model. *Environ. Plan. A Econ. Space* **2005**, *37*, 1311–1328. [CrossRef]
36. Xiao, J.; Shen, Y.; Ge, J.; Tateishi, R.; Tang, C.; Liang, Y.; Huang, Z. Evaluating urban expansion and land use change in Shijiazhuang, China, by using GIS and remote sensing. *Landsc. Urban Plan.* **2006**, *75*, 69–80. [CrossRef]
37. Pastorino, M.; Gallo, F.; Di Febbraro, A.; Moser, G.; Sacco, N.; Serpico, S.B. Multimodal Fusion of Mobility Demand Data and Remote Sensing Imagery for Urban Land-Use and Land-Cover Mapping. *Remote Sens.* **2022**, *14*, 3370. [CrossRef]
38. Hoymann, J. Quantifying demand for built-up area—A comparison of approaches and application to regions with stagnating population. *J. Land Use Sci.* **2012**, *7*, 67–87. [CrossRef]
39. Jiang, H.; Guo, H.; Sun, Z.; Xing, Q.; Zhang, H.; Ma, Y.; Li, S. Projections of urban built-up area expansion and urbanization sustainability in China's cities through 2030. *J. Clean. Prod.* **2022**, *367*, 133086. [CrossRef]
40. Erb, K.-H. Actual land demand of Austria 1926–2000: A variation on Ecological Footprint assessments. *Land Use Policy* **2004**, *21*, 247–259. [CrossRef]
41. Jackson, L.E.; Bird, S.L.; Matheny, R.W.; O'Neill, R.V.; White, D.; Boesch, K.C.; Koviach, J.L. A Regional Approach to Projecting Land-Use Change and Resulting Ecological Vulnerability. *Environ. Monit. Assess.* **2004**, *94*, 231–248. [CrossRef]
42. Lee, C.; Lee, S. Analyzing spatiotemporal land use change using an urban growth model based on multilevel logistic regression and future land demand scenarios. *Appl. Geogr.* **2023**, *160*, 103099. [CrossRef]
43. Ustaoglu, E.; Batista e Silva, F.; Lavalle, C. Quantifying and modelling industrial and commercial land-use demand in France. *Environ. Dev. Sustain.* **2020**, *22*, 519–549. [CrossRef]
44. Wang, L.; Wang, K.; Zhang, J.; Zhang, D.; Wu, X.; Zhang, L. Multiple objective oriented land supply for sustainable transportation: A perspective from industrial dependence, dominance and restrictions of 127 cities in the Yangtze River Economic Belt of China. *Land Use Policy* **2020**, *99*, 105069. [CrossRef]
45. Sun, H.; Chen, Y.Y.; Fan, Z.Q. Study the Residential Land Demand by Ridge Regression and Multiple Linear Regression. *Key Eng. Mater.* **2011**, *467–469*, 1250–1255. [CrossRef]
46. Samardžić-Petrović, M.; Kovačević, M.; Bajat, B.; Dragićević, S. Machine Learning Techniques for Modelling Short Term Land-Use Change. *ISPRS Int. J. Geo-Inf.* **2017**, *6*, 387. [CrossRef]
47. Pijanowski, B.C.; Brown, D.G.; Shellito, B.A.; Manik, G.A. Using neural networks and GIS to forecast land use changes: A Land Transformation Model. *Comput. Environ. Urban Syst.* **2002**, *26*, 553–575. [CrossRef]
48. Aburas, M.M.; Ahamad, M.S.S.; Omar, N.Q. Spatio-temporal simulation and prediction of land-use change using conventional and machine learning models: A review. *Environ. Monit. Assess.* **2019**, *191*, 205. [CrossRef] [PubMed]
49. Rogan, J.; Chen, D. Remote sensing technology for mapping and monitoring land-cover and land-use change. *Prog. Plan.* **2004**, *61*, 301–325. [CrossRef]
50. Hu, X.; Li, Z.; Cai, Y.; Wu, F. Urban construction land demand prediction and spatial pattern simulation under carbon peak and neutrality goals: A case study of Guangzhou, China. *J. Geogr. Sci.* **2022**, *32*, 2251–2270. [CrossRef]
51. Almeida, C.M.D.; Monteiro, A.M.V.; Câmara, G.; Soares-Filho, B.S.; Cerqueira, G.C.; Pennachin, C.L.; Batty, M. GIS and remote sensing as tools for the simulation of urban land-use change. *Int. J. Remote Sens.* **2005**, *26*, 759–774. [CrossRef]
52. Warth, G.; Braun, A.; Assmann, O.; Fleckenstein, K.; Hochschild, V. Prediction of Socio-Economic Indicators for Urban Planning Using VHR Satellite Imagery and Spatial Analysis. *Remote Sens.* **2020**, *12*, 1730. [CrossRef]
53. Needham, B.; Louw, E.; Metzemakers, P. An economic theory for industrial land policy. *Land Use Policy* **2013**, *33*, 227–234. [CrossRef]
54. Zhou, L.; Tian, L.; Cao, Y.; Yang, L. Industrial land supply at different technological intensities and its contribution to economic growth in China: A case study of the Beijing-Tianjin-Hebei region. *Land Use Policy* **2021**, *101*, 105087. [CrossRef]
55. Miller, P.; de Barros, A.G.; Kattan, L.; Wirasinghe, S.C. Public transportation and sustainability: A review. *KSCE J. Civ. Eng.* **2016**, *20*, 1076–1083. [CrossRef]
56. Chen, Y.; Zhang, S.; Huang, D.; Li, B.-L.; Liu, J.; Liu, W.; Ma, J.; Wang, F.; Wang, Y.; Wu, S. The development of China's Yangtze River Economic Belt: How to make it in a green way. *Sci. Bull.* **2017**, *62*, 648–651. [CrossRef] [PubMed]

57. Moher, D.; Liberati, A.; Tetzlaff, J.; Altman, D.G. Preferred Reporting Items for Systematic Reviews and Meta-Analyses: The PRISMA Statement. *Ann. Intern. Med.* **2009**, *151*, 264–269. [CrossRef] [PubMed]
58. Matthew, J.P.; Joanne, E.M.; Patrick, M.B.; Isabelle, B.; Tammy, C.H.; Cynthia, D.M.; Larissa, S.; Jennifer, M.T.; Elie, A.A.; Sue, E.B.; et al. The PRISMA 2020 statement: An updated guideline for reporting systematic reviews. *BMJ* **2021**, *372*, n71. [CrossRef] [PubMed]
59. Wong, F.S. Time series forecasting using backpropagation neural networks. *Neurocomputing* **1991**, *2*, 147–159. [CrossRef]
60. Guan, Q.; Wang, L.; Clarke, K.C. An Artificial-Neural-Network-based, Constrained CA Model for Simulating Urban Growth. *Cartogr. Geogr. Inf. Sci.* **2005**, *32*, 369–380. [CrossRef]
61. Ju-Long, D. Control problems of grey systems. *Syst. Control. Lett.* **1982**, *1*, 288–294. [CrossRef]
62. Ding, C. Land policy reform in China: Assessment and prospects. *Land Use Policy* **2003**, *20*, 109–120. [CrossRef]
63. Zander, P.; Kächele, H. Modelling multiple objectives of land use for sustainable development. *Agric. Syst.* **1999**, *59*, 311–325. [CrossRef]
64. Mao, X.; Huang, X.; Song, Y.; Zhu, Y.; Tan, Q. Response to urban land scarcity in growing megacities: Urban containment or inter-city connection? *Cities* **2020**, *96*, 102399. [CrossRef]
65. Ren, Y.; Tian, Y.; Xiao, X. Spatial effects of transportation infrastructure on the development of urban agglomeration integration: Evidence from the Yangtze River Economic Belt. *J. Transp. Geogr.* **2022**, *104*, 103431. [CrossRef]
66. Jin, F.; Ding, J.; Wang, J.E.; Liu, D.; Wang, C. Transportation development transition in China. *Chin. Geogr. Sci.* **2012**, *22*, 319–333. [CrossRef]

land

MDPI

Article

Machine Learning for Criteria Weighting in GIS-Based Multi-Criteria Evaluation: A Case Study of Urban Suitability Analysis

Lan Qing Zhao, Alysha van Duynhoven and Suzana Dragićević *

Spatial Analysis and Modeling Laboratory, Department of Geography, Simon Fraser University, 8888 University Drive, Burnaby, BC V5A 1S6, Canada; lan_qing_zhao@sfu.ca (L.Q.Z.); alyshav@sfu.ca (A.v.D.)
* Correspondence: suzanad@sfu.ca

Abstract: Geographic Information System-based Multi-Criteria Evaluation (GIS-MCE) methods are designed to assist in various spatial decision-making problems using spatial data. Deriving criteria weights is an important component of GIS-MCE, typically relying on stakeholders' opinions or mathematical methods. These approaches can be costly, time-consuming, and prone to subjectivity or bias. Therefore, the main objective of this study is to investigate the use of Machine Learning (ML) techniques to support criteria weight derivation within GIS-MCE. The proposed ML-MCE method is explored in a case study of urban development suitability analysis of the City of Kelowna, Canada. Feature importance values drawn from three ML techniques–Random Forest (RF), Extreme Gradient Boosting (XGB), and Support Vector Machine (SVM)–are used to derive criteria weights. The suitability scores obtained using the ML-MCE methodology are compared with Equal-Weights (EW) and the Analytical Hierarchy Process (AHP) approach for criteria weighting. The results indicate that ML-derived criteria weights can be used in GIS-MCE, where RF and XGB techniques provide more similar values for criteria weights than those derived from SVM. The similarities and differences are confirmed with Kappa indices obtained from comparing pairs of suitability maps. The proposed new ML-MCE methodology can support various decision-making processes in urban land-use planning.

Keywords: multi-criteria evaluation (MCE); machine learning (ML); geographic information science (GIS); random forest (RF); extreme gradient boosting (XGB); support vector machine (SVM); urban suitability analysis

1. Introduction

The Multi-Criteria Evaluation (MCE) method is often used in decision-making for solving problems related to suitability analysis where multiple criteria, often conflicting, are considered simultaneously [1,2]. When integrated with Geographic Information Systems (GIS), MCE can be used with geospatial data to address a wide range of spatial problems, such as land suitability analysis [3], environmental management [4], ecological capacity evaluations [5], or disaster risk assessment [6], to name a few. This integration also supports informed decision-making related to urban land-use developments by facilitating structured analysis of complex spatial problems [7]. By including several criteria, such as land-use zoning, environmental quality, transportation accessibility, or demographics, GIS-MCE can be used to obtain the suitability of locations for various land-uses, thereby aiding urban planning processes. Studies have demonstrated GIS-MCE in urban land-use suitability analysis, with effective applications for identifying locations of new industrial areas [8], green spaces [9], managing urban developments [10,11], and planning infrastructure placement [12].

A GIS-MCE procedure generally consists of six main steps: (1) defining the spatial decision problem; (2) selecting criteria; (3) collecting relevant spatial data; (4) designing and applying suitability functions to normalize criteria data layers; (5) assigning criteria weights to reflect their relative importance in the overall decision-making problem; and (6) calculating the aggregated suitability scores that can be then presented as a suitability map. One of the key steps of GIS-MCE is eliciting criteria weights because they directly influence the output suitability scores [13,14]. Criteria weights in GIS-MCE are typically determined using opinions from experts and stakeholders or via mathematical approaches such as the Analytical Hierarchy Process (AHP) [15], the Technique for Order of Preference by Similarity to Ideal Solution (TOPSIS) [16], or the Entropy Method [17]. Despite the usefulness of GIS-MCE, determining criteria weights remains challenging. For example, obtaining expert and stakeholder opinions is costly and time-consuming [18]. In contrast, mathematical approaches offer a more structured and efficient way to derive criteria weights. However, these methods are partially dependent on subjective opinions of spatial analysts, stakeholders, or decision-makers that consequently introduce biases [19].

Proposed by Saaty in 1977, the AHP approach is the most common MCE criteria weighting technique used in GIS software such as TerrSet [20] and ArcMap [21]. The AHP involves making pairwise comparisons of all criteria to establish the hierarchy of criteria. As such, the derived weights reflect the relative importance of each criterion to the suitability problem. For example, urban development suitability might consider criteria such as land-uses, demographics, or transportation accessibility. The decision-maker can determine the weights for these criteria through pairwise comparisons, thereby quantifying the relative importance of each criterion. However, the AHP approach relies on subjective judgments from those who establish the relative importance of the criteria. These judgments can vary among individuals and may be influenced by biases or incomplete information, potentially leading to inconsistent or unreliable MCE suitability outputs [22]. Furthermore, the pairwise comparison process inherent to AHP becomes more complex and difficult to manage when larger numbers of criteria are considered [23]. To manage these challenges, the potential of Machine Learning (ML) techniques was recently explored due to their capacity to handle and analyze large amounts of data [24].

In geospatial applications, ML is primarily used to classify Remote Sensing (RS) or geospatial data [25,26] or to estimate new states of geospatial phenomena [27,28]. For example, ML was applied in various geospatial applications such as estimating habitat [29] or agricultural suitability [30], guiding infrastructure placement [31,32], predicting urban growth suitability [33], or calculating flood susceptibility [34,35]. These studies directly used values outputted by ML techniques to establish suitability or susceptibility scores. Rather than using direct outputs of ML techniques, other recent studies have explored leveraging feature importance analyses to derive criteria weights in GIS-MCE methods. For instance, Singh et al. [36] used the feature importance analyses associated with a Random Forest (RF) classifier to derive criteria weights in a GIS-MCE method in an agroforestry suitability assessment. In that study, the relative importance of each factor involved in training the classifier is used to derive factor significance directly from available geospatial data. The relative importance values were standardized to become criteria weights for a traditional GIS-MCE method. While this study has integrated ML techniques with GIS-MCE, there is a need to further investigate the capacity of different ML techniques for deriving criteria weights, which remains unexplored in different spatial contexts, including urban land-use change and residential development suitability analyses.

For these reasons, the primary objectives of this research are to: (1) present a novel methodology involving three ML techniques–Random Forest (RF), Extreme Gradient Boosting (XGB), and Support Vector Machines (SVMs)–to derive criteria weights for GIS-MCE to assist in an urban suitability decision problem and (2) compare the resulting suitability scores of the ML-MCE method with those obtained using neutral Equal-Weights (EW) and AHP approaches. The proposed methodology is applied to real-world datasets for the City of Kelowna, British Columbia (B.C.), Canada. The main innovation of this research is the use of ML techniques to generate criteria weights for GIS-MCE in urban suitability analysis. The presented methodology can be applied in urban planning and land use management.

2. Methodology

An overview of the proposed methodology is shown in Figure 1. This study addresses the problem of finding suitable locations for new urban developments in the City of Kelowna, Canada. To begin the MCE analysis, spatial criteria are selected, necessary geospatial data are collected, and criteria layers are derived. Next, criteria layers are normalized using developed suitability functions before spatial criteria can be integrated. The selection of criteria and the specification of suitability functions are based on scientific literature, as it was not possible to involve stakeholders and experts in this study. Criteria weights are derived first using a neutral approach, such as EW, then the AHP approach. These two sets of criteria weights are compared with those obtained through the proposed ML-MCE method, which leverages the feature importance assessments of three ML techniques (RF, XGB, and SVM). ML techniques were trained under two different regimes to highlight the effects of training data configuration on derived criteria weights. The derived weights are then aggregated via the Weighted Linear Combination (WLC) method to compute suitability scores for each approach. The most common aggregation method in GIS-MCE, the WLC method, involves multiplying each normalized criterion layer by its assigned weight and summing the results to obtain a final suitability score for each raster cell in the study area [37]. Finally, the resulting suitability scores are classified into four suitability classes, and suitability maps are generated for comparisons and as a starting point for a decision-making process.

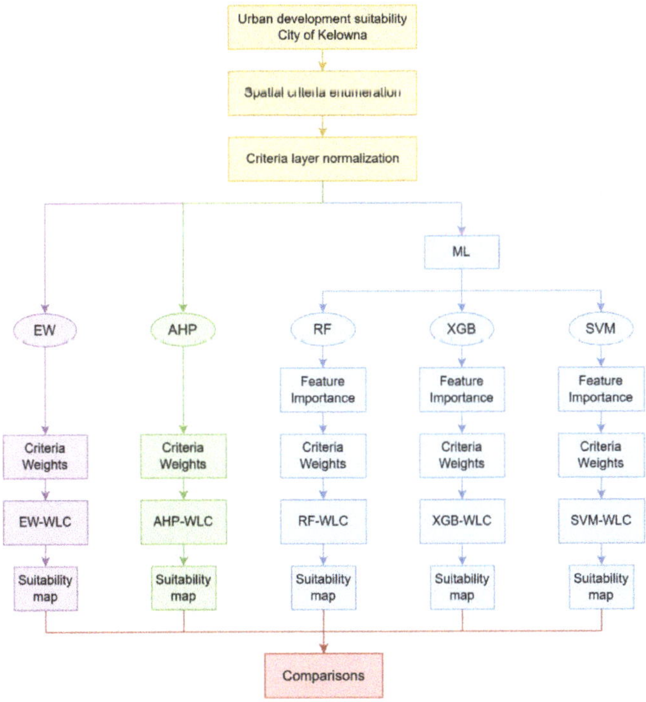

Figure 1. General overview of the proposed methodology.

2.1. Study Area and Datasets

The chosen study area for this research is the City of Kelowna, B.C., Canada (Figure 2), due to its fast urban development. The City of Kelowna is situated along the eastern shore of Okanagan Lake and is surrounded by a diverse landscape that includes urban areas, agricultural lands, forests, waterbodies, and First Nation land. The city experienced significant population growth of 14% between 2016 and 2021, making it one of the fastest-growing cities in Canada [38]. As a result, the municipality faces significant challenges related to new urban developments due to its surrounding natural environment and agricultural lands.

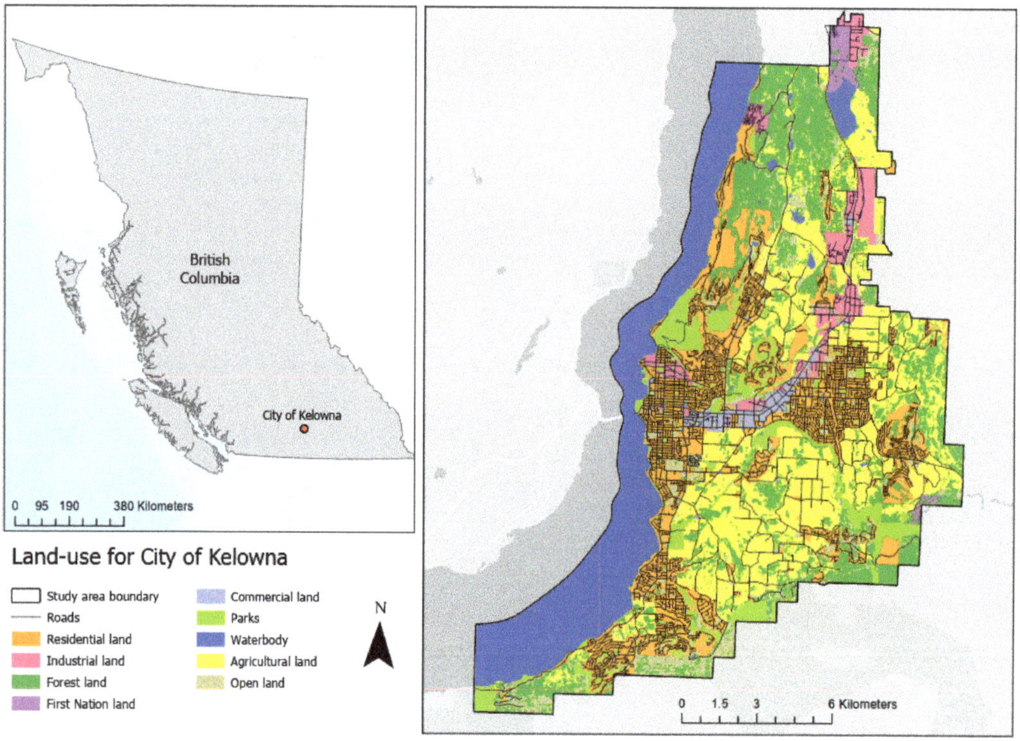

Figure 2. Study area: City of Kelowna, with major land-uses for the year 2015.

The geospatial data collected for this research are obtained from various sources, including City of Kelowna Open Data [39], B.C. Open Data [40], and Open Government Canada [41–43]. This comprehensive dataset encompasses information on road networks, zoning, elevation, topography, infrastructure distribution, demographic statistics for the year 2016, as well as land-use and land cover (LULC) data. The LULC data for the years 2015 and 2020 are used exclusively for the needs of training and testing ML techniques. All geospatial data are rasterized to a 30m spatial resolution and are set in the Transverse Mercator projection with the NAD 1983 UTM Zone 11 N coordinate system.

2.2. Selection of Criteria and Suitability Functions

Eleven criteria presented in Table 1 are selected as the most important factors contributing to new urban development in general and within the City of Kelowna. Using the Euclidean distance method [44], proximity to roads, commercial areas, residential areas, industrial areas, parks, waterbodies, forest, and agricultural lands are derived from the 2015 LULC data. Next, the "Elevation" and "Slope" layers are obtained from a Digital Elevation Model (DEM). Lastly, "Proximity to high population density" is derived from a population density data layer. Specifically, population densities are calculated from the 2016 Canada Census data at the Dissemination Area (DA) level using the dasymetric mapping method [45,46], which refines population distribution within each DA to residential land-use types. Based on the calculated average population density, each raster cell has been assigned a value of 1 if the population density is higher than the average and a value of 0 if it is lower than the average. In addition to the eleven criteria, there are key constraint areas where new urban development is not expected to occur that are excluded from the suitability analysis, along with existing urban areas, road networks, waterbodies, First Nation land, agricultural land reserves, and parks.

Table 1. List of the criteria with respective suitability functions in breakpoint notation and the rationales for the urban development suitability analysis.

Criteria	Suitability Functions and Units	Rationale
Proximity to roads	Crit (Proximity to roads) = {(30,0), (30,1), (120,1), (900,0)} [m]	Within 30 m of roads poses safety risks. The 30 m to 120 m zone balances reduced negative impacts with good road access. Beyond 900 m from roads is considered inaccessible [47].
Proximity to commercial	Crit (Proximity to commercial) = {(600,1), (3600,0)} [m]	Commercial services need to be accessible and within easy walking distance (600 m). Suitability decreases beyond this distance until 3600 m, which is considered beyond a walkable distance [48]
Proximity to residential	Crit (Proximity to residential) = {(300,1), (900,0)} [m]	New residential development typically happens at locations within 300 m of existing residential areas, with isolated areas being less preferred for new housing. Beyond 900 m is considered too far [49].
Proximity to industrial	Crit (Proximity to industrial) = {(600,0), (3600,1)} [m]	Residential developments are generally planned at least 600 m away from industrial zones to minimize exposure to pollution and noise. It is considered that at 3600 m, industrial areas are beyond a walkable distance [50]
Proximity to parks	Crit (Proximity to parks) = {(300,1), (1500,0)} [m]	To ensure parks are in short walking distance from new housing, the breakpoints are chosen from the city's current planning [51]
Proximity to waterbodies	Crit (Proximity to waterbodies) = {(30,0), (30,1), (600,1), (3000,0)} [m]	Within 30 m of waterbodies poses safety risks. A distance between 30 m and 600 m is considered appropriate for enjoying scenery and leisure activities [52]. The maximum distance from waterbodies within the study area is 3000 m.
Proximity to forest	Crit (Proximity to forest) = {(0,0), (300,1)} [m]	Residential developments should maintain more than 300 m from forests to ensure forest and biodiversity conservation [53].
Proximity to agricultural	Crit (Proximity to agricultural) = {(30,0), (900,1)} [m]	Agricultural land in the City of Kelowna is protected and preserved. New residential development should start at least from 30m which is an equivalent to the width size of a minor road [50].
Elevation	Crit (Elevation) = {(500,1), (860,0)} [m]	The minimum elevation in Kelowna is 335 m and the maximum is 860 m. The suitable range for residential development is usually within 100 m to 200 m of the minimum elevation. It is difficult to construct residential development at higher elevations [54].
Slope	Crit (Slope) = {(20,1), (40,0)} [degree]	Slopes less than 20 degrees are considered suitable for new developments, while slopes beyond 40 degrees are considered too steep and unsafe for housing developments [55].
Proximity to high population density	Crit (Proximity to high population density) = {(600,1), (3000,0)} [m]	Residential area development tends to occur near existing high-density population zones. Typically, areas distant from these centers are less likely to experience new development due to lack of infrastructure and services [56].

After criteria data layers are prepared, suitability functions are developed and applied to normalize each criterion. The suitability functions in breakpoint notation and their rationale are presented in Table 1. The choice of suitability function breaks is determined based on information obtained from City of Kelowna policies and scientific literature. For example, the criterion "Proximity to roads" is represented by the suitability function {(30,0), (30,1), (120,1), (900,0)} [m]. This indicates that the suitability value remains at a minimum of 0 when locations lie within 0 to 30 m from roads. Between 30 m and 120 m, the suitability value linearly increases to reach a maximum value of 1. From 120 m to 900 m, suitability declines linearly and remains 0 at locations beyond 900 m from existing roads. A value of 1 represents full satisfaction while a value of 0 represents no satisfaction. Each criteria data layer is normalized with a corresponding suitability function such that all the values range from 0 to 1 (Figure 3). The normalized data layers are then weighted and aggregated to calculate the overall suitability scores for each raster cell.

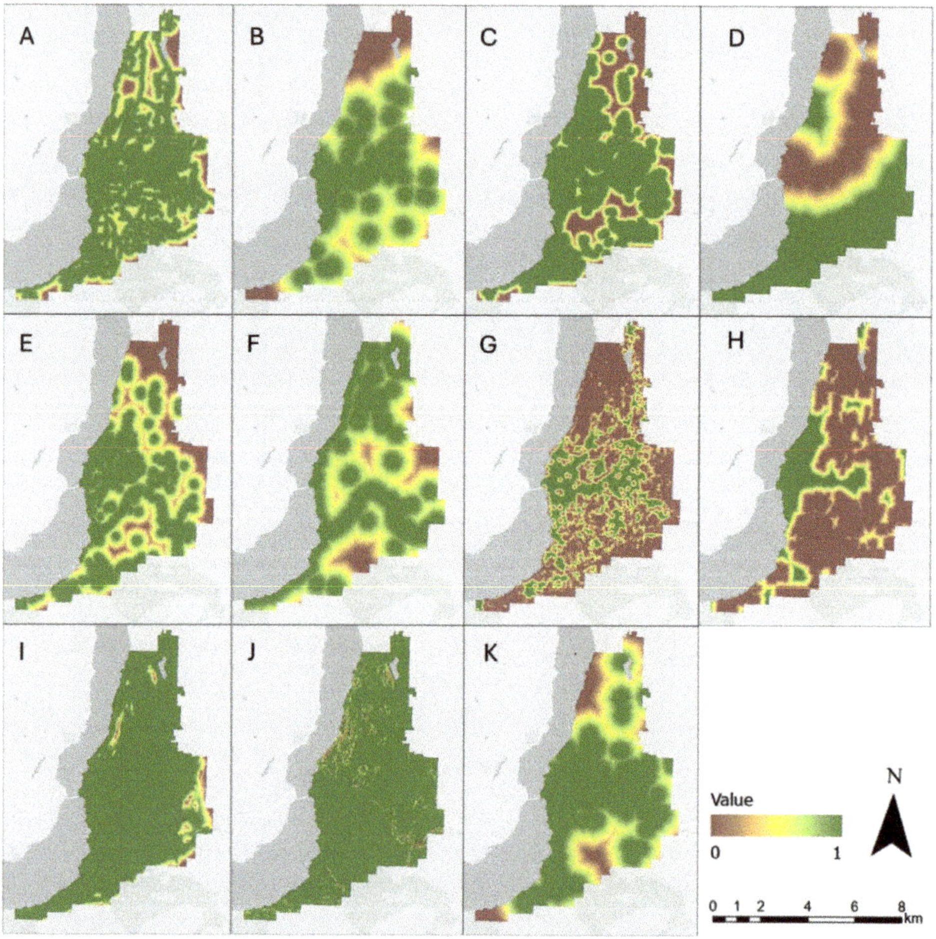

Figure 3. Maps presenting the normalized criteria data layers for the City of Kelowna are (**A**) proximity to roads; (**B**) proximity to commercial; (**C**) proximity to residential; (**D**) proximity to industrial; (**E**) proximity to parks; (**F**) proximity to waterbodies; (**G**) proximity to forest; (**H**) proximity to agricultural; (**I**) elevation; (**J**) slope; and (**K**) proximity to high population density.

Multicollinearity of Criteria

Before proceeding with the next steps of GIS-MCE analysis, it is important to examine the interrelationships among criteria and identify possible correlations or redundancies. A multicollinearity analysis involving the calculation of Variance Inflation Factors (VIFs) [57] was applied using the following formula:

$$VIF = \frac{1}{1 - R^2} \quad (1)$$

where R^2 measures how well a given criterion can be estimated using the other criteria. A VIF equal to 1 indicates no correlation or collinearity between a given criterion and any others, while a VIF exceeding 5 suggests a high level of collinearity. The results of the VIF analysis using the normalized criteria layers data are presented in Table 2. Since no criteria exceeded a value of 5, it was concluded that no multicollinearity is present, so all criteria were included in the GIS-MCE analysis.

Table 2. The VIF values for the normalized data layers.

Criteria	VIF
Proximity to roads	1.605
Proximity to commercial	3.275
Proximity to residential	2.149
Proximity to industrial	1.633
Proximity to parks	2.022
Proximity to waterbodies	1.175
Proximity to forest	1.146
Proximity to agricultural	1.489
Elevation	1.347
Slope	1.140
Proximity to high population density	2.365

2.3. *Criteria Weighting Approaches*

This section presents in detail the approaches used for deriving criteria weights.

2.3.1. The Equal-Weights Approach

The simplest way of assigning criteria weights in MCE is by applying an EW approach, where criteria are assigned equal importance [58]. This method serves as a baseline and neutral criteria weighting approach that can be useful for procedures like sensitivity analysis [59] and facilitating suitability map comparisons [60]. In this research, the EW approach is used to enable comparisons with all other approaches. As this study considers eleven criteria, each criterion is assigned approximately 9.09% (rounding to two decimal places), summing to a total of 100%.

2.3.2. The Analytical Hierarchy Process Approach

The AHP approach involves conducting pairwise comparisons of all criteria that are relevant to a spatial problem and organizing them in a hierarchical structure [61]. In the AHP, the criteria are compared pairwise using a scale from 1 to 9, where values reflect the relative importance of one element over another. For example, a value of 1 assigned between two criteria indicates that they are considered "equally important". Conversely, a value of 9 indicates that one criterion is "extremely important" compared to the other. The pairwise comparison is repeated for all criteria combinations to create a comparison matrix. The sum of each column of the resulting matrix is then calculated and used as a divisor for each cell in the matrix to generate new values for each matrix cell. The weight for each criterion is then calculated by taking the average of each row in the matrix. Furthermore, the Consistency Index (CI) within the AHP approach measures the consistency of the pairwise comparison judgments made by the spatial analyst or decision-maker [15,19]. Typically, a CI value less than 0.1 indicates an acceptable level of consistency in the pairwise comparisons, suggesting that the judgments are reliable. Since the relative importance of the criteria in the AHP is assigned manually, either by experts, decision-makers, or spatial analysts, it is prone to subjective judgments, biases, or incomplete information [22]. This can potentially lead to inconsistent or unreliable results for criteria weights. For this study, the relative importance of the criteria is informed by the existing literature. Table 3 presents the relative importance assigned to the criteria in the AHP pairwise comparison matrix and the derived criteria weights for the purpose of urban suitability analysis. The obtained CI value for the AHP comparisons is 0.03, indicating that the judgments involved in the pairwise comparisons are within an acceptable level of consistency.

2.3.3. The Proposed ML Approach

The proposed ML approach involves deriving criteria weights based on feature importance values from three ML techniques, namely RF, XGB, and SVM. Each technique is defined for a classification task where the training target is a binary label, with 0 indicating non-urban and 1 indicating urban locations. In this study, the impact of two training regimes on feature importance values is also explored. The training regimes and procedure for deriving criteria weights are described in the following sections.

Description of ML Training Regimes

The first training regime follows a traditional approach where the training label is derived from one map at a single timestamp, representing land use [32,34]. However, this setup tends to emphasize features related to longstanding urban developments rather than recognizing factors and conditions that were important for setting up new urban developments. As such, ML techniques may potentially be overfitted to specific features. To address this issue, a second training regime is proposed to obtain feature importance values specifically related to siting new urban developments. By including two maps to supply information about locations that have become urban, each ML technique learns to detect patterns and important features related to where new developments are most likely to occur. The proposed two-map training regime also enables each ML classifier to capture recent decision-maker priorities without the time-consuming task of acquiring expert opinions. Likewise, the two-map training regime assumes that decision-makers are better informed about the historical changes in a study area, thus allowing the ML method to capture which factors they have prioritized. The results from both training regimes are included in further steps of the MCE suitability analysis.

Evaluating Fit of ML Techniques

While feature importance values obtained from the RF, XGB, and SVM classifiers are used to derive criteria weights, direct outputs of each classifier are used to evaluate the fit of each ML technique. As such, the effects of one-map and two-map training regimes are examined for each ML type with respect to the outputs and feature importance values. The one-map training regime utilizes the eleven criteria outlined in Table 1 as the training data features, while urban and non-urban areas drawn from the 2015 LULC data supply the training target or label. In the one-map training regime, the ML techniques determine the likelihood of urban area presence given the input feature values of each training data sample. In contrast, the two-map training regime uses the same input features, while the training labels are derived from the differences between LULC data available for 2015 and 2020. In this approach, the objective of the two-map training regime is to determine the likelihood of new urban areas given the input feature values of each training data sample. This method allows ML techniques to learn the relationship between the criteria and the newly developed urban areas to derive feature importance values.

Table 3. AHP pairwise comparison matrix and the weights obtained for the criteria with obtained value for Consistency Index of 0.03.

	Proximity to Roads	**Proximity to Commercial**	**Proximity to Residential**	**Proximity to Industrial**	**Proximity to Parks**	**Proximity to Waterbody**	**Proximity to Forest**	**Proximity to Agricultural**	**Elevation**	**Slope**	**Proximity to High Pop. Density**	**AHP Criteria Weight (%)**
Proximity to Roads	1	4	3	5	6	6	6	7	7	8	3	**29.8**
Proximity to Commercial	1/4	1	1/3	2	2	2	2	3	3	4	1/2	**8.9**
Proximity to Residential	1/3	3	1	4	5	5	5	6	6	7	1	**19.5**
Proximity to Industrial	1/5	1/2	1/4	1	1	1	1	2	2	3	1/2	**5.6**
Proximity to Parks	1/6	1/2	1/5	1	1	1	1	2	2	3	1/2	**5.4**
Proximity to Waterbodies	1/6	1/2	1/5	1	1	1	1	2	2	3	1/2	**5.4**
Proximity to Forest	1/6	1/2	1/5	1	1	1	1	2	2	3	1/2	**5.4**
Proximity to Agricultural	1/7	1/3	1/6	1/2	1/2	1/2	1/2	1	1	2	1/3	**3.2**
Elevation	1/7	1/3	1/6	1/2	1/2	1/2	1/2	1	1	2	1/3	**3.2**
Slope	1/8	1/4	1/7	1/3	1/3	1/3	1/3	1/2	1/2	1	1/4	**2.2**
Proximity to High Pop. Density	1/3	2	1	2	2	2	2	3	3	4	1	**11.3**

Deriving Criteria Weights Using ML Feature Importance Analysis

Once each ML technique is trained, a feature importance analysis is conducted, and resulting values are converted to criteria weights. In this study, feature importance values help in understanding the contribution of each feature to the estimation of the target label. Feature importance values are determined using properties inherent to the scikit-learn and the XGBoost classifier implementations to support the identification of which features contributed most to each classifier's learned decision mechanisms. After feature importance values are obtained from the respective ML techniques under each training regime, each set of feature importance values is normalized so that the importance of all features sums to 1. Normalized feature importance values can then be directly used in place of criteria weights and further integrated into MCE analysis. The details of setting up the ML techniques and evaluations are presented in the following subsections.

2.3.4. Operationalizing the Proposed ML Approach

ML Techniques

The ML techniques implemented in this study include RF [62], XGB [63], and SVM [64] classifiers. The RF ensemble algorithm is a robust decision tree method capable of handling high-dimensional data. The RF technique leverages a number of decision trees that contribute to a majority vote used to obtain a final outputted value. The XGB technique is also an ensemble tree-based algorithm, instead relying on a concept called *gradient boosting* that involves refining a collection of weak learners at each iteration of training. XGB involves training multiple decision trees sequentially, with each tree focusing on correcting the errors of the previous one. In contrast, the SVM technique is based on a max-margin approach that aims to delineate a hyperplane to maximize separation between positive and negative training samples. To delineate the hyperplane, input features are first mapped to higher-dimension feature spaces using a kernel function. This underlying procedure enables SVM to generalize to unseen data, reducing the risk of overfitting. The RF technique benefited from the default *scikit-learn* parameters of the *Random Forest Classifier*, which creates 100 decision trees in the ensemble. Next, the number of XGB estimators was set to 100 and the maximum tree depth was set to 5. Lastly, a linear kernel was used for SVM and the regularization parameter (C) was set to 1.

Preparing ML Training Datasets

The next step requires the preparation of a training dataset. A training dataset consists of two components: training features and a training label. Training features are the independent variables or attributes provided as inputs to ML techniques, while the training label is the dependent variable or output. To form the training dataset for the one-map training regime, training features are derived from the eleven criteria data layers. The training label denotes the presence or absence of urban land-use for the year 2015, where a binary value of 0 indicates "non-urban" and 1 indicates "urban". To configure the training dataset for the two-map training regime, the same training features are maintained from the one-map training regime. However, the LULC data from 2015 and 2020 are used to identify new and recent urban developments. In the two-map training regime, the training label instead denotes persistent non-urban areas with a value of 0, while locations that have undergone development in the interval are assigned a value of 1. Details regarding the training configuration for both methods are listed in Table 4.

Next, a stratified random sampling approach is employed to form a balanced training dataset such that the ML techniques are not biased toward majority samples. The sampling approach is based on previous studies that mitigated imbalances by ensuring there are equal numbers of positive and negative targets for the ML techniques to learn [35,65]. Specifically, 2000 points are sampled from the study area, where 50% of the training dataset labels have a value of 0 and the latter have a value of 1. This is to ensure the training dataset is balanced to mitigate ML biases toward the majority label in both the one-map and two-map methods [35]. The training samples are subsequently split into training and testing sets, with 70% of the points being used to fit the ML techniques and the remaining 30% reserved for testing [66,67].

Table 4. The configuration for the two ML training regimes.

	One-Map Training Regime	Two-Map Training Regime
Training features	Eleven criteria data layers	Eleven criteria from data layers
Training label	2015 LULC	2015 and 2020 LULC
Training label values	0: non-urban areas in 2015 1: urban areas in 2015	0: persistent non-urban areas 1: areas that became urban between 2015 and 2020

Evaluation of ML Techniques

To evaluate the ML classifier parameterization and quality of fit, assessments were conducted with respect to the withheld testing dataset. The assessment involves using both the Receiver Operating Characteristic Area Under Curve (ROC-AUC) and F1 score with their respective components, as demonstrated in previous studies [29,35,68,69]. The emphasis of the assessment is to confirm that the ML techniques trained with one-map and two-map training regimes are operational for the respective tasks as opposed to comparing or optimizing predictive accuracy or performance.

The ROC-AUC score relies on the True Positive Rate (TPR) and the False Positive Rate (FPR). The TPR, also known as "recall", quantifies the proportion of true positive (TP) predictions or correctly classified positives relative to all TP predictions and false negative predictions (FN) [35]. Meanwhile, the FPR represents the proportion of false positive predictions (FP) relative to all actual negatives, which includes FP and True Negative (TN). The ROC curve depicts the relationship between the TPR and FPR of ML outcomes, serving as a graphical representation for assessing the performance of a binary ML classifier. Values obtained from calculating the area under the ROC curve (ROC-AUC) range from 0 to 1, with higher values indicating better ML performance. An ROC-AUC of 0.5 implies that the ML technique performs no better than random guessing, and a value of 1.0 signifies perfect classification [70]. The F1 score is determined using precision and recall. Precision equals the ratio of TP to all positive results (TP+FP), which includes those that were misidentified. The F1 score provides a single measure that balances precision and recall. Resulting values similarly adhere to a range of 0 to 1, where 1 indicates perfect precision and 0 indicates that the ML classifier could not classify any data sample correctly [71].

2.4. Deriving Suitability Scores and Suitability Maps

After the criteria weights are established with the various weighting approaches, the criteria represented by normalized data layers are aggregated using the WLC method to derive the final suitability score for each raster cell. The calculation of output suitability scores $S_{(r_i)}$ for a spatial problem r_i has been accomplished using the following WLC formula [37]:

$$S_{(r_i)} = \sum_{j=1}^{n} w_j f_j(r_i) \quad (2)$$

where w_j is the weight of importance assigned to criterion j and $f_j(r_i)$ is the suitability function for r_i for the jth criterion. The output suitability scores range from value 0 to 1, where 0 represents locations that are completely unsuitable while 1 represents the most suitable locations for the decision problem. In this study, a value of 1 indicates areas that are the most suitable for urban development. The resulting suitability scores are classified into several classes to produce suitability maps depicting locations that range from poor to excellent suitability for the new urban development in the study area. Suitability maps are derived for each weighting method for the purpose of comparison.

3. Results and Discussion

To prepare and normalize the data layers for this study, the ArcGIS Pro (version 3.2) software is used [72]. Next, the AHP criteria weights are developed using Microsoft Excel [73]. The ArcGIS Pro software was used to prepare the ML training datasets and implement the proposed ML weighing approach. Specifically, an ArcGIS Python Notebook and *ArcPy* [72] were used. An ArcGIS Python Notebook provides an interactive means of writing and running Python code within ArcGIS Pro. The ML approach is implemented using functions from the *ArcPy*, *scikit-learn* [74], and *xgboost* [63] Python libraries to prepare training datasets, to train each classifier, to perform the feature importance analysis, and to obtain the criteria weights.

3.1. Evaluating the ML Techniques

Figure 4 presents the obtained ROC curves and the AUC values for each ML technique and training regime. The high AUC values indicate that the ML techniques and their parameters have been effectively configured. As the ML techniques in this research are not intended to produce any predictions, the ROC values were determined with respect to data withheld for testing purposes presented in Section 2.3.4.

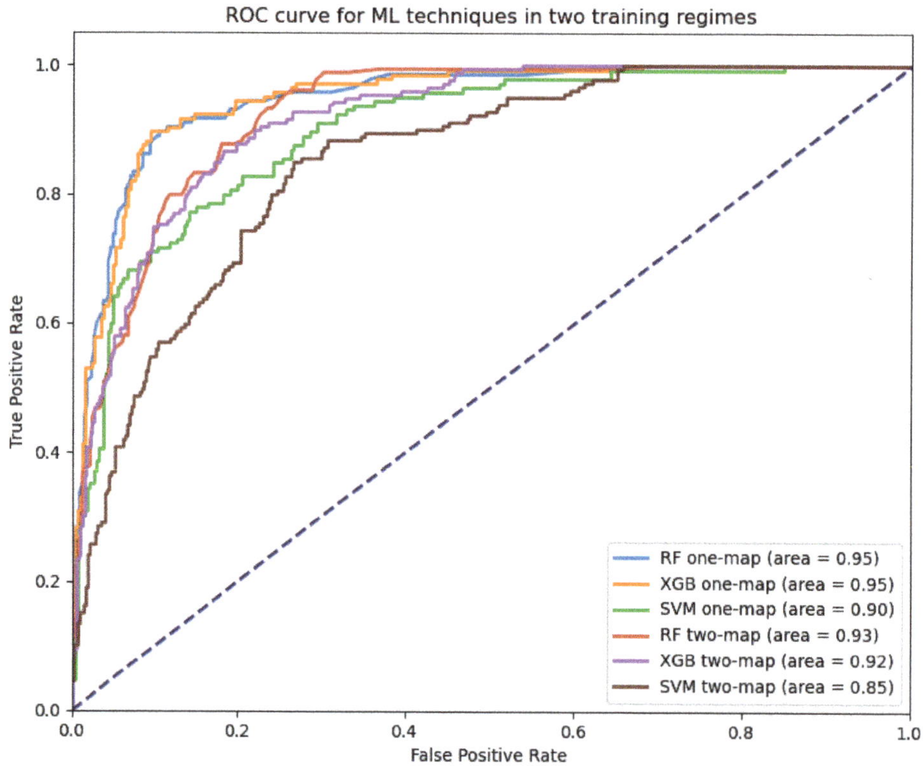

Figure 4. ROC-AUC for ML techniques in both one-map and two-map training regimes.

Table 5 presents the evaluation results for the ML techniques using the two different training regimes described in Section 2.3.3. Regardless of the training data setup used, RF and XGB exhibit high recall, precision, and F1 scores. Conversely, SVM's performance under both training regimes shows lower recall scores.

Table 5. The evaluation results from one-map and two-map training regimes.

Training Regime	Technique	Recall	Precision	F1 Score
One-map training	RF	80.69%	84.17%	0.82
	XGB	84.83%	82.00%	0.83
	SVM	68.28%	79.84%	0.74
Two-map training	RF	74.30%	76.00%	0.75
	XGB	74.30%	76.88%	0.76
	SVM	52.51%	71.21%	0.60

3.2. Comparing Derived Criteria Weights under Each Training Regime

The criteria weights obtained from all weighting approaches, including the RF, XGB, and SVM techniques under both training regimes, are summarized in Table 6a. Additionally, Table 6b summarizes the ranking of the criteria by each approach along with the average ranking for each criterion.

Comparing the criteria rankings of the AHP and ML approaches, "Proximity to roads" consistently ranks among the top three criteria affecting urban development. Additionally, "Elevation" and "Slope" are identified as the least important criteria in the AHP, and the RF technique was trained with both one-map and two-map regimes. The overall criteria rankings from the AHP closely align with those obtained using the RF and XGB techniques under the two-map training regime. However, the SVM technique presents a different weight distribution compared to the AHP under both training regimes.

Regarding the criteria weights obtained from the proposed ML approach, RF and XGB yield very similar criteria weight outcomes while SVM presents different results in both training regimes. Under the one-map training regime, "Proximity to agricultural" is ranked as the most important criterion of both RF and XGB techniques, with a weight of over 25%. "Proximity to forest" was also allocated a high criteria weight based on its larger contribution within the RF and XGB techniques. In contrast, "Slope" was determined as the most important criterion of the SVM technique, with a normalized weight of 25.8%. The most important criteria determined with the SVM technique can indicate a different stakeholder viewpoint that prioritizes certain criteria as being the most important. For example, "Slope" is often regarded as one of the least important criteria while "Proximity to roads" is typically considered one of the most important criteria by stakeholders and decision-makers [75–77]. Evidently, the feature importance values obtained from ML techniques trained under the one-map training regime are influenced by longstanding urban locations in the City of Kelowna. As such, the one-map training regime may have limited applicability regarding uncovering factors associated with favorable new urban development locations in real-world decision-making processes.

In contrast, the goal of the two-map training regime was to obtain feature importance values that take new urban developments into consideration (Table 4). By accounting for the actual recent study area changes that occurred between two timesteps, the obtained values can be seen as reflections of the choices of criteria weights developed by stakeholders and decision-makers. As such, the "Proximity to roads" criterion was identified as the most important criterion of both the RF and XGB techniques (Table 6). "Proximity to high population density" is also ranked higher in both techniques trained using the two-map training regime. The criteria weights associated with the SVM technique also are more realistic regarding "Proximity to roads," which ranked as the second most important, while "Slope" was allocated less weight. By comparing the criteria weights obtained under the two-map training regime with those specified manually in previous studies, the proposed ML approach reduces the difficulties of criteria weight derivation.

Table 6. (a) List of the criteria weights obtained by each weighting approach, with the top three criteria and weights being presented in bold. Due to rounding, some summations may not be exactly 100%. (b) Ranking of the criteria weights obtained by each weighting approach and the average ranking of the criteria, with the top three average ranking criteria being presented in bold.

(a)			One-Map Training Regime			Two-Map Training Regime		
Criteria	EW (%)	AHP (%)	RF (%)	XGB (%)	SVM (%)	RF (%)	XGB (%)	SVM (%)
Proximity to roads	9.09	**29.8**	**10.75**	**18.17**	**23.7**	**18.1**	**23.7**	**18.4**
Proximity to commercial	9.09	8.9	9.63	3.13	1.8	10.7	6.6	1.8
Proximity to residential	9.09	**19.5**	4.37	5.82	1.4	5.6	7.7	0.3
Proximity to industrial	9.09	5.6	6.54	3.54	3.3	**10.9**	**11.9**	9.2
Proximity to parks	9.09	5.4	4.66	4.71	1.5	9.4	5.1	7.8
Proximity to waterbodies	9.09	5.4	6.65	3.63	0.6	9.0	5.2	7.7
Proximity to forest	9.09	5.4	**21.92**	**14.13**	**15.8**	10.0	**13.4**	**32.9**
Proximity to agricultural	9.09	3.2	**27.11**	**26.89**	15.0	9.5	8.4	9.9
Elevation	9.09	3.2	1.05	9.77	6.4	3.3	6.8	**10.4**
Slope	9.09	2.3	2.00	6.84	**25.8**	2.5	3.1	1.2
Proximity to high pop. density	9.09	**11.3**	5.32	3.38	4.7	**11.0**	8.1	0.4
Total	100	100	100	100	100	100	100	100

(b)		One-Map Training regime			Two-Map Training regime			Average Criteria Ranking
Criteria	AHP	RF	XGB	SVM	RF	XGB	SVM	
Proximity to roads	1	3	2	2	1	1	2	**1.7**
Proximity to commercial	4	4	11	8	4	8	8	6.7
Proximity to residential	2	9	6	10	9	6	11	7.6
Proximity to industrial	5	6	9	7	3	3	5	5.4
Proximity to parks	6	8	7	9	7	10	6	7.6
Proximity to waterbodies	6	5	8	11	8	9	7	7.7
Proximity to forest	6	2	3	3	5	2	1	**3.1**
Proximity to agricultural	9	1	1	4	6	4	4	**4.1**
Elevation	10	11	4	5	10	7	3	7.1
Slope	11	10	5	1	11	11	9	8.3
Proximity to high pop. density	3	7	10	6	2	5	10	6.1

After comparing the criteria weights attributed to each ML technique under both training regimes, the two-map training regime is considered to be a better-informed approach for training ML techniques, obtaining feature importance values and deriving criteria weights for the task of determining suitable locations for new urban developments. Therefore, the criteria weights associated with the two-map ML techniques are further compared with the traditional approaches in the subsequent sections.

3.3. Comparing Effects of Sample Size on Derived Criteria Weights

Additional tests have been conducted to determine if the number of training points influences the outputs of the ML techniques. For example, the RF technique was trained using the two-map regime with 2000, 4000 and 6000 training points. The results demonstrate consistent criteria weight distributions and rankings (Table 7) regardless of the choice of number of training points. This is consistent with findings from [78], which found that lower numbers of training data samples even improved outcomes. Therefore, the presented results following this section are produced using 2000 data samples to train each ML technique.

Table 7. List of the criteria weights and rankings obtained by RF with 2000, 4000, and 6000 training points. The top three average ranking criteria are bolded.

Criteria	RF 2000 (%)	Criteria Ranking	RF 4000 (%)	Criteria Ranking	RF 6000 (%)	Criteria Ranking	Average Ranking
Proximity to roads	18.1	1	19.1	1	18.6	1	**1.0**
Proximity to commercial	10.7	4	10.8	2	11.5	3	**3.0**
Proximity to residential	5.6	9	5.5	9	5.5	9	9.0
Proximity to industrial	10.9	3	10.2	4	11.2	4	3.7
Proximity to parks	9.4	7	9.7	6	9.0	6	6.3
Proximity to waterbodies	9.0	8	8.6	8	8.7	7	7.7
Proximity to forest	10.0	5	10.0	5	8.5	8	6.0
Proximity to agricultural	9.5	6	9.4	7	9.2	5	6.0
Elevation	3.3	10	3.7	10	3.4	10	10.0
Slope	2.5	11	2.6	11	2.7	11	11.0
Proximity to high pop. density	11.0	2	10.4	3	11.7	2	**2.3**
Total	100		100		100		

3.4. Analyzing Obtained Suitability Maps

After integrating the criteria weights using the WLC aggregation method, the obtained suitability scores are classified into four classes using the equal interval classification method. The classes are defined as follows: excellent (0.76–1.00), good (0.51–0.75), moderate (0.26–0.50), and poor (0.0–0.25) suitability for urban development. Figure 5 presents suitability maps created using criteria weights derived from the EW, AHP, and respective ML techniques trained with the two-map regime in the WLC aggregation method. Areas classified as having excellent suitability are interpreted as having the highest potential for future urban development. At the other extreme, "poor" suitability indicates areas unsuitable for new urban developments within the study area. For comparison purposes, suitability maps produced using the criteria weights obtained from the RF-WLC approach under both one-map and two-map training regimes are presented in Figure 6 to support visual assessments of the effects of the training data methods.

Additionally, the surface area (in km^2) of derived suitability for each class is summarized in Figure 7 and Table 8 to quantify the differences between the various weighting approaches. The EW approach identified the smallest amount of "poor suitability" area and a moderate amount of area classified as excellent. Conversely, the suitability map created with the AHP-WLC approach shows the largest area as having excellent suitability, almost double that of the RF-WLC and XGB-WLC approaches. The RF-WLC and XGB-WLC approaches demonstrate similar values for surfaces under the same suitability classes. On the other hand, the SVM-WLC approach presents a distinct outcome compared to the other approaches. The outcome generated using the SVM-WLC approach depicts the least area as having "excellent" suitability, almost close to zero, with most of the area classified as having "moderate" suitability.

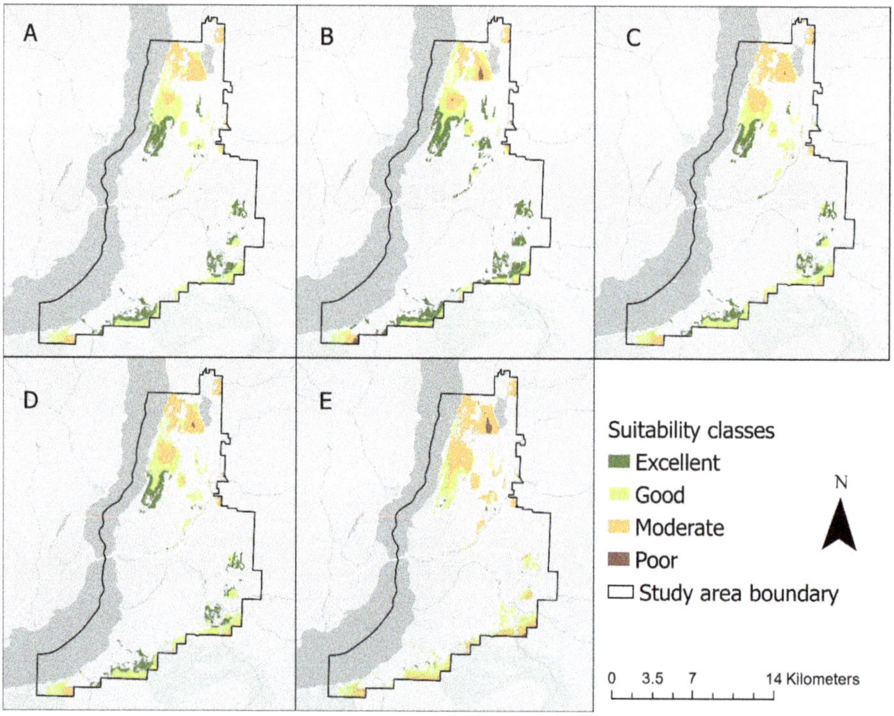

Figure 5. Output suitability maps for the City of Kelowna with four suitability classes derived from the five approaches: (**A**) EW-WLC; (**B**) AHP-WLC; (**C**) RF-WLC; (**D**) XGB-WLC; (**E**) SVM-WLC.

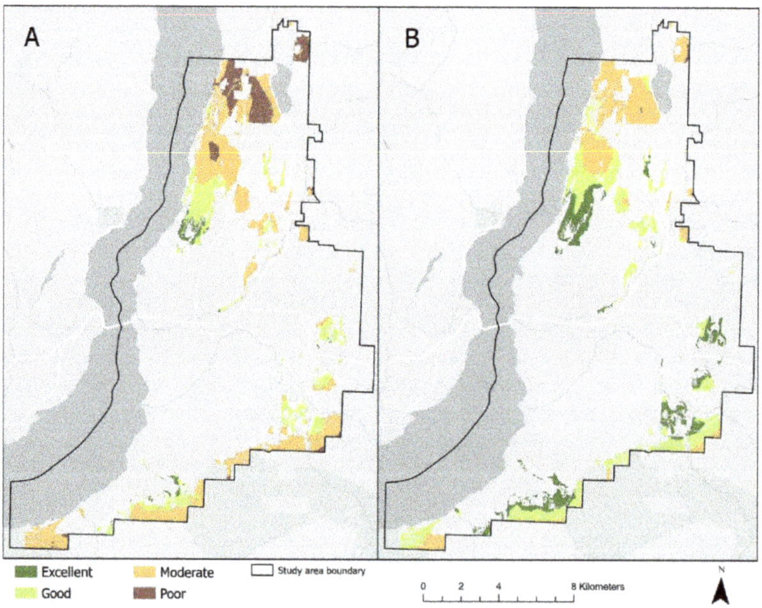

Figure 6. Output suitability maps for RF-WLC obtained with (**A**) one-map training and (**B**) two-map training regimes.

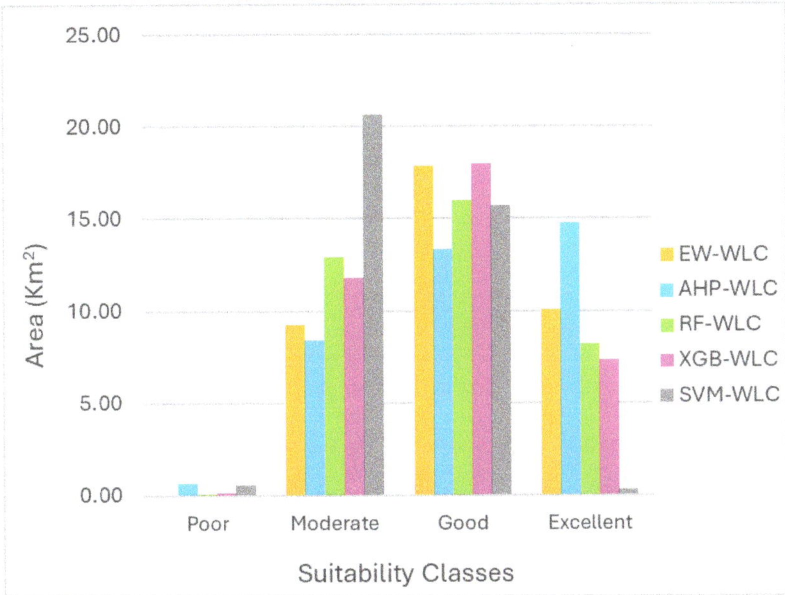

Figure 7. Surface area associated with each suitability class from each criteria weighting approach.

Table 8. Surface area and proportion of study region with urban development potential assigned to each suitability class, where bold values indicate highest surface area under specific suitability class.

	Criteria Weighting Technique									
	EW-WLC		**AHP-WLC**		**RF-WLC**		**XGB-WLC**		**SVM-WLC**	
Surface area for each class	**km^2**	**%**	**km^2**	**%**	**km^2**	**%**	**km^2**	**%**	**km^2**	**%**
Excellent	10.09	27.11	**14.76**	**39.65**	8.20	22.02	7.33	19.69	0.31	0.84
Good	**17.87**	**48.01**	13.37	35.91	**16.00**	**42.98**	**17.98**	**48.31**	15.67	42.11
Moderate	9.26	24.87	8.43	22.65	12.92	34.70	11.79	31.67	**20.66**	**55.50**
Poor	0	0	0.67	1.79	0.11	0.30	0.12	0.33	0.58	1.56
Total	37.22	100	37.22	100	37.22	100	37.22	100	37.22	100

3.5. Comparing Suitability Maps

Results from GIS-MCE analysis, such as suitability scores and maps, are typically used in further deliberations of stakeholders to guide decision-making processes. Given that developing suitability maps is not equivalent to making predictions of new land-use states, the techniques for validating suitability maps are limited. Instead, suitability maps serve as a starting point for stakeholder discussions that would lead to final decisions about where new urban developments should be sited. Nevertheless, to compare the suitability maps developed with the various approaches, the respective outcomes are examined using visual and areal cross-comparison techniques, the Kappa Index of Agreement (KIA), and a visual comparison with recent real-world imagery obtained for the City of Kelowna.

3.5.1. Suitability Map Cross-Comparison

The cross-comparison technique [79] was used to generate the agreement/disagreement maps with respect to the "excellent" suitability class for all criteria weighting methods. Focusing on comparing the "excellent" class is important because potential stakeholders and urban developers will be the most interested in locations identified as having the highest suitability for new urban developments. The EW-WLC map is used as a neutral suitability map by which to base the comparisons. Figure 8 presents the cross-comparison maps obtained from each weighting approach with the WLC method, indicating locations where the highest suitability scores obtained overlap or differ. The areas in agreement are shown in red, while disagreeing areas are shown in blue on the maps. Overall, it can be observed that SVM-WLC generates the smallest area with "excellent" suitability and shows the most disagreement with all other approaches. Conversely, the AHP-WLC map depicts the largest area under the "excellent" suitability class (Figure 7, Table 8) while simultaneously showing a significant amount of locational disagreement with other approaches (Figure 8). Regardless, the RF-WLC and XGB-WLC show the most agreement with this suitability class among all MCE approaches.

With respect to the neutral baseline (EW-WLC), both the RF-WLC and XGB-WLC approaches exhibit high agreement between "excellent" areas. The similarities between EW-WLC, RF-WLC, and XGB-WLC can be attributed to the selection of criteria and several local trends observed in transformed values of factors (Figure 3). For example, "Proximity to Roads" and "Proximity to Residential" display similarly high values in some areas, thus contributing to the observation that the EW-WLC approach has assigned "Excellent" suitability classes to several similar areas identified via the RF-WLC and XGB-WLC approaches. Meanwhile, both the RF-WLC and XGB-WLC approaches assign "Proximity to Roads" one of the highest criteria weights. Due to the numerous combinations of criteria weights that can lead to an "excellent" class being assigned to certain areas by the EW-WLC approach, this demonstrates the importance of comparing criteria weights with the existing literature (Section 3.2) and the need for inspections of the local factor values contributing to high suitability values. Despite the observed local similarities between the RF-WLC and XGB-WLC approaches with the neutral baseline, the agreement with other suitability classes deviates, as observed in Figure 5.

3.5.2. Quantifying Agreement between Suitability Maps

Following the visual and areal cross-comparisons of each suitability map, the similarity between the suitability class locations in each pair of maps is quantified using the Kappa Index of Agreement (KIA) [52,80,81] given that suitability maps are not the result of a space-time process. KIA values are presented in Table 9 and, when approaching 1, suggest high agreement between suitability classes present in the two maps being compared. Conversely, KIA values approaching 0 indicate low alignment of suitability classes at most locations compared across the study region.

Table 9. Summary of Kappa Index of Agreement (KIA) values obtained by comparing each pair of suitability maps.

vs.	AHP-WLC	RF-WLC	XGB-WLC	SVM-WLC
EW-WLC	0.75	0.84	0.83	0.40
AHP-WLC	-	0.75	0.70	0.32
RF-WLC	-	-	0.90	0.54
XGB-WLC	-	-	-	0.55

Across each pair of compared suitability maps, the KIA values confirm previous evaluations. Additionally, the KIA values further confirm the similarities between the suitability maps attributed to the RF and XGB techniques. For example, RF-WLC and XGB-WLC exhibit the highest agreement of any pair of suitability maps based on the KIA value of 0.90 (Table 9). In contrast the SVM-WLC map is associated with the lowest KIA values obtained from all suitability map comparisons. The minimum KIA value (0.32) is associated with the comparison between AHP-WLC and SVM-WLC, indicating that these two maps exhibit the least similarity of suitability class locations across the study region. Considering the KIA values attributed to the overall comparisons of EW-WLC and

RF-WLC (0.84), and EW-WLC and XGB-WLC (0.83), the deviating locational agreement between suitability classes other than "excellent" as previously discussed (Section 3.5.1) are also confirmed.

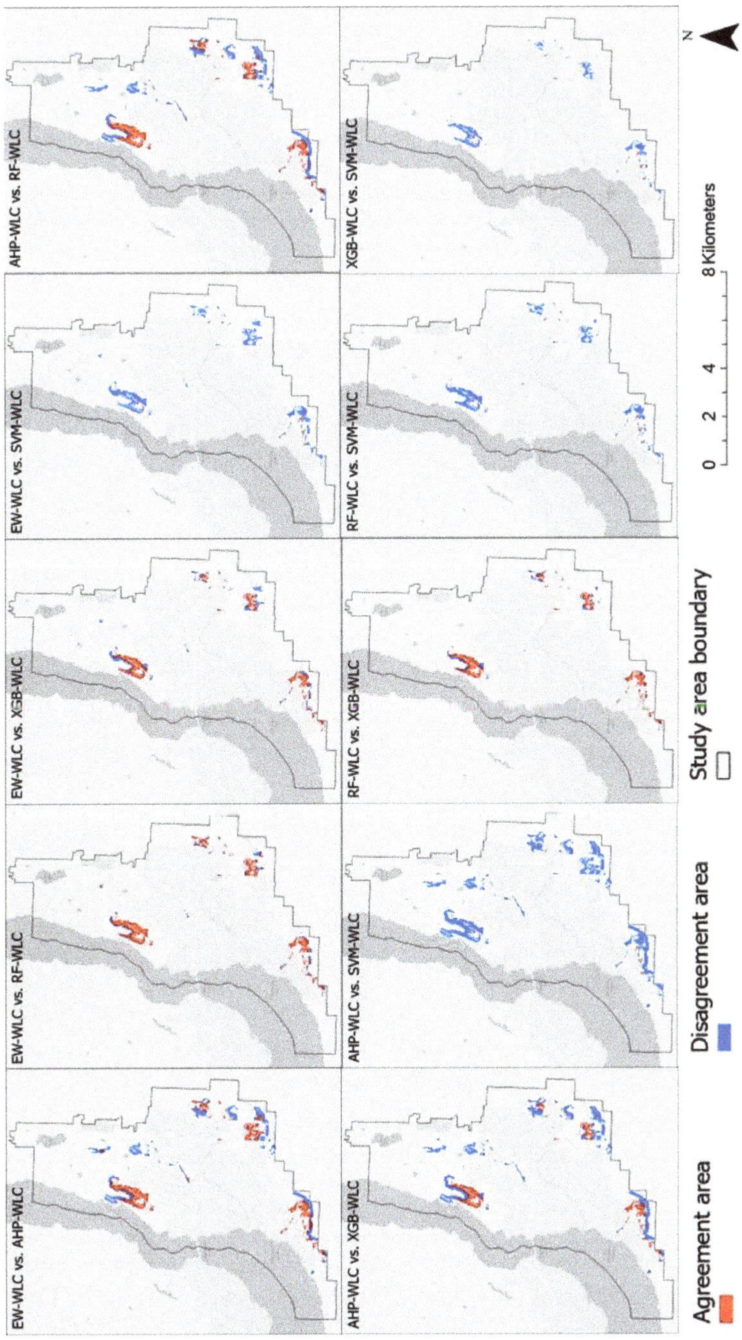

Figure 8. Cross-comparison maps for excellent suitability class obtained by each MCE-WLC approach.

3.5.3. Examining Suitability Maps versus Real-World Images

In order to confirm the applicability of the proposed ML-MCE method and obtained suitability scores, a manual visual verification with an alternative source of "field observations" is conducted with respect to recent Google Earth satellite imagery. As the RF-WLC yields very similar results to XGB-WLC and attained the highest ROC-AUC score among the ML techniques trained under the two-map regime, the RF-WLC suitability map was compared to areas with and without recent urban developments in the study area. Figure 9 presents a comparison of the RF-WLC suitability map with real-world satellite imagery from 2023, focusing on areas of "moderate" and "poor" suitability. The highlighted locations identified to have low suitability show that no urban development has occurred since. This also suggests that the RF-WLC suitability map concurs with knowledge about the criteria. For example, areas determined to be less accessible (Figure 3A), more isolated from current urban developments (Figure 3C), and closer to industrial zones (Figure 3D) are appropriately identified as less suitable for new urban developments.

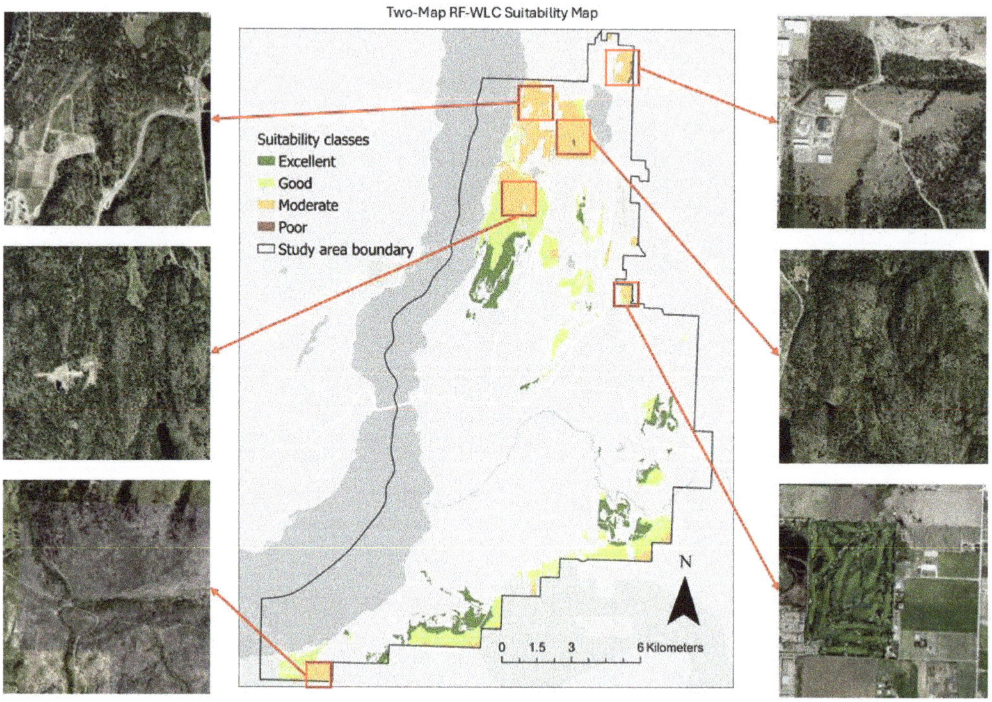

Figure 9. Comparisons of obtained "moderate" and "poor" suitability classes for City of Kelowna using the RF-WLC approach with Google Earth imagery for year 2023.

Next, Figure 10 presents a comparison of the suitability maps obtained with the RF-WLC approach trained under both the one- and two-map regimes versus real-world satellite imagery for the year 2023. Focusing on the areas of "excellent" suitability, the visual assessment indicates that areas identified as highly suitable for new urban developments using the two-map RF-WLC approach corresponded with some recent new urban developments in the study area. Overall, the visual assessment further confirms that the two-map training regime for ML techniques is well suited for deriving criteria weights in the proposed ML-MCE method. In contrast, the RF-WLC suitability maps attributed to the one-map training regime show that many of the selected locations do not have the highest suitability scores. This suggests that urban developers and planners in the City of Kelowna are indeed considering the areas with characteristics or criteria identified with the two-map ML-MCE methodology proposed in this study.

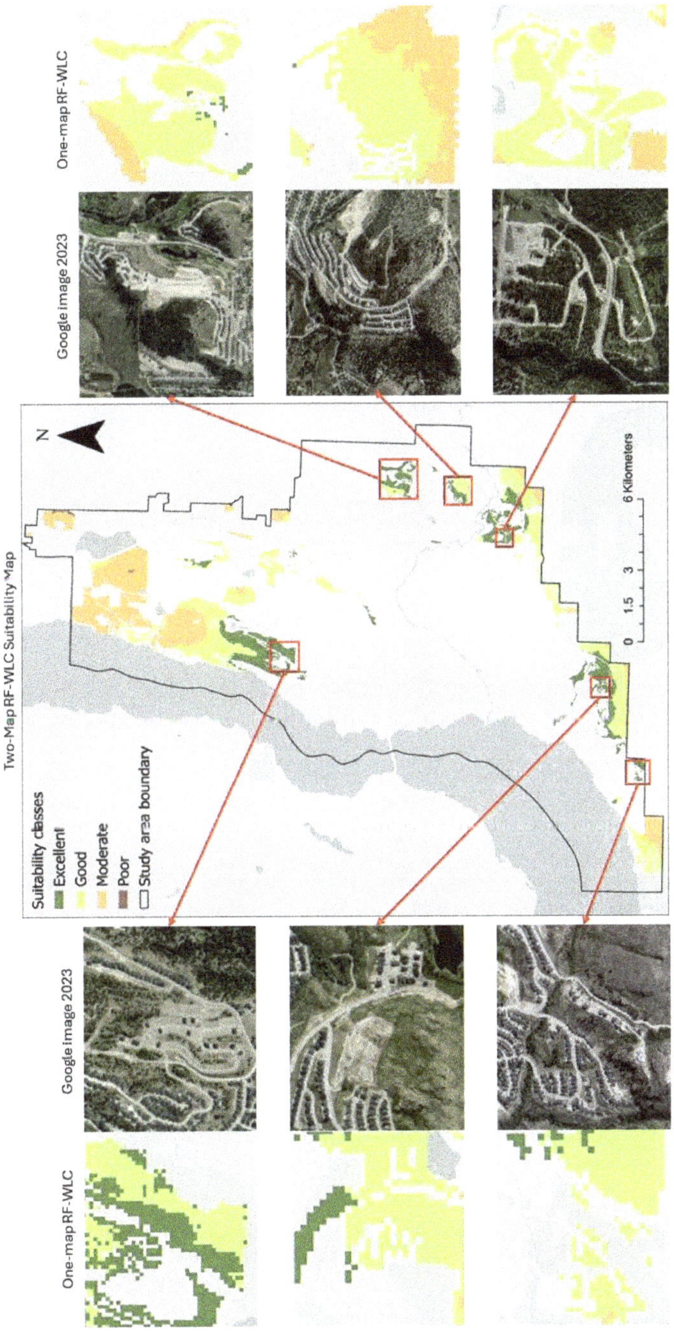

Figure 10. Comparisons of areas identified to have "excellent" suitability in the City of Kelowna using the RF-WLC approach versus Google Earth imagery obtained for the year 2023.

3.6. Limitations and Future Directions

Although the comparison of ML techniques and training regimes showed the value of deriving criteria weights with two timesteps for urban development suitability, there remain several limitations and opportunities for improvement. In real-world implementation, ML techniques and software may not be widely accessible to urban planners, as the proper setup, training, and testing of ML techniques may be a challenge for municipal planners. Next, while the four-class equal interval breaks used to assign suitability classes are easier to visualize and understand, the real variation of the suitability values obtained from each approach require additional comparisons at finer-level classes or on a continuous scale in future studies. Another limitation of this work is that the WLC method used in this study is based on a Simple Additive Scoring (SAS) method, which takes the mean of all the components in the analysis [75]. When the number of criteria increases, the importance of criteria decreases, oversimplifying the decision-making process of many complex spatial problems like urban land-use suitability analysis. To overcome this issue in future studies, a more complex MCE method can be applied, such as the Ordered Weighted Averaging (OWA) or Logic Scoring of Preference (LSP) methods. Another limitation of this study is that it only considered eleven criteria, while complex urban planning spatial decision problems may require a much larger number of criteria, especially for larger study areas and urban metropolitan regions. Investigating the limits of the number of criteria remains unexplored with the ML-MCE method and could be explored in future work. Lastly, the ML techniques in this study are trained using the same normalized criteria data layers used for the EW and AHP approaches. Future studies should explore and compare the effects of integrating domain knowledge in data layer pre-processing procedures versus standard ML data normalization strategies.

4. Conclusions

This research explores the integration of ML with GIS-MCE for determining criteria weights and with the aim to analyze urban development suitability. Focusing specifically on a case study in the City of Kelowna, results indicate that approaches such as the neutral EW and the traditional AHP produce different criteria weights than weights obtained via feature importance analyses of the three ML techniques. Among the ML techniques explored, the criteria weights derived from RF feature importance values closely aligned with those of XGB, while SVM shows markedly different criteria weights. Additionally, this study examined the differences in criteria weights, distributions, and rankings that result from different training regimes. A traditional one-map training regime for ML techniques is compared with the proposed two-map training regime for deriving feature importance related to new urban developments specifically. The results demonstrate that the two-map training regime is more suitable for this case study and enables criteria weights that are more likely to align with perspectives of potential stakeholders and urban planners. After comparing the suitability maps created using the RF-WLC approach with the 2023 Google Earth satellite images, the visual assessments confirmed the effectiveness of the proposed ML-MCE method in real-world urban planning in the City of Kelowna.

Overall, the proposed ML-MCE method provides a novel, data-driven, and systematic framework for combining geospatial data and ML techniques for urban development suitability analysis. This framework has the potential to assist in city planning and urban land-use management by supporting decision-makers.

Author Contributions: Conceptualization, Formal Analysis, Investigation, Methodology, Original draft, Review and Editing L.Q.Z., A.v.D. and S.D.; Funding Acquisition, Supervision, Suzana Dragicevic; Software, L.Q.Z. and A.v.D. All authors have read and agreed to the published version of the manuscript.

Funding: This research was funded by the Natural Sciences and Engineering Research Council (NSERC) of Canada Graduate Scholarships Master's Program (CGSM) scholarship and Discovery Grant (RGPIN-2023-04052) awarded to the first and third authors, respectively.

Data Availability Statement: Publicly available datasets were used in this study. The data can be found at https://opendata.kelowna.ca/ (accessed on 3 July 2023) (City of Kelowna Zoning data, Road Networks data), https://open.canada.ca/data/en/dataset/4e615eae-b90c-420b-adee-2ca35896caf6 (accessed on 3 July 2023) (Canada Land-use and Land Cover data 2015, 2020), https://open.canada.ca/data/en/dataset/d0cdef71-9343-46c3-b2e7-c1ded5907686 (accessed on 3 July 2023) (Canada Lakes and Rivers Boundaries data), https://open.canada.ca/data/en/dataset/7f245e4d-76c2-4caa-951a-45d1d2051333 (accessed on 3 July 2023) (Canadian DEM data), https://www12.statcan.gc.ca/census-recensement/2021/geo/index-eng.cfm (accessed on 3 July 2023) (Canadian Census data), and https://catalogue.data.gov.bc.ca/dataset/alc-alr-polygons/resource/d35b18b0-ecfa-468b-b31a-2bf16a459a7c (accessed on 3 July 2023) (BC Agricultural Land Reserve data).

Acknowledgments: The authors are grateful for the full support of the Natural Sciences and Engineering Research Council of Canada (NSERC). The authors are also thankful for the constructive comments provided by two anonymous reviewers.

Conflicts of Interest: The authors have no conflicts of interest to declare.

References

1. Malczewski, J. GIS-based Multicriteria Decision Analysis: A Survey of the Literature. *Int. J. Geogr. Inf. Sci.* **2006**, *20*, 703–726. [CrossRef]
2. Malczewski, J.; Jankowski, P. Emerging Trends and Research Frontiers in Spatial Multicriteria Analysis. *Int. J. Geogr. Inf. Sci.* **2020**, *34*, 1257–1282. [CrossRef]
3. Chen, J. GIS-Based Multi-Criteria Analysis for Land Use Suitability Assessment in City of Regina. *Environ. Syst. Res.* **2014**, *3*, 13. [CrossRef]
4. Gelan, E. GIS-Based Multi-Criteria Analysis for Sustainable Urban Green Spaces Planning in Emerging Towns of Ethiopia: The Case of Sululta Town. *Environ. Syst. Res.* **2021**, *10*, 13. [CrossRef]
5. Masoudi, M.; Aboutalebi, M.; Asrari, E.; Cerdà, A. Land Suitability of Urban and Industrial Development Using Multi-Criteria Evaluation (MCE) and A New Model by GIS in Fasa County, Iran. *Land* **2023**, *12*, 1898. [CrossRef]
6. Abid, S.K.; Sulaiman, N.; Chan, S.W.; Nazir, U.; Abid, M.; Han, H.; Ariza-Montes, A.; Vega-Muñoz, A. Toward an Integrated Disaster Management Approach: How Artificial Intelligence Can Boost Disaster Management. *Sustainability* **2021**, *13*, 12560. [CrossRef]
7. Masoudi, M.; Centeri, C.; Jakab, G.; Nel, L.; Mojtahedi, M. GIS-Based Multi-Criteria and Multi-Objective Evaluation for Sustainable Land-Use Planning (Case Study: Qaleh Ganj County, Iran) "Landuse Planning Using MCE and Mola". *Int. J. Environ. Res.* **2021**, *15*, 457–474. [CrossRef]
8. Dai, L.; Zhao, X.; He, H.S.; Deng, H.; Yu, D.; Zhou, L.; Wu, S. Evaluating Land-Use Suitability of an Industrial City in Northeast China. *Int. J. Sustain. Dev. World Ecol.* **2008**, *15*, 378–382. [CrossRef] [PubMed]
9. Abebe, M.T.; Megento, T.L. Urban Green Space Development Using GIS-Based Multi-Criteria Analysis in Addis Ababa Metropolis. *Appl. Geomat.* **2017**, *9*, 247–261. [CrossRef]
10. Liu, R.; Zhang, K.; Zhang, Z.; Borthwick, A.G.L. Land-Use Suitability Analysis for Urban Development in Beijing. *J. Environ. Manag.* **2014**, *145*, 170–179. [CrossRef] [PubMed]
11. Zhang, X.; Fang, C.; Wang, Z.; Ma, H. Urban Construction Land Suitability Evaluation Based on Improved Multi-Criteria Evaluation Based on GIS (MCE-GIS): Case of New Hefei City, China. *Chin. Geogr. Sci.* **2013**, *23*, 740–753. [CrossRef]
12. Caprioli, C.; Bottero, M. Addressing Complex Challenges in Transformations and Planning: A Fuzzy Spatial Multicriteria Analysis for Identifying Suitable Locations for Urban Infrastructures. *Land Use Policy* **2021**, *102*, 105147. [CrossRef]
13. Chen, Y.; Yu, J.; Khan, S. Spatial Sensitivity Analysis of Multi-Criteria Weights in GIS-Based Land Suitability Evaluation. *Environ. Model. Softw.* **2010**, *25*, 1582–1591. [CrossRef]
14. Feick, R.; Hall, B. A Method for Examining the Spatial Dimension of Multi-Criteria Weight Sensitivity. *Int. J. Geogr. Inf. Sci.* **2004**, *18*, 815–840. [CrossRef]
15. Saaty, T.L. A Scaling Method for Priorities in Hierarchical Structures. *J. Math. Psychol.* **1977**, *15*, 234–281. [CrossRef]
16. Hwang, C.-L.; Yoon, K. *Multiple Attribute Decision Making*; Lecture Notes in Economics and Mathematical Systems; Springer: Berlin/Heidelberg, Germany, 1981; Volume 186, ISBN 978-3-540-10558-9.
17. Shannon, C.E. A Mathematical Theory of Communication. *Bell Syst. Tech. J.* **1948**, *27*, 379–423. [CrossRef]
18. Greene, R.; Luther, J.E.; Devillers, R.; Eddy, B. An Approach to GIS-Based Multiple Criteria Decision Analysis That Integrates Exploration and Evaluation Phases: Case Study in a Forest-Dominated Landscape. *For. Ecol. Manag.* **2010**, *260*, 2102–2114. [CrossRef]
19. Saaty, T.L. The Analytic Hierarchy Process: Decision Making in Complex Environments. In *Quantitative Assessment in Arms Control: Mathematical Modeling and Simulation in the Analysis of Arms Control Problems*; Avenhaus, R., Huber, R.K., Eds.; Springer: Boston, MA, USA, 1984; pp. 285–308, ISBN 978-1-4613-2805-6.
20. Clark Labs TerrSet. Available online: https://clarklabs.org/terrset/ (accessed on 10 September 2023).

21. Esri ArcMap. Available online: https://www.esri.com/en-us/arcgis/products/arcgis-desktop/resources (accessed on 10 September 2023).
22. Liu, Y.; Eckert, C.M.; Earl, C. A Review of Fuzzy AHP Methods for Decision-Making with Subjective Judgements. *Expert Syst. Appl.* **2020**, *161*, 113738. [CrossRef]
23. Dodevska, Z.; Radovanović, S.; Petrović, A.; Delibašić, B. When Fairness Meets Consistency in AHP Pairwise Comparisons. *Mathematics* **2023**, *11*, 604. [CrossRef]
24. Choi, Y. GeoAI: Integration of Artificial Intelligence, Machine Learning, and Deep Learning with GIS. *Appl. Sci.* **2023**, *13*, 3895. [CrossRef]
25. Chaturvedi, V.; de Vries, W.T. Machine Learning Algorithms for Urban Land Use Planning: A Review. *Urban Sci.* **2021**, *5*, 68. [CrossRef]
26. Huang, M.; Gong, D.; Zhang, L.; Lin, H.; Chen, Y.; Zhu, D.; Xiao, C.; Altan, O. Spatiotemporal Dynamics and Forecasting of Ecological Security Pattern under the Consideration of Protecting Habitat: A Case Study of the Poyang Lake Ecoregion. *Int. J. Digit. Earth* **2024**, *17*, 2376277. [CrossRef]
27. Chang, Z.; Du, Z.; Zhang, F.; Huang, F.; Chen, J.; Li, W.; Guo, Z. Landslide Susceptibility Prediction Based on Remote Sensing Images and GIS: Comparisons of Supervised and Unsupervised Machine Learning Models. *Remote Sens.* **2020**, *12*, 502. [CrossRef]
28. Janiec, P.; Gadal, S. A Comparison of Two Machine Learning Classification Methods for Remote Sensing Predictive Modeling of the Forest Fire in the North-Eastern Siberia. *Remote Sens.* **2020**, *12*, 4157. [CrossRef]
29. Garzón, M.B.; Blazek, R.; Neteler, M.; de Dios, R.S.; Ollero, H.S.; Furlanello, C. Predicting Habitat Suitability with Machine Learning Models: The Potential Area of *Pinus Sylvestris* L. in the Iberian Peninsula. *Ecol. Model.* **2006**, *197*, 383–393. [CrossRef]
30. Taghizadeh-Mehrjardi, R.; Nabiollahi, K.; Rasoli, L.; Kerry, R.; Scholten, T. Land Suitability Assessment and Agricultural Production Sustainability Using Machine Learning Models. *Agronomy* **2020**, *10*, 573. [CrossRef]
31. Al-Ruzouq, R.; Shanableh, A.; Yilmaz, A.G.; Idris, A.; Mukherjee, S.; Khalil, M.A.; Gibril, M.B.A. Dam Site Suitability Mapping and Analysis Using an Integrated GIS and Machine Learning Approach. *Water* **2019**, *11*, 1880. [CrossRef]
32. Al-Ruzouq, R.; Abdallah, M.; Shanableh, A.; Alani, S.; Obaid, L.; Gibril, M.B.A. Waste to Energy Spatial Suitability Analysis Using Hybrid Multi-Criteria Machine Learning Approach. *Environ. Sci. Pollut. Res.* **2022**, *29*, 2613–2628. [CrossRef] [PubMed]
33. Gharaibeh, A.A.; Jaradat, M.A.; Kanaan, L.M. A Machine Learning Framework for Assessing Urban Growth of Cities and Suitability Analysis. *Land* **2023**, *12*, 214. [CrossRef]
34. Ali, S.A.; Parvin, F.; Vojteková, J.; Costache, R.; Linh, N.T.T.; Pham, Q.B.; Vojtek, M.; Gigović, L.; Ahmad, A.; Ghorbani, M.A. GIS-Based Landslide Susceptibility Modeling: A Comparison between Fuzzy Multi-Criteria and Machine Learning Algorithms. *Geosci. Front.* **2021**, *12*, 857–876. [CrossRef]
35. Mahdizadeh Gharakhanlou, N.; Perez, L. Flood Susceptible Prediction through the Use of Geospatial Variables and Machine Learning Methods. *J. Hydrol.* **2023**, *617*, 129121. [CrossRef]
36. Singh, R.; Behera, M.D.; Das, P.; Rizvi, J.; Dhyani, S.K.; Biradar, Ç.M. Agroforestry Suitability for Planning Site-Specific Interventions Using Machine Learning Approaches. *Sustainability* **2022**, *14*, 5189. [CrossRef]
37. Malczewski, J. On the Use of Weighted Linear Combination Method in GIS: Common and Best Practice Approaches. *Trans. GIS* **2000**, *4*, 5–22. [CrossRef]
38. Government of Canada. Census Profile, 2021 Census of Population. Available online: https://www12.statcan.gc.ca/census-recensement/2021/dp-pd/prof/index.cfm?Lang=E (accessed on 3 July 2023).
39. City of Kelowna. Open Kelowna. Available online: https://opendata.kelowna.ca/ (accessed on 3 July 2023).
40. Government of BC Agricultural Land Reserve. Available online: https://catalogue.data.gov.bc.ca/dataset/alc-alr-polygons/resource/d35b18b0-ecfa-468b-b31a-2bf16a459a7c (accessed on 3 July 2023).
41. Government of Canada Land Cover of Canada. Available online: https://open.canada.ca/data/en/dataset/4e615eae-b90c-420b-adee-2ca35896caf6 (accessed on 3 July 2023).
42. Government of Canada Lakes and Rivers Boundary Files. Available online: https://open.canada.ca/data/en/dataset/d0cdef71-9343-46c3-b2e7-c1ded5907686 (accessed on 3 July 2023).
43. Government of Canada Canadian Digital Elevation Model. Available online: https://open.canada.ca/data/en/dataset/7f245e4d-76c2-4caa-951a-45d1d2051333 (accessed on 3 July 2023).
44. Danielsson, P.-E. Euclidean Distance Mapping. *Comput. Graph. Image Process.* **1980**, *14*, 227–248. [CrossRef]
45. Deichmann, U.; Balk, D. *Transforming Population Data for Interdisciplinary Usages: From Census to Grid*; Center for International Earth Science Information Network: Washington, DC, USA, 2001.
46. Dulal, C.R.; Thomas, B. A Grid-Based Approach for Refining Population Data in Rural Areas. *J. Geogr. Reg. Plan.* **2014**, *7*, 47–57. [CrossRef]
47. City of Kelowna. *Pedestrian Protection: Requirements for Construction Sites*; City of Kelowna: Kelowna, BC, Canada, 2019.
48. Michelle, B.; Margaret, G.; Joanne, P.; Moreno, R.; Lyle, W.; Greg, Y. Translink Transit-Oriented Communities Design Guidelines Creating More Livable Places around Transit in Metro Vancouver. 2012. Available online: https://www.translink.ca/-/media/translink/documents/plans-and-projects/managing-the-transit-network/transit-oriented-communities/transit_oriented_communities_design_guidelines.pdf (accessed on 10 December 2023).

49. Hatch, K.; Dragićević, S.; Dujmović, J. Logic Scoring of Preference and Spatial Multicriteria Evaluation for Urban Residential Land Use Analysis. In *Geographic Information Science*; Duckham, M., Pebesma, E., Stewart, K., Frank, A.U., Eds.; Lecture Notes in Computer Science; Springer International Publishing: Cham, Switzerland, 2014; Volume 8728, pp. 64–80, ISBN 978-3-319-11592-4.
50. City of Kelowna. *Proposed Zoning Bylaw No. 12375*; City of Kelowna: Kelowna, BC, Canada, 2022.
51. City of Kelowna. *Urban Centres Roadmap: Final Report*; City of Kelowna: Kelowna, BC, Canada, 2016.
52. Shen, S.; Dragićević, S.; Dujmović, J. GIS-Based Logic Scoring of Preference Method for Urban Densification Suitability Analysis. *Comput. Environ. Urban Syst.* **2021**, *89*, 101654. [CrossRef]
53. City of Kelowna. *Urban Forest Strategy*; City of Kelowna: Kelowna, BC, Canada, 2011.
54. Al-Ghorayeb, A.; Al-Shaar, W.; Elkordi, A.; Faour, G.; Al-Shaar, M.; Attalah, Y. Land Suitability Analysis for Sustainable Urban Development: A Case of Nabatiyeh Region in Lebanon. *J* **2023**, *6*, 267–285. [CrossRef]
55. City of Kelowna. *Hillside Development Guidelines*; City of Kelowna: Kelowna, BC, Canada, 2009.
56. Mouratidis, K. Is Compact City Livable? The Impact of Compact versus Sprawled Neighbourhoods on Neighbourhood Satisfaction. *Urban Stud.* **2018**, *55*, 2408–2430. [CrossRef]
57. Daoud, J.I. Multicollinearity and Regression Analysis. *J. Phys. Conf. Ser.* **2017**, *949*, 012009. [CrossRef]
58. Munda, G. Social Multi-Criteria Evaluation of Policy Options. In *Multicriteria Decision Aiding Interventions: Applications for Analysts*; Norese, M.F., De Vicente y Oliva, M.A., Abi-Zeid, I., Eds.; Springer International Publishing: Cham, Switzerland, 2023; pp. 217–233, ISBN 978-3-031-28465-6.
59. Vogel, R. Methodology and Software Solutions for Multicriteria Evaluation of Floodplain Retention Suitability. *Cartogr. Geogr. Inf. Sci.* **2016**, *43*, 301–320. [CrossRef]
60. Eastman, J.R. Multi-Criteria Evaluation and GIS. In *Multi-Criteria Evaluation and GIS*; John Wiley & Sons: Hoboken, NJ, USA, 1999; pp. 493–502, ISBN 978-0-471-98635-7.
61. Darko, A.; Chan, A.P.C.; Ameyaw, E.E.; Owusu, E.K.; Pärn, E.; Edwards, D.J. Review of Application of Analytic Hierarchy Process (AHP) in Construction. *Int. J. Constr. Manag.* **2019**, *19*, 436–452. [CrossRef]
62. Breiman, L. Random Forests. *Mach. Learn.* **2001**, *45*, 5–32. [CrossRef]
63. Chen, T.; Guestrin, C. XGBoost: A Scalable Tree Boosting System. In Proceedings of the 22nd ACM SIGKDD International Conference on Knowledge Discovery and Data Mining, San Francisco, CA, USA, 13–17 August 2016; ACM: New York, NY, USA; pp. 785–794.
64. Boser, B.E.; Guyon, I.M.; Vapnik, V.N. A Training Algorithm for Optimal Margin Classifiers. In Proceedings of the Fifth Annual Workshop on Computational Learning Theory, Pittsburgh, PA, USA, 27–29 July 1992; ACM: New York, NY, USA; pp. 144–152.
65. Rather, T.; Kumar, S.; Khan, J. Using Machine Learning to Predict Habitat Suitability of Sloth Bears at Multiple Spatial Scales. *Ecol. Process.* **2021**, *10*, 48. [CrossRef]
66. Nguyen, Q.H.; Ly, H.-B.; Ho, L.S.; Al-Ansari, N.; Le, H.V.; Tran, V.Q.; Prakash, I.; Pham, B.T. Influence of Data Splitting on Performance of Machine Learning Models in Prediction of Shear Strength of Soil. *Math. Probl. Eng.* **2021**, *2021*, 4832864. [CrossRef]
67. Singh, V.; Pencina, M.; Einstein, A.J.; Liang, J.X.; Berman, D.S.; Slomka, P. Impact of Train/Test Sample Regimen on Performance Estimate Stability of Machine Learning in Cardiovascular Imaging. *Sci. Rep.* **2021**, *11*, 14490. [CrossRef]
68. Avand, M.; Moradi, H.; Lasboyee, M.R. Spatial Modeling of Flood Probability Using Geo-Environmental Variables and Machine Learning Models, Case Study: Tajan Watershed, Iran. *Adv. Space Res.* **2021**, *67*, 3169–3186. [CrossRef]
69. Mohsin, M.; Ali, S.A.; Shamim, S.K.; Ahmad, A. A GIS-Based Novel Approach for Suitable Sanitary Landfill Site Selection Using Integrated Fuzzy Analytic Hierarchy Process and Machine Learning Algorithms. *Environ. Sci. Pollut. Res.* **2022**, *29*, 31511–31540. [CrossRef] [PubMed]
70. Chen, Y.; Chen, W.; Janizadeh, S.; Bhunia, G.S.; Bera, A.; Pham, Q.B.; Linh, N.T.T.; Balogun, A.-L.; Wang, X. Deep Learning and Boosting Framework for Piping Erosion Susceptibility Modeling: Spatial Evaluation of Agricultural Areas in the Semi-Arid Region. *Geocarto Int.* **2022**, *37*, 4628–4654. [CrossRef]
71. Chicco, D.; Jurman, G. The Advantages of the Matthews Correlation Coefficient (MCC) over F1 Score and Accuracy in Binary Classification Evaluation. *BMC Genom.* **2020**, *21*, 6. [CrossRef] [PubMed]
72. Esri ArcGIS Pro. Available online: https://www.esri.com/en-us/arcgis/products/arcgis-pro/overview (accessed on 1 April 2024).
73. Microsoft. Microsoft 365. Available online: https://www.microsoft.com/en-ca/microsoft-365/get-started-with-office-2021 (accessed on 23 June 2024).
74. Pedregosa, F.; Varoquaux, G.; Gramfort, A.; Michel, V.; Thirion, B.; Grisel, O.; Blondel, M.; Prettenhofer, P.; Weiss, R.; Dubourg, V.; et al. Scikit-Learn: Machine Learning in Python. *J. Mach. Learn. Res.* **2011**, *12*, 2825–2830.
75. Montgomery, B.; Dragićević, S. Comparison of GIS-Based Logic Scoring of Preference and Multicriteria Evaluation Methods: Urban Land Use Suitability. *Geogr. Anal.* **2016**, *48*, 427–447. [CrossRef]
76. Shuaibu, J.A.; Kara, C. Evaluating Suitability for Sustainable Urban Growth of Abuja by Using MCE and GIS. *Int. J. Adv. Appl. Sci.* **2019**, *6*, 68–76. [CrossRef]
77. Ustaoglu, E.; Aydınoglu, A.C. Suitability Evaluation of Urban Construction Land in Pendik District of Istanbul, Turkey. *Land Use Policy* **2020**, *99*, 104783. [CrossRef]
78. Rienow, A.; Mustafa, A.; Krelaus, L.; Lindner, C. Modeling Urban Regions: Comparing Random Forest and Support Vector Machines for Cellular Automata. *Trans. GIS* **2021**, *25*, 1625–1645. [CrossRef]

79. Roy Chowdhury, P.K.; Bhaduri, B.L.; McKee, J.J. Estimating Urban Areas: New Insights from Very High-Resolution Human Settlement Data. *Remote Sens. Appl. Soc. Environ.* **2018**, *10*, 93–103. [CrossRef]
80. Congalton, R.G. A Review of Assessing the Accuracy of Classifications of Remotely Sensed Data. *Remote Sens. Environ.* **1991**, *37*, 35–46. [CrossRef]
81. Romano, G.; Dal Sasso, P.; Trisorio Liuzzi, G.; Gentile, F. Multi-Criteria Decision Analysis for Land Suitability Mapping in a Rural Area of Southern Italy. *Land Use Policy* **2015**, *48*, 131–143. [CrossRef]

land

MDPI

Article

Simulation of Urban Growth Boundary under the Guidance of Stock Development: A Case Study of Wuhan City

Yang Zhang [1,2,3,4], Xiaojiang Xia [3,*], Jiandong Li [5], Luge Xing [3], Chengchao Yang [3], Haofeng Wang [6,7], Xiaoai Dai [3,4,8] and Jue Wang [4,9]

1 School of Architecture, Southeast University, Nanjing 210096, China; zhangyang2021@seu.edu.cn
2 Key Laboratory of Digital Mapping and Land Information Application, Ministry of Natural Resources, Wuhan 430079, China
3 College of Geography and Planing, Chengdu University of Technology, Chengdu 610059, China; ruger_xing@stu.cdut.edu.cn (L.X.); yangchengchao@stu.cdut.edu.cn (C.Y.); daixiaoa@cdut.edu.cn (X.D.)
4 Digital Hu Huanyong Line Research Institute, Chengdu University of Technology, Chengdu 610059, China; wangjue19@cdut.edu.cn
5 School of Economics and Management, Southwest Petroleum University, Chengdu 610500, China; 202321000593@stu.swpu.edu.cn
6 Information Center of Natural Resources and Planning of Wuhan City, Wuhan 430014, China; hhfwang2009@whu.edu.cn
7 Key Laboratory of City Simulation, Ministry of Natural Resources, Wuhan 430014, China
8 State Key Laboratory of Geohazard Prevention and Geoenvironment Protection, Chengdu 610059, China
9 Business School, Chengdu University of Technology, Chengdu 610059, China
* Correspondence: xiaxiaojiang@cdut.edu.cn

Abstract: The implementation of an urban growth boundary (UGB) can effectively control urban sprawl and promote efficient land use, which is crucial for future urban development. However, most of existing studies overlook the reuse of existing idle and inefficient land within the city in the delineation of UGBs. With China's urban construction shifting from incremental development to stock development, this study focuses on Wuhan and presents a set of technical approaches for delineating UGBs with a stock development orientation. First, a built-up area composite index (POI&ISA) is constructed based on point of interest (POI) kernel density analysis and impervious surface index extraction to evaluate constructive levels in 2010 and 2020 and identify the urban vitality zone. Then, we combine the current land use status and control policies to divide the urban spatial development potential into five categories: urban vitality land, urban non-vitality land, other vitality land, other non-vitality land, and restricted development land. Finally, the PLUS model is applied in the analysis of the driving forces of land use change in Wuhan, simulating the UGBs in three stages of incremental development (2020–2030), incremental and stock development (2030–2040), and stock development (2040–2050). Finally, the PLUS model simulation projects the UGB areas to be 436.436 km^2, 474.617 km^2, and 520.396 km^2 for the years 2030, 2040, and 2050, respectively. The predicted timespan of urban development extends up to 30 years, serving as a reliable reference for Wuhan's long-term and near-term planning.

Keywords: stock development; POI&ISA index; urban growth boundary; PLUS model; Wuhan city

1. Introduction

The rapid development of technology and globalization have led to an unprecedented growth in the global urbanization process. From 1985 to 2015, the global urban area increased from 362,700 km^2 to 653,400 km^2, nearly doubling. While this swift urbanization has presented significant economic and population development opportunities, it has also given rise to challenges such as excessive internal urban pressure and difficulties in urban expansion [1,2]. To address these challenges, organizations such as the United Nations Conference on Housing and Sustainable Urban Development have stressed the importance of unlocking urban potential and promoting sustainable urban development [3]. This is crucial to prevent chaotic urban expansion and marginalization.

China, as one of the most populous countries globally, has undergone rapid urbanization. However, the prevailing development paradigm prioritizing "quantity over quality" has led to a myriad of issues, including resource depletion, environmental degradation, and inefficient land utilization [4,5]. The trajectory of urbanization is gradually shifting from the "incremental era" to the "stock era". This transition underscores that the conventional approach of solely advocating urban expansion and spatial development is no longer conducive to current urban progress. Instead, a more sustainable approach is needed. Stock development, which emphasizes the optimal utilization of existing urban resources over mere urban boundary expansion, involves comprehensive investment and transformation of cities to modernize urban functions, repurpose urban spaces, and rejuvenate urban development [6–8]. This change in the concept of urban development undoubtedly puts more emphasis on sustainability [9].

The urban growth boundary (UGB) is a crucial tool for managing urban expansion [10]. Its origins can be traced back to the Garden City theory proposed by British scholar Howard in his seminal work *Garden Cities of To-Morrow* in 1898 [11]. The theory aimed to curb urban sprawl by establishing a robust "green belt" between the central city and surrounding towns to segregate them and prevent indiscriminate urban sprawl. In the 1930s, the United States government began establishing new towns in rural areas with development potential to alleviate the "urban disease" in large cities. These new towns were surrounded by farmland and trees, forming a barrier that restricted the disorderly expansion of urban areas. They were called green belt towns. In1938, the British government enacted the Green Belt Act, marking the beginning of large-scale green belt construction in London. Subsequently, in 1976, the city of Salem, USA, defined the UGB as the delineation between urban and rural areas, beyond which urban development is restricted [12]. Over time, its definition has expanded to encompass various regulatory aspects, including local government oversight of expansion [13]. In China, since the incorporation of delineating UGBs in the "Urban Planning Formulation Means" in 2006, numerous scholars have undertaken related research, and many large cities have carried out the delineation of UGBs [14–17]. The UGB holds significant importance in steering prudent urban development, conserving the ecological environment, and augmenting urban quality [18,19].

In prior research, scholars have extensively explored UGBs and proposed various methods for their delineation, both domestically and internationally. These methods include dynamic simulation approaches, backward deduction methods, and comprehensive methodologies [10]. The dynamic simulation method is qualitative and treat the city as a complex dynamic system that evolves with societal changes and economic progress [20,21]. To employ this method, it is essential to consider a multitude of factors influencing urban development and constraints on urban expansion, such as population size, GDP, and land demand [22]. Ultimately, geographic simulation models yield simulated results for UGBs. However, this method is highly subjective and heavily reliant on researchers' experience. The backward deduction method is quantitative and begins with policies and laws related to urban expansion and ecological preservation to demarcate areas restricted or prohibited for development [23–25]. This approach establishes clear rigid boundaries for future urban expansion or internal renewal. Nonetheless, it can only delineate rigid boundaries, lacking flexibility to respond to various changes with uncertainty. The comprehensive method integrates dynamic simulation and backward deduction methods, considering both future urban development needs and delineating rigid boundaries during urban expansion. It reflects the dynamic change in urban construction space more scientifically, and has been increasingly applied in the delineation practice of UGBs [26,27].

To accurately delineate UGBs, it is crucial to conduct dynamic simulation of future urban development. One commonly used method is the cellular automata (CA) model, which is a kind of discrete grid dynamical model that considers time, space, and state as discrete variables. It also takes into local spatial interactions and causal relationships over time, making it an effective tool for simulating the simulate the spatiotemporal evolution process of complex systems and is widely utilized for delineating UGBs [28]. Several expanded iterations of the CA model have been developed for delineating UGBs, including the conversion of land use and its effects at small regional extent (CLUE-S) [29,30], slope–land use–exclusion–urban–transportation–hillshade (SLEUTH) [31], artificial neural network cellular automata (ANN-CA) [32], and future land-use simulation (FLUS) [33]. Among these models, the CLUE-S model stands out for its ability to comprehensively consider both natural and human factors. By using iterative method, it can integrate spatial and non-spatial analyses to better simulate land use change scenarios at small scales [34]. The SLEUTH model can combine large-scale spatial databases and various resolution remote sensing data to simulate and predict changes in urban land use from micro to macro and from 10 years to 100 years on spatial and temporal scales [35]. ANN-CA is a simple model that does not require manual determination of model structure, conversion rules, and model parameters. Instead, it uses neural networks to replace conversion rules and automatically obtains model weight parameters through sample training of the neural network [36]. The FLUS model is a simulation and prediction model that can effectively deal with the complexity and uncertainty of mutual conversion between various land use types within the national territory under the influence of human–land relationships [37]. However, in the transition phase of most models, the competition and interaction among different land types are often neglected. The patch-generating land use simulation (PLUS) model, a derivative of the CA model, effectively tackles the complexity and uncertainty of urban growth by considering the combined impact of natural and human activities. This approach enhances simulation accuracy and enables the delineation of UGBs [38–40].

Although many cities in China have experienced land urbanization faster than population urbanization, the current delineation of UGB only takes the incremental expansion of cities into account. This delineation process often overlooks the reuse of existing idle and inefficient land within the city, resulting in a tendency to overestimate the determined scale of urban development. In the case of urbanization shifting from incremental development to stock development, it is necessary to coordinate the redevelopment within the city and the external expansion, delineate the UGBs at different stages scientifically, and ensure the orderly and efficient development of urban space.

This paper presents technical approaches to delineate UGBs in a manner that promotes intensive development and reduces the extensive use of land. This study focuses on Wuhan City, Hubei Province, China and aims to achieve three objectives: (1) identify urban vitality zone and non-vitality zones; (2) divide the urban spatial development potential; and (3) simulate future UGBs for different periods. The remainder of this paper is structured as follows: Section 2 provides a detailed description of the study area and data sources. Section 3 presents a set of technical approaches and main processes in detail for delineating UGBs, including the POI&ISA index, classification of urban spatial development potential, and PLUS model. Section 4 shows the delineation results of UGBs in 2030, 2040, and 2050. Finally, Section 5 summarizes the main contributions of this research.

2. Study Area and Data

2.1. Overview of the Study Area

Wuhan is located in the eastern part of Hubei Province, in the central region of China (Figure 1). The terrain is mostly flat, with hills and low mountains situated in the north and south. The western and southern regions have higher elevations, while the eastern and northern regions are lower. Wuhan situated at the confluence of the Yangtze River and the Han River, and the urban area has numerous lakes. The city has a typical subtropical monsoon climate, with four distinct seasons, ample sunshine, abundant rainfall, and plentiful water and heat resources.

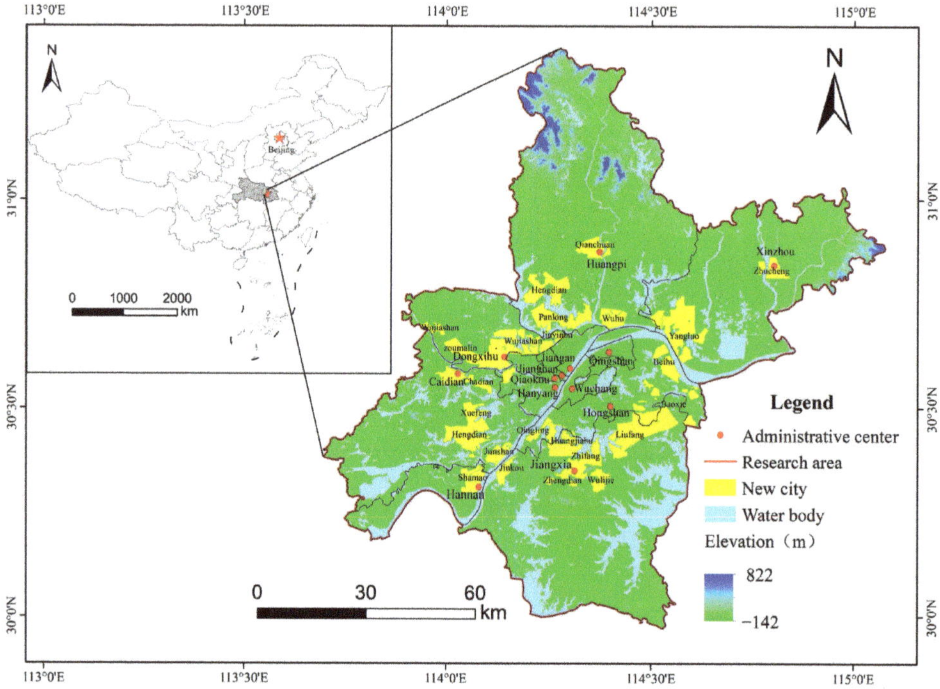

Figure 1. Study area.

Since the late 1970s, the Wuhan government implemented a strategy to expand the city in all directions. This involved integrating neighboring towns and cities into the scope of administrative division, eventually forming 13 districts. It includes seven central urban districts—Hongshan District, Wuchang District, Jianghan District, Jiang'an District, Qingshan District, Qiaokou District, and Hanyang District—and six new urban districts: Jiang xia District, Xinzhou District, Hannan District, Caidian District, Dongxihu District, and Huangpi District. For a long time, the urban area of Wuhan has been expanding mainly within the boundaries of the seven central urban districts, along the banks of the Yangtze River and the Han River. However, in recent years, due to the rapid development of the economy and society and the acceleration of urbanization, not only has the traditional urban area of Wuhan continued to expand, but the urban scale of six new urban districts has also greatly increased. From 1988 to 2011, the urban built-up area of Wuhan increased rapidly [38].

Despite the rapid urban expansion in Wuhan, its internal land use efficiency is suboptimal, resulting in the wastage of different types of land. Additionally, the significance of the old urban areas is gradually diminishing. The Wuhan City Master Plan (2010–2020) emphasizes the importance of transforming old urban areas, enhancing their vitality, and promoting the intensive and economical use of land to achieve sustainable development and ensure the long-term growth of Wuhan. At the same time, the plan proposes to actively promote the development of new cities and build 24 new cities and new city clusters. Therefore, it is necessary to scientifically delineate the UGB of Wuhan and guide the sustainable use of land.

2.2. Data Sources

This study utilized data from four main sources: remotely sensed imagery data, land use and control line data, point of interest (POI) data, and additional impact factors data (Table 1). Remotely-sensed imagery data of Landsat 5 in 2010 and Landsat 8 in 2020 were used for Wuhan. Land use data from 2010 to 2020 were acquired from Chinese Academy of Sciences Resource and Environmental Science Data Platform. Land control line data included the ecological red line and basic farmland preservation area from relevant planning schemes such as Wuhan Land Use Master Plan and 1:2000 Ecological Control Line Plan. The POI data used in this study were acquired via the application programming interface provided by Baidu Maps. These data included businesses, restaurants, attractions, commercial services, and public services, which represented real geographic entities. Each data segment includes information such as name, address, category, and latitude and longitude coordinates.

Table 1. Data classification and sources.

Classification	Data Name	Data Source
Remotely sensed imagery data	Wuhan City Landsat 5 and Landsat 8 data	Geospatial Data Cloud (https://www.gscloud.cn/) (accessed on 17 May 2022)
Land use and control line data	Wuhan City land use data for 2010 and 2020	Chinese Academy of Sciences Resource and Environment Science Data Center (https://www.resdc.cn/) (accessed on 1 February 2022)
	Ecological red line data Basic farmland data	Overall land use planning of Wuhan City and 1:2000 ecological control line planning
Points of interest (POI) data	POI	Baidu Maps (https://map.baidu.com/) (accessed on 22 September 2022)
Additional impact factors data	DEM	Geospatial Data Cloud (https://www.gscloud.cn/) (accessed on 17 May 2022)
	Slope	Extraction of DEM data
	PD GDP Administrative boundary	Chinese Academy of Sciences Resource and Environment Science Data Center (https://www.resdc.cn/) (accessed on 31 December 2021)
	Expressways Primary roads Secondary roads Railways	Open Street Map (https://www.openstreetmap.org/) (accessed on 10 March 2023)

The additional impact factors data mainly involve traffic network data, digital elevation model (DEM), population density (PD), and gross domestic product (GDP). The Wuhan traffic network data were sourced from the Open Street Map platform and encompasses highways, primary roads, secondary roads, and railways. DEM, PD, and GDP are all from Chinese Academy of Sciences Resource and Environmental Science Data Platform.

To ensure consistency with the study area's extent, two remote sensing images for each year collected were mosaiced using ENVI 5.3, and the other acquired data underwent clipping and masking using ArcGIS 10.8. Simultaneously, the spatial coordinates were set to WGS_1984_EASE-Grid_2.0_Global, with 4440 rows and 5106 columns, maintaining a uniform resolution of 30 $\times$ 30 m, as required.

3. Methods and Processes

The research process is divided into three main steps: Firstly, we construct a POI&ISA index of the study area to identify urban vitality zones in both 2010 and 2020. Secondly, we divide the urban spatial development potential based on the current land use status and control policies. Finally, we apply the PLUS model to simulate the UGBs oriented towards stock development in different periods. The research framework is illustrated in Figure 2.

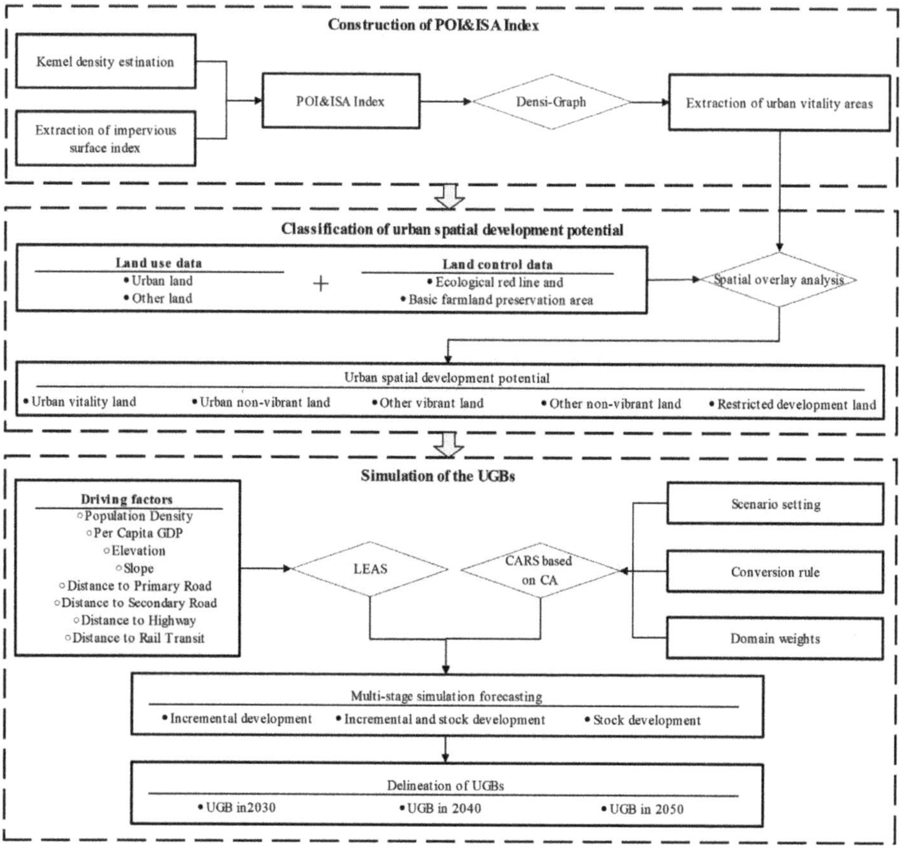

Figure 2. Research processes.

3.1. POI&ISA Index

3.1.1. Kernel Density Estimation

The kernel density estimation is based on a regular circle with a specific bandwidth around any point of interest in the study area. The weights are determined by the distance from the center point, with closer distances receiving greater weight. The estimated density of each point is the weighted density of all the points in the area. The calculation formula is as follows:

$$Q_i = \frac{1}{m\pi D^2} \times \sum_{j=1}^{m} \left[E_j \left(1 - \frac{L^2{}_{ij}}{D^2} \right)^2 \right]$$

where Qi is the kernel density of point i, D is the bandwidth, Ej is the weight of research object j, L_{ij} is the Euclidean distance between point i and research object j, and m is the number of research object j within the bandwidth.

In general, the density of interest points in urban areas is higher than in rural areas. Therefore, this study conducts kernel density analysis on five major types of points of interest within the administrative districts of Wuhan, namely, transport facilities, public facilities, healthcare, education, and business services, during the two periods of 2010 and 2020. The kernel density analysis was performed using ArcGIS 10.8 software, with a bandwidth of 1000 m and a kernel size of 30 m, to meet the requirement of the impervious surface index pixel. The resulting kernel density distribution of POIs in Wuhan is depicted in Figure 3.

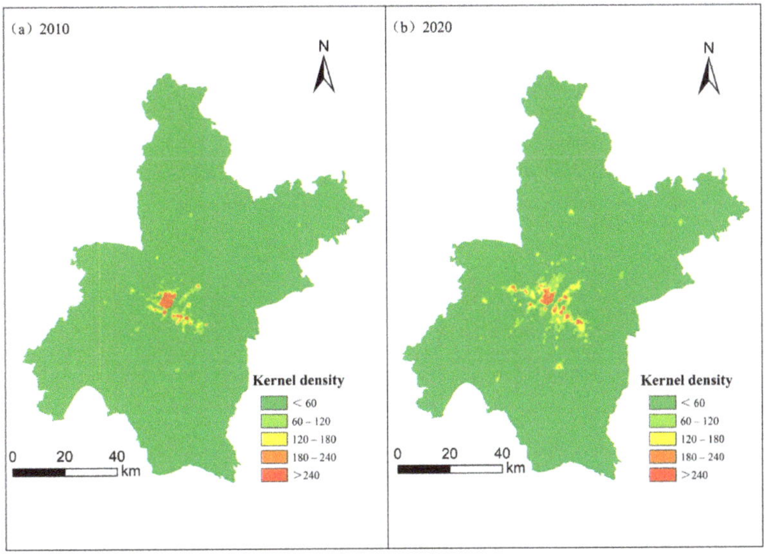

Figure 3. Kernel density of POIs in Wuhan. (**a**) 2010, (**b**) 2020.

The figure illustrates the POI kernel density values in Wuhan City for the years 2010 and 2020. It is evident that the areas with high kernel density in Wuhan are primarily situated within the seven central urban districts, especially in the regions along the Yangtze and Han rivers. In contrast, the six new districts exhibit high kernel density only in the areas where the respective district governments are located and, in some parts, adjacent to the central urban districts. However, these areas are less concentrated and relatively smaller in scale.

A comparison of the results from the two years clearly indicates that over the decade, the high kernel density areas in Wuhan have expanded to some extent. The added high-density zones in the city center are mainly concentrated along the Han River and Yangtze River, while in the new urban areas, expansion is primarily around district administrative centers.

3.1.2. Extraction of Impervious Surface Index

This study adopts the V-I-S model proposed by Ridd [41], recognizing the varying spectral reflectance of features across different bands in remote sensing images. After performing radiometric calibration, atmospheric correction, and mosaic cropping on the remote sensing images, we identified four spectral feature combinations: high albedo, low albedo, bare soil, and vegetation. We then applied a linear spectral mixture decomposition model to extract the impervious surface index of Wuhan in 2010 and 2020 by summing up the abundance of the high albedo end-members and the low albedo end-members.

The linear spectral mixture decomposition model expresses the reflectance of a single image element in each spectral band as a linear combination of its unit component reflectances and their respective abundances. This is calculated using the following formula:

$$L_b = \sum_{j=1}^{M} h_i L_{i,b} + a_b$$

where L_b the spectral reflectance in band b; M is the number of end-elements; h_i is the proportional weight of end-element i in the image element; $L_{i,b}$ is the reflectance of end-element i in the b-band; and a_b is the value of the unmodelled residual error.

To ensure accurate extraction results of the impervious surface index, this study removes the water body through a mask before applying the linear spectral mixture decomposition model. The resulting impervious surface index of Wuhan in 2010 and 2020 is presented in the Figure 4.

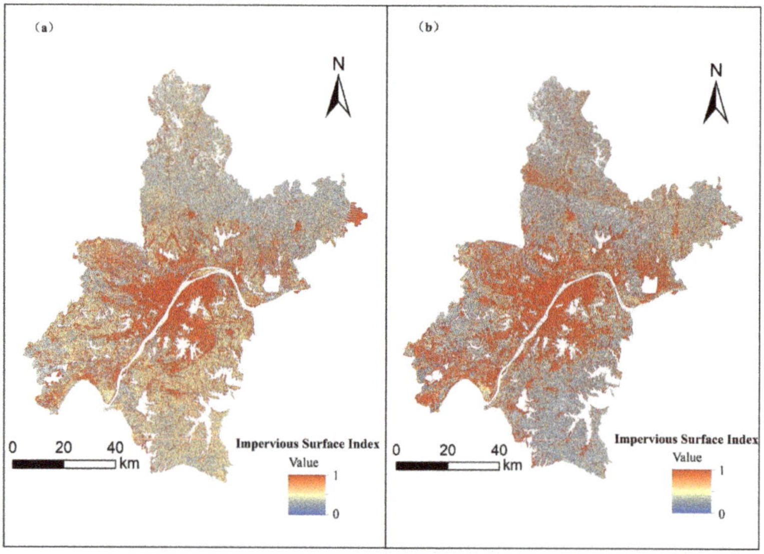

Figure 4. Impervious surface index in Wuhan. (**a**) 2010, (**b**) 2020.

The impervious surface index ranges from 0 to 1, with higher values indicating a higher proportion of artificial surface and lower values indicating a lesser proportion, as illustrated in the figure. In terms of regional distribution, areas with high impervious surface index in Wuhan are mainly concentrated in the seven central urban districts. Furthermore, other areas with high impervious surface index are concentrated around the district offices of each new urban district, as well as some new cities and new city clusters.

A comparison of the impervious surface results between the two periods reveals that the extent of impervious areas in Wuhan has increased considerably over the decade, reflecting the rapid development of urban construction. The expansion trend is categorized into two forms. The first form is characterized by high impervious surface index areas in the central urban districts, which radiate in all directions. For example, impervious surfaces have expanded along the Han River towards and the Yangtze River. The second form is based on the sites of each new district government or other areas with high impervious surface index, expanding into suitable development areas. For instance, the area with high impervious surface index in Huangpi District has expanded in all directions.

3.1.3. POI&ISA Index Construction

It is important to note that using only POI kernel density or impervious surface index to extract urban vitality zones can lead to inaccurate results. For instance, some villages with complete public service facilities may have high POI kernel density values, but they do not necessarily belong to the category of urban vitality zones. Similarly, some mining and industrial areas may have high impervious surface indexes, but they do not meet the requirements of urban vitality zones and should not be classified as such. To minimize these shortcomings, this paper proposes using the POI kernel density and impervious surface index to construct the built-up area integrated index (POI&ISA index). This approach avoids the limitations of using only the POI kernel density and impervious surface index to extract urban vitality zones, while also reducing errors in the extraction process. To construct the POI&ISA index, this paper sets the weights to 0.5 and combines the two types of data using the geometric mean. This eliminates the influence of extreme values and differences in data magnitude. The formula for calculating the POI&ISA index is as follows:

$$POIISA_i = \sqrt{POI_i \times ISA_i}$$

$POI\&ISA_i$ in the formula is POI&ISA index, POI_i is the kernel density value at point i, and ISA_i is the impervious surface index at point i.

In ArcGIS 10.8, the POI kernel density data and impervious surface index data were imported. The raster calculator tool was then used to obtain the POI&ISA index. The results are shown below (Figure 5).

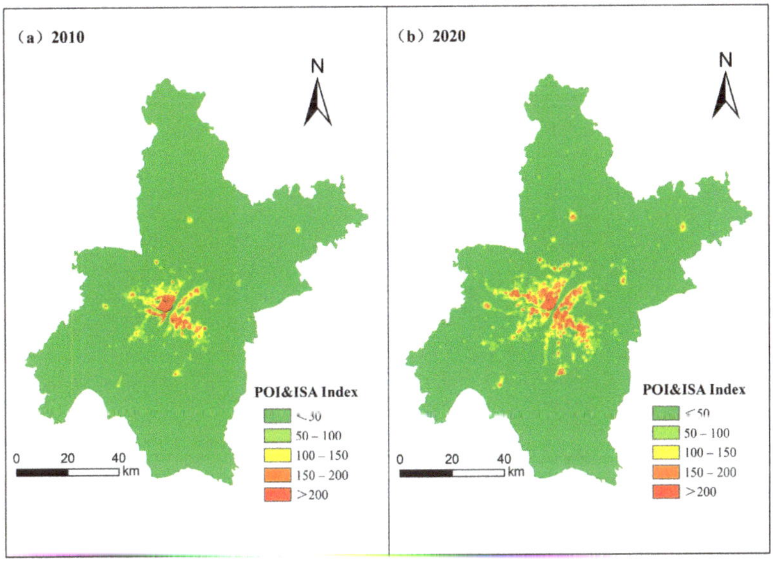

Figure 5. POI&ISA index in Wuhan. (**a**) 2010, (**b**) 2020.

The figure illustrates that the seven central urban districts of Wuhan have high POI&ISA index in both 2010 and 2020, indicating their strong foundation for development. These districts are situated adjacent to the Yangtze and Han Rivers, providing them with prominent location advantages. With flat terrain, well-developed infrastructure, high road network density, and a relatively prosperous economy, these areas are conducive to both production and living, resulting in large populations, high densities of artificial surface, well-developed secondary and tertiary industries, and high urban vitality. Surrounding these high-value areas are sub-high-value zones distributed in gradients, which are crucial regions for Wuhan's urban expansion. In the six new urban districts, regions with higher POI&ISA index are classified into two main categories. The first category includes areas where new urban district government offices are located, such as Xinzhou and Huangpi District. The second category are mainly some new cities such as Zhifang, Yangluo, and Wujiashan. Although these areas have relatively low levels of development compared to the central urban districts, they have great potential for future growth due to their location advantages and government support.

A comparison of the POI&ISA index of Wuhan between the two periods reveals that newly expanded areas with high POI&ISA index are primarily developed by spreading from the original high POI&ISA index areas. These areas have faster economic development, higher population density, and better infrastructure. The POI&ISA index of some new cities and new city clusters like Wujiashan, Caidian, Yangluo, Zhifang, Changfu, Zhuancheng, Qianchuan, and Beihu are increasing significantly, leading to an expansion of areas with high POI&ISA index.

3.1.4. Extraction of Urban Vitality Zones

Urban vitality zones are identified based on the obtained POI&ISA index by determining a threshold value. This is achieved by selecting several different POI&ISA index using the ArcGIS10.8 reclassification tool. The generated zones are then equated into a regular circle and a Densi-Graph folded line graph is constructed by taking the corresponding radius increment Δd of the circle as the Y-axis according to the area formula of the circle. The critical point between the high and low POI&ISA index is considered as the point of sudden change of slope, which indicates the boundary of the urban vitality zones. The resulting Densi-Graph folding line is shown below (Figure 6).

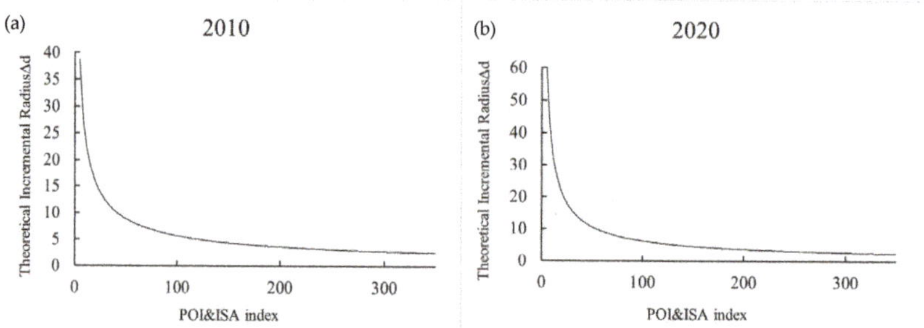

Figure 6. Densi-Graph diagram of Wuhan. (**a**) 2010, (**b**) 2020.

The figure illustrates a sudden change in slope for the POI&ISA index in 2010 and 2020 at approximately 25. Therefore, the POI&ISA index of 25 are selected as the demarcation points between urban vitality and non-vitality zones for the two periods, as shown in Figure 7.

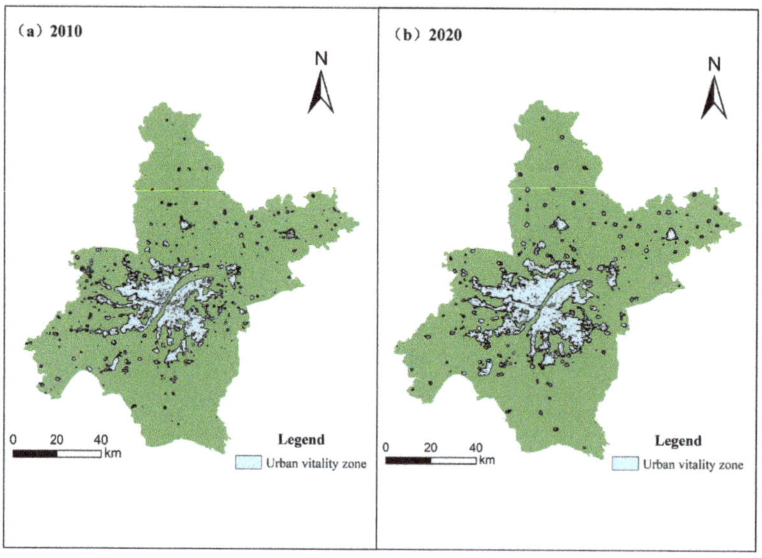

Figure 7. Urban vitality zone extraction results in Wuhan. (**a**) 2010, (**b**) 2020.

The total area of urban vitality zones increased from 978.94 km^2 in 2010 to 1171.076 km^2 over the following ten years, representing a 19.62% cumulative increase, or 192.136 km^2. Urban vitality zones are concentrated in seven central urban areas. In the six new urban districts, urban vitality zones are classified into three types: those distributed along rivers and lakes, those located in the streets where the district governments are situated, and discrete areas in each district that are more developed.

When comparing the results of urban vitality zones between the two periods, three main trends in the expansion of Wuhan's urban vitality zone were identified. The first trend is the continuous improvement of the original urban vitality zones by filling in the central urban districts, allowing for their ongoing development. The second trend is the expansion of the urban vitality zones along the Yangtze River to Wujiashan, Caidian, and Yangluo, as well as Changfu, Zhifang, and Panlong. The third trend involves expanding the areas surrounding the district government offices in each new urban districts or the better-developed areas to include surrounding regions. For example, this could include the expansion of former areas such as Qianchuan in Huangpi District, Zhucheng in Xinzhou District, and Shamao in Hannan District.

3.2. Classification of Urban Spatial Development Potential

The extracted urban vitality zones should be combined with Wuhan's land use data, ecological protection zones, basic farmland protection zones, and water body data. These combined data can then be used to identify five types of land: urban vitality land, urban non-vitality land, other vitality land, other non-vitality land, and restricted development land (Table 2). These types of land can be identified using tools such as "intersection inverse", "merge", and "reclassification" in ArcGIS 10.8. Figure 8 shows the results of identifying five types of land.

Table 2. Identification of the five land types.

Land Types	Specific Description	Development Potential Description
Urban vitality land	Belong to urban land, with high POI&ISA index	Low development potential, should be preserved for the future
Urban non-vitality land	Belong to urban land, but with low POI&ISA index, and not within ecological protection areas or basic farmland protection areas	It serves as the primary source of land for future urban expansion
Other vitality land	Not classified as urban land, ecological protection areas, or basic farmland protection areas, but with high POI&ISA index	In the future development process, it is more likely to be developed into urban areas
Other non-vitality land	Not classified as urban land, ecological protection areas, or basic farmland protection areas, and with low POI&ISA index	In the future development process, the likelihood of development into urban areas is low
Restricted development land	This type of land includes ecological protection areas, basic farmland protection areas, and water bodies in Wuhan	It constitutes the impassable, rigid boundaries during the future urban development process

The concentration of urban vitality land in Wuhan can be observed in Jiang'an, Hannan, Qiaokou, Hanyang, Qingshan, and Wuchang Districts, and the western and southern parts of Hongshan. The urban non-vitality land is primarily situated in the eastern part of Qingshan District. The other vitality land is mainly distributed in the periphery of urban vitality land, and other non-vitality land is generally in the periphery of other vitality land.

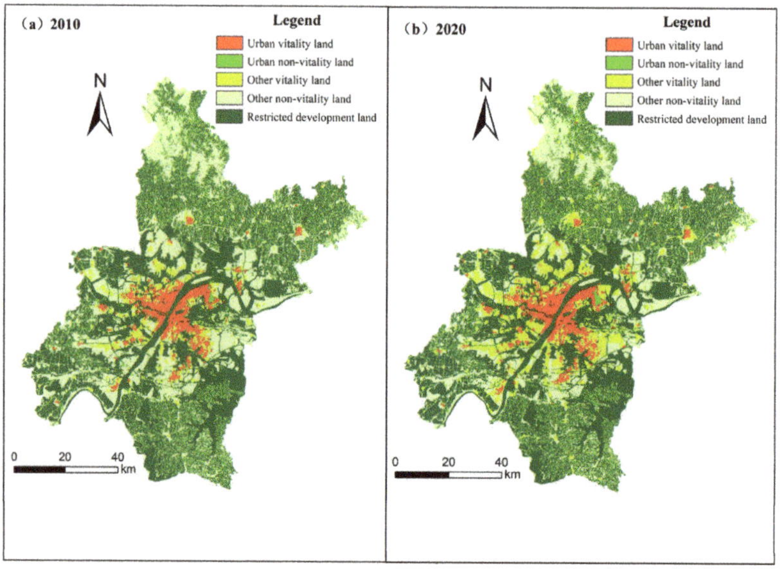

Figure 8. Land types of urban spatial development potential in Wuhan. (**a**) 2010, (**b**) 2020.

When comparing the land type classification results of urban spatial development potential in 2010 and 2020, Wuhan's urban vitality land development trend can be divided into two scenarios. Firstly, the land within the urban vitality zones has gradually improved, increasing the internal vitality of these zones. Secondly, the increasing efficiency of land use in the surrounding areas gradually converts the non-vitality land into urban vitality land. The trend in Wuhan's development of other vitality land is primarily to transform some of the non-vitality zones into vitality zones. The number of patches and the ratio to the total area of each land type in Wuhan are presented in Table 3.

Table 3. Area and percentage of patches of various types of land in 2010 and 2020.

Land Types	Year	Area (km^2)	Number of Patches	Proportion
Urban vitality land	2010 2020	357.2343 383.3595	396,927 425,955	4.16% 4.47%
Urban non-vitality land	2010 2020	38.8152 25.0497	43,128 27,833	0.45% 0.30%
Other vitality land	2010 2020	475.7229 1052.7021	52,8581 1,169,669	5.54% 12.27%
Other non-vitality land	2010 2020	3092.1705 2499.9192	3,435,745 2,777,688	36.04% 29.14%
Restricted development land	2010 2020	4614.2568 4617.3312	5,126,952 5,130,368	53.81% 53.82%

The table shows that Wuhan's urban vitality land increased by 26.1252 km^2 between 2010 and 2020, while urban non-vitality land decreased by 13.7655 km^2. In addition, other vitality land increased by 576.9722 km^2, and other non-vitality land decreased by 592.2513 km^2. It is worth noting that the area of urban vitality land and other vitality land significantly increased. The most significant change is in the area of other vitality land, which has grown to 2.21 times its size in 2010. This reflects the effectiveness of the Wuhan government's efforts to develop low-vitality areas.

3.3. PLUS Model

The PLUS model, developed by Liang Xun et al. [33], is based on the FLUS model. It comprises two primary components: the land expansion analysis strategy (LEAS) and the multi-class random patch seeding model (CARS) based CA.

3.3.1. LEAS

The driving factors in this paper are selected based on the principles of data accessibility, relevance, and accuracy. According to the study area's natural geographic and socioeconomic profiles, eight indicators from socioeconomic, natural environment, and traffic network were selected as driving factors (Table 4). Figure 9 show the spatial distribution of these driving factors.

Table 4. Driving factors and their meanings and units.

Type of Driving Factor	Driving Factor	Meaning	Unit
Socioeconomic factor	Population density	Population density of each grid	People/km^2
	Per capita GDP	Per capita GDP of each grid	Ten thousand yuan/km^2
Natural environment factor	Elevation	Elevation value of each grid	Meters
	Slope	Slope of each grid	Degrees
Transportation network factor	Distance to primary Roads	Distance from the center of each grid to primary roads	Meters
	Distance to secondary roads	Distance from the center of each grid to secondary roads	Meters
	Distance to highways	Distance from the center of each grid to highways	Meters
	Distance to rail transit	Distance from the center of each grid to rail transit	Meters

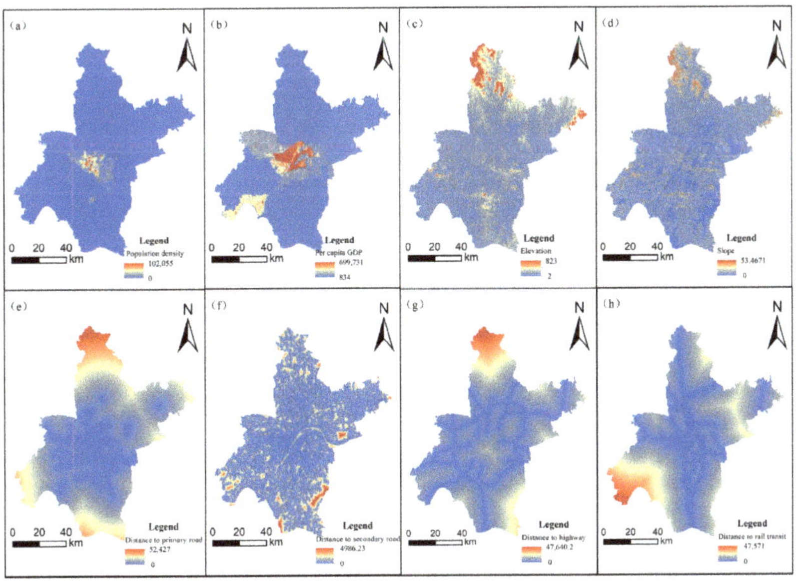

Figure 9. Spatial distribution of driving factors. (**a**) PD, (**b**) per capita GDP, (**c**) elevation, (**d**) slope, (**e**) primary road, (**f**) secondary road, (**g**) highway, (**h**) rail transit.

3.3.2. Simulation Parameter Settings of CARS

In the PLUS model, a binary image with only 0 and 1 is required to specify the conversion of land types. Here, 0 indicates that the land use type cannot be converted to other land use types, while 1 indicates that it can be converted as required. This study aligns with the requirements of Wuhan's Territorial Spatial Master Plan (2021–2035) and designates basic farmland, water bodies, and ecological reserves as non-convertible areas in the simulation of the future.

The conversion rule specifies whether a land type can be converted during the simulation process. To indicate whether a particular type of land can be converted, the land type conversion matrix assigns a value of 1, while a value of 0 indicates it cannot be converted. In this paper, we present a matrix (Table 5) that is based on the relevant literature and policy requirements.

Table 5. Land type transfer matrix.

Land Types	Urban Vitality Land	Urban Non-Vitality Land	Other Vitality Land	Other Non-Vitality Land	Restricted Development Land
Urban vitality land	1	0	0	0	0
Urban non-vitality land	1	1	0	0	0
Other vitality land	1	1	1	0	0
Other non-vitality land	1	1	1	1	0
Restricted development land	0	0	0	0	1

It specifies that urban vitality land can only be converted into other urban vitality land and not into any other type of land. Urban non-vitality land can be converted into urban vitality land and other vibrant sites, but not into other land types. Other vitality land can be converted into urban vitality land, urban non-vitality land, and other vitality land, but not into other types of land. Other non-vitality land can be converted into other land types except for restricted development land. Restricted development land cannot be converted to any other land type.

Domain weights have been set to determine the expansion intensity of a specific land type, ranging from 0 to 1. A higher value indicates easier expansion and lower chances of conversion to other land types. The weight is determined based on experience or expansion of each land type. This paper presents the results of various experiments conducted to determine the optimal domain weights in different contexts (Table 6).

Table 6. Domain weights.

Land Types / Domain Weigth	Urban Vitality Land	Urban Non-Vitality Land	Other Vitality Land	Other Non-Vitality Land	Restricted Development Land
2030	0.5	0.15	0.15	0.25	0.3
2040	0.5	0.25	0.25	0.35	0.3
2050	0.5	0.45	0.5	0.55	0.3

To analyze land use expansion, it was crucial to convert the 2010 and 2020 land use data of Wuhan into the "unsigned char" format required by the PLUS model. Then, we overlayed the land use data from both years to identify the changed raster cells. The resulting files were imported into the LEAS module of the PLUS model to determine the probability of various land types. The land type data and the driving factor file were used to assess the likelihood of different land types occurring. The parameters for the random forest regression were set as follows: a sampling rate of 0.01, meaning approximately 1% of the pixels are used for training, with 20 regression trees and a maximum of 8 features, not exceeding the number of driving factors. The running parameters were set to 6 to ensure the module ran efficiently. This process yielded the development probabilities for five land types (Figure 10).

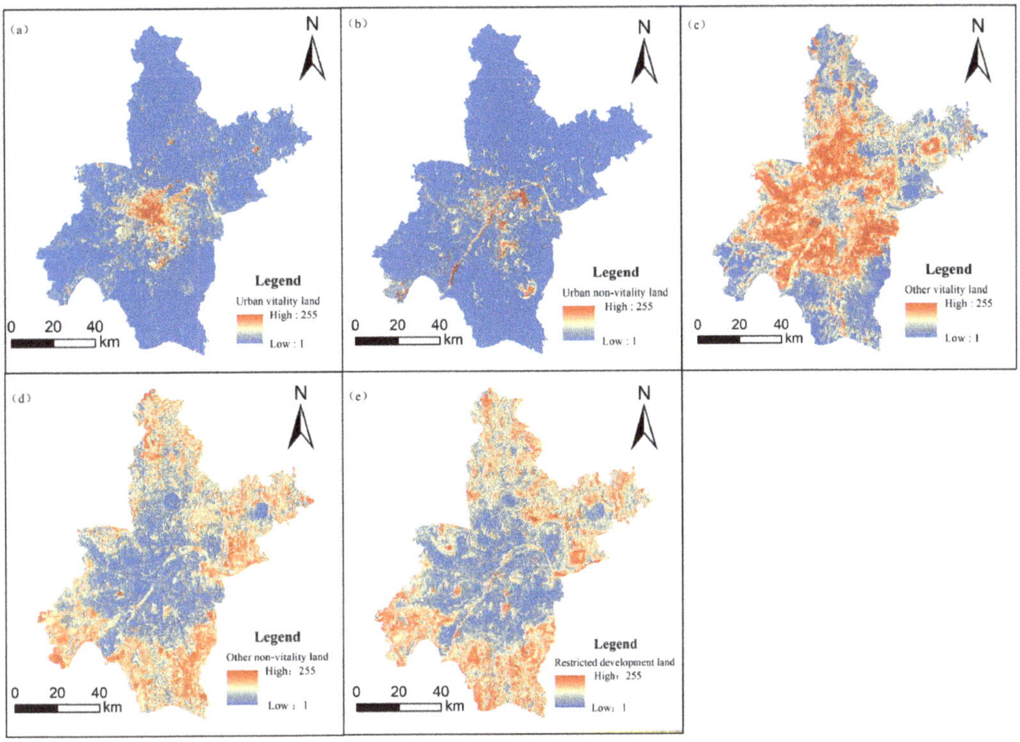

Figure 10. Development probabilities of five land types. (**a**) Urban vitality land, (**b**) urban non-vitality land, (**c**) other vitality land, (**d**) other non-vitality land, (**e**) restricted development land.

When calculating the development probability of each type of land, the PLUS model can also determine the contribution of each driving factor to different land types.

Figure 11 illustrates the factors that contribute to the expansion of urban vibrant land. PD, GDP, rail transit, and primary roads have the highest impact, while elevation and slope have the least impact. On the other hand, the expansion of urban non-vitality land is mainly influenced by GDP, highways, slope, and PD, while rail transit and secondary roads have minimal impact. Regarding the expansion of other vitality land, GDP rail transit and primary roads have the highest impact, while elevation and slope have the least impact. Similarly, for the expansion of other non-vitality land, GDP, primary roads, and secondary roads contribute the most, while rail transit and slope have minimal impact. In terms of restricted development land, primary roads and PD have the most significant impact, while elevation and slope have minimal impact.

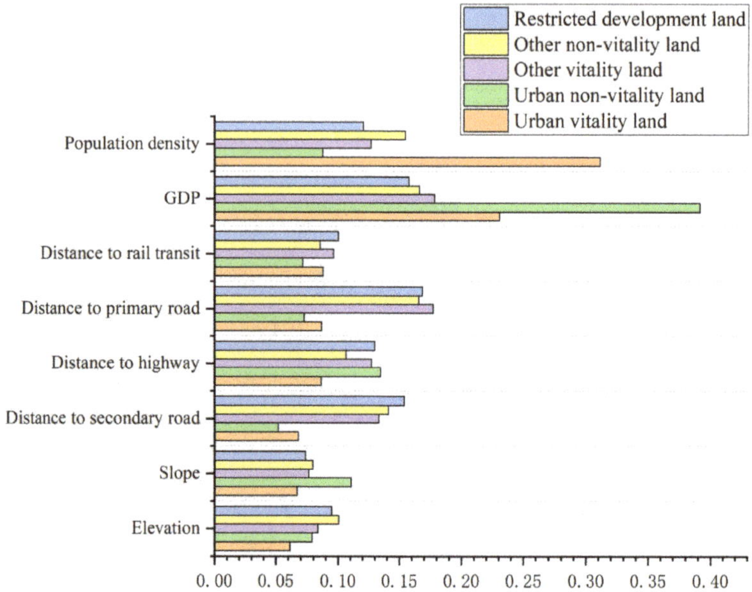

Figure 11. Contribution of each driving factor to different land types.

4. Results and Analysis

4.1. Multi-Stage Simulation Forecasting

4.1.1. Modelling Projections of Urban Growth Boundaries in Three Phases

This study outlines the simulation of Wuhan's future growth boundary, divided into three phases: incremental development (2020–2030), incremental and stock development (2030–2040), and stock development (2040–2050).

During the incremental development stage, the expansion strategy is based on incremental expansion, focusing on developing other vitality land to expand the urban boundary. This study assumes that the change rule of land types in this stage is consistent with the previous one, where restricted development land is prohibited from development. The table below shows the number of image elements for each type of land in 2030, obtained through Markov chain plate simulation of the PLUS model (Table 7).

Table 7. Incremental development-phase land requirements.

Land Use Types	Urban Vitality Land	Urban Non-Vitality Land	Other Vitality land	Other Non-Vitality Land	Restricted Development Land
2020	425,955	27,833	1,169,669	2,777,688	5,130,368
2030	461,510	23,020	1,599,169	2,317,446	5,130,368

During the incremental and stock development phase, the expansion strategy is to develop vitality land while upgrading non-vitality land in the city. This study assumes that the rule for changing land types in this stage remains consistent with the previous one, and that restricted development land is still prohibited from development. The table below shows the number of image elements for each type of land type in 2040, obtained through the PLUS model Markov chain plate simulation (Table 8).

During the stock development stage, there is relatively low demand for incremental development. Instead, the focus is on developing non-vitality land to enhance the expansion characteristics of urban vitality. This study assumes that the rule for changing land types in this stage remains consistent with the previous stage. The development of land for rigid edges is restricted. Using the PLUS model Markov chain plate simulation, the number of similar elements of different land types in 2050 was determined. The results are presented in the Table 9 below.

Table 8. Demand for land in both incremental and stock development phases.

Land Use Types	Urban Vitality Land	Urban Non-Vitality Land	Other Vitality Land	Other Non-Vitality Land	Restricted Development Land
2030	461,510	23,020	1,599,169	2,317,446	5,130,368
2040	506,834	19,049	1,941,799	1,933,463	5,130,368

Table 9. Demand for land at stock development stage.

Land Use Types	Urban Vitality Land	Urban Non-Vitality Land	Other Vitality Land	Other Non-Vitality Land	Restricted Development Land
2040	506,834	19,049	1,941,799	1,933,463	5,130,368
2050	560,287	16,923	2,210,832	1,613,103	5,130,368

4.1.2. Dynamic Analysis of Simulation Results

The probabilities for the development of each type of land use were determined using land use data from 2010 and 2020. These parameters were then used in the PLUS model to simulate land use in the three phases mentioned above (Figure 12). The simulation produced results for each phase, and the attribute table tool was used to calculate the area of each type of land use in different phases and create a table (Table 10).

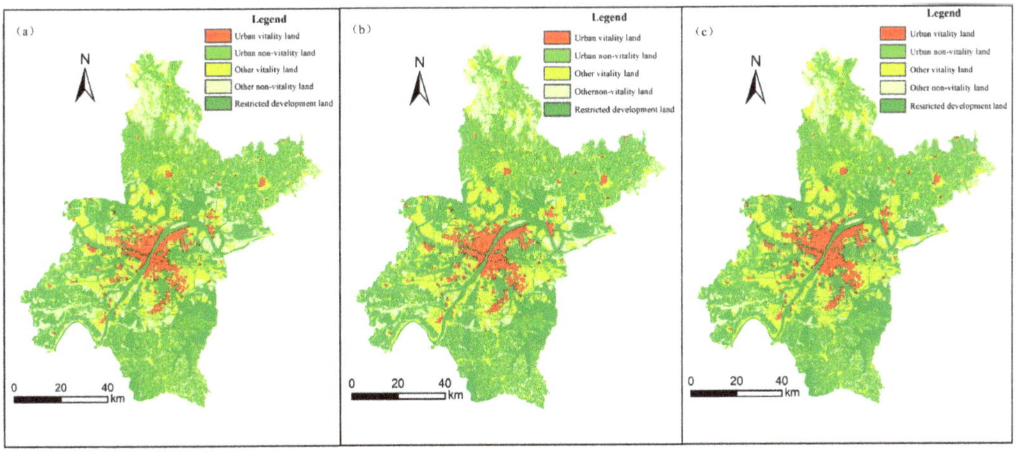

Figure 12. Simulation results of land use in Wuhan at different stages. (**a**) Incremental development stage from 2020 to 2030, (**b**) incremental and stock development stage from 2030 to 2040, (**c**) stock development stage from 2040 to 2050.

After several comparative analyses, it was concluded that the simulation results align with the general law of urban development and can better predict future urban development. The provided figures and tables illustrate that from 2020 to 2050, the area of urban vitality land will continue to expand, while the area of non-vitality land will decrease. The area of restricted development land will remain unchanged due to its rigid boundaries, which cannot be overcome during the city's future expansion. The following sections describe the specifics of each of the three phases.

During the incremental development stage, the area of each land type increased or decreased by varying amounts. Specifically, the area of urban vitality land increased by

31.9995 km^2, while the area of urban non-vitality land decreased by 4.3317 km^2. The area of other vitality land increased by 386.55 km^2, while the area of other non-vitality land decreased by a larger amount, which was 414.2178 km^2. It is important to note that these changes are in comparison to the year 2020. The simulation results indicate that urban vitality land is expanding through infill, primarily from surrounding urban vitality zones. Expansion from urban non-vitality land and other non-vitality land is minimal. The increase in other non-vitality land comes from within that category. The new increase in other vitality land is mainly distributed in the Wujiashan and Changfu areas.

Table 10. Area of five land types at different stages.

Land Types	2020	Incremental Development (2030)	Both Incremental Development and Stock Development (2040)	Stock Development (2050)
Urban vitality land	383.3595	415.359	456.1506	504.2583
Urban non-vitality land	25.0497	20.718	17.7606	15.2307
Other vitality land	1052.7021	1439.2521	1747.0026	1989.7488
Other non-vitality land	2499.9192	2085.7014	1740.1167	1451.7927
Restricted development land	4617.3312	4617.3312	4617.3312	4617.3312

During the phases of incremental and stock development, the area of each land type increased by 40.7916 km^2 compared to 2030. The area of urban vitality land increased, while the area of urban non-vitality land decreased by 2.9574 km^2. Additionally, the area of other vitality land increased by 307.7505 km^2, and the area of other non-vitality land decreased by 345.5847 km^2. The expansion of urban vitality land is primarily concentrated in its original base and the surrounding areas. Qiaokou and Hongshan District have experienced the most significant expansion.

During the stock development stage, the area of each site increases by 48.1007 km^2 compared to 2040. The area of urban vitality land also increases by 48.1007 km^2, while the area of urban non-vitality land decreases by 2.5299 km^2. Additionally, the area of other vitality land increases by 242.7462 km^2, and the area of other non-vitality land decreases by 288.324 km^2. The analysis of the simulation results map reveals that urban vitality land is expanding in all directions and filling its interior.

4.2. Delineation of UGBs

This study considers the urban vitality land and urban non-vitality land within the simulation results of each stage as the scope of the UGB. Based on the delineation of the UGB, this study concludes that the UGBs of Wuhan will expand to 436.463 km^2, 474.617 km^2, and 520.396 km^2 in 2030, 2040, and 2050, respectively. As shown in Figure 13, a1, a2, a3, and a4 are the enlarged maps of several major urban built-up areas in 2030, while b1, b2, b3, and b4 as well as c1, c2, c3, and c4 correspond to 2040 and 2050, respectively.

In terms of overall distribution, Wuhan's UGB exhibits a pattern of "one center and multiple clusters". The "one center" refers to the old urban area of Wuhan, main within the seven central districts. The "multiple clusters" mainly consist of the core areas of some new cities and new city clusters. In terms of spatial form, there are some gaps within the main patches of Wuhan's UGB in 2030. However, by 2040 and 2050, the gaps that are not restricted for development are filled and each patch gradually expands outward and aggregates with each other. Overall, as Wuhan's urban development shifts from incremental to stock development mode, the shape of its UGB is also optimized.

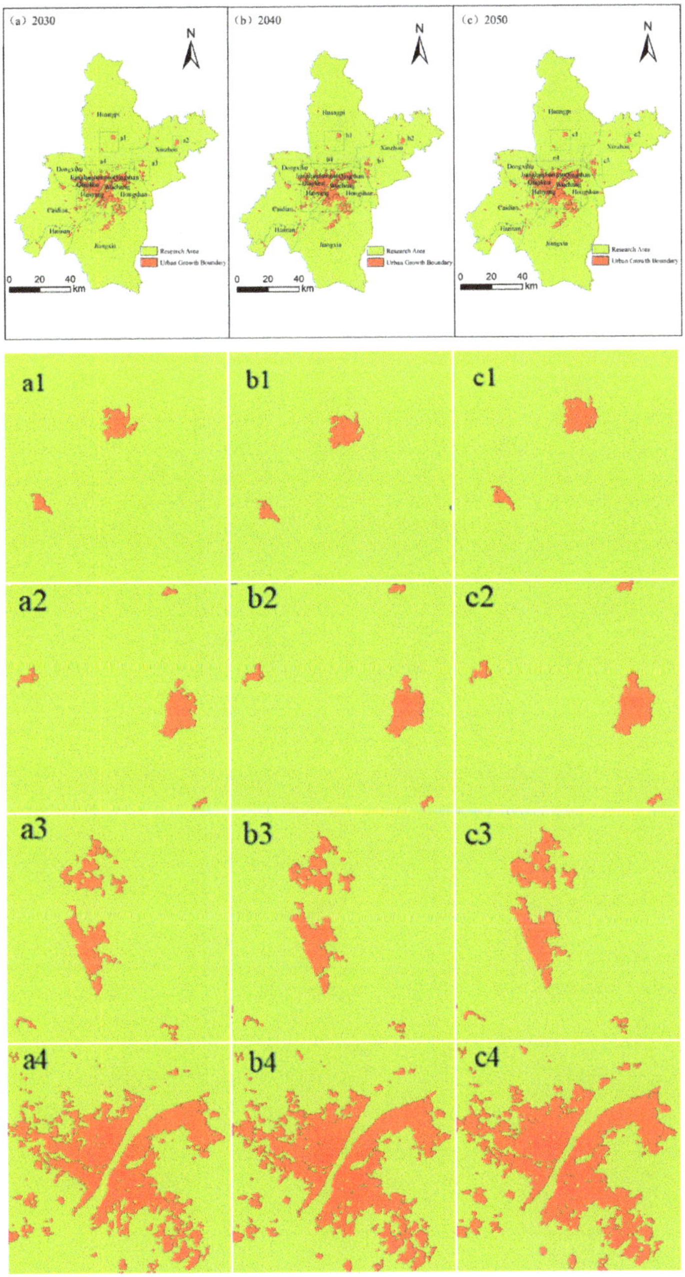

Figure 13. UGBs of Wuhan at different stages. (**a**) UGB in 2030, (**b**) UGB in 2040, (**c**) UGB in 2050.

5. Conclusions and Discussion

This study presents a scientific and rational prediction of the future UGB in the study area of Wuhan. This study uses a built-up area composite index (POI&ISA index) constructed by superimposing POI kernel density and impervious surface index. This index can be applied to extract vitality zones in Wuhan and further classify urban spatial development potential into five categories based on current land use status: urban vitality land, urban non-vitality land, other vitality land, other non-vitality land, and restricted development land in 2010 and 2020. The LEAS is then used to identify the development potential of different land types and the contribution of various driving factors by extracting land type expansion. Then, the future demand for each type of land is predicted through Markov chain. Based on these analyses, different expansion strategies oriented towards stock development are set to obtain the simulation results of the UGBs of Wuhan for the periods of 2030, 2040, and 2050. The UGB delineation process proposed in this paper considers the excavation of inefficient and idle land in different stages of the future, which is undoubtedly more in line with the future trend of urban spatial development under the background of stock development and will guide smart growth of cities and efficient utilization of urban land. The delineation of UGBs in three periods for Wuhan can also provide certain references for controlling the increasingly imbalanced urban expansion of Wuhan.

Compared to previous studies, this study has three innovative contributions: (1) When evaluating urban vitality, a comprehensive evaluation is carried out by constructing a built-up area composite index (POI&ISA index). This method can provide a more integrated perspective to extract urban vitality zones and urban non-vitality zones more accurately, reducing the error generated by using only a single factor to extract urban vitality zones. It also provides a better data basis for subsequent land type classification and simulation. (2) This study predicts the urban expansion scenarios of Wuhan in three different periods in the future, spanning up to 30 years. As China's urbanization enters the middle and later stages of development, urban development is gradually shifting from large-scale incremental construction to stock quality improvement. The results provide a reference for near-term urban development planning and can also aid in the city's long-term planning. (3) The PLUS model is employed to simulate future UGBs expansion. This model is relatively new and has broad application prospects. It integrates the rule mining method based on land expansion analysis and the CA model based on multiple types of random patch seeds. This integration is effective in mining the driving factors of land expansion and predicting the patch-level evolution of the land use landscape.

Although this study has made significant research contributions, it still has some limitations. Firstly, the development process of the city is impacted by multiple factors and facets. However, the PLUS model selects driving factors from three categories: natural environment factors, socioeconomic factors, and traffic network factors, with a total of eight types of data. Thus, this study may have shortcomings in the selection of driving factors. Secondly, the parameter settings in the PLUS model will have a direct impact on the final simulation results. This study relies heavily on the existing results of other scholars and repeated experiments to improve parameter setting. While efforts have been made to minimize its impact on the simulation results, some subjectivity may present. In addition, this study sets three urban development scenarios for the periods of 2020–2030, 2030–2040, and 2040–2050, respectively, focusing on incremental development, balanced development between incremental and stock, and stock development. Although the gradual transition of urban development is considered, the proportion of incremental and stock development in actual urban development is often difficult to accurately determine, especially considering that different cities are at different stages of development. Therefore, the results of the UGB delineation in this study are mainly aimed at guiding the direction of urban development.

Author Contributions: Y.Z., X.X. and J.W. conceived and designed this study. Funding acquisition was led by Y.Z. and H.W.; J.L., H.W. and X.D. contributed to data acquisition. J.L. and X.D. contributed to methodology. J.L., L.X. and C.Y. were involved in drafting the initial manuscript. C.Y. contributed to visualization. Y.Z. and X.X. contributed to the final manuscript revision. All authors have read and agreed to the published version of the manuscript.

Funding: This study was supported by the Key Program of National Natural Science Foundation of China (42230107), and the Open Research Fund Program of Key Laboratory of Digital Mapping and Land Information Application, Ministry of Natural Resources (ZRZYBWD202201).

Data Availability Statement: The data presented in this study are available on request from the corresponding author.

Conflicts of Interest: The authors declare no conflicts of interest.

References

1. Poudel, D.P.; Blackburn, S.; Manandhar, R.; Adhikari, B.; Ensor, J.; Shrestha, A.; Timsina, N.P. The Urban Political Ecology of 'Haphazard Urbanisation' and Disaster Risk Creation in the Kathmandu Valley, Nepal. *Int. J. Disaster Risk Reduct.* **2023**, *96*, 103924. [CrossRef]
2. Almulhim, A.I.; Cobbinah, P.B. Can Rapid Urbanization Be Sustainable? The Case of Saudi Arabian Cities. *Habitat Int.* **2023**, *139*, 102884. [CrossRef]
3. Buckley, R.M.; Simet, L. An Agenda for Habitat III: Urban Perestroika. *Environ. Urban.* **2016**, *28*, 64–76. [CrossRef]
4. Wang, L.; Li, C.; Ying, Q.; Cheng, X.; Wang, X.; Li, X.; Hu, L.; Liang, L.; Yu, L.; Huang, H.; et al. China's Urban Expansion from 1990 to 2010 Determined with Satellite Remote Sensing. *Chin. Sci. Bull.* **2012**, *57*, 2802–2812. [CrossRef]
5. Bai, X.; Shi, P.; Liu, Y. Society: Realizing China's Urban Dream. *Nature* **2014**, *509*, 158–160. [CrossRef]
6. Cao, K.; Deng, Y.; Song, C. Exploring the Drivers of Urban Renewal through Comparative Modeling of Multiple Types in Shenzhen, China. *Cities* **2023**, *137*, 104294. [CrossRef]
7. Cao, K.; Harris, R.; Liu, S.; Deng, Y. How Does Urban Renewal Affect Residential Segregation in Shenzhen, China? A Multi-Scale Study. *Sustain. Cities Soc.* **2024**, *102*, 105228. [CrossRef]
8. Zhang, W.; Zhang, X.; Wu, G. The Network Governance of Urban Renewal: A Comparative Analysis of Two Cities in China. *Land Use Policy* **2021**, *106*, 105448. [CrossRef]
9. Guo, R.; Ding, Y.; Shang, L.; Wang, D.; Cao, X.; Wang, S.; Bonatz, N.; Wang, L. Sustainability-Oriented Urban Renewal and Low-Impact Development Applications in China: Case Study of Yangpu District, Shanghai. *J. Sustain. Water Built Environ.* **2018**, *4*, 05017006. [CrossRef]
10. He, F.; Yang, J.; Zhang, Y.; Sun, D.; Wang, L.; Xiao, X.; Xia, J. Offshore Island Connection Line: A New Perspective of Coastal Urban Development Boundary Simulation and Multi-Scenario Prediction. *GISci. Remote Sens.* **2022**, *59*, 801–821. [CrossRef]
11. Howard, E. *Garden Cities of To-Morrow*; MIT Press: Cambridge, MA, USA, 1965; ISBN 978-0-262-58002-1.
12. Wang, Y.; Gu, C.; Li, X. Research Progress on Urban Growth Boundaries in China and Abroad. *Urban Plan. Int.* **2014**, *29*, 1–11.
13. Bengston, D.N.; Fletcher, J.O.; Nelson, K.C. Public Policies for Managing Urban Growth and Protecting Open Space: Policy Instruments and Lessons Learned in the United States. *Landsc. Urban Plan.* **2004**, *69*, 271–286. [CrossRef]
14. Huang, D.; Huang, J.; Liu, T. Delimiting Urban Growth Boundaries Using the CLUE-S Model with Village Administrative Boundaries. *Land Use Policy* **2019**, *82*, 422–435. [CrossRef]
15. Liu, X.; Wei, M.; Li, Z.; Zeng, J. Multi-Scenario Simulation of Urban Growth Boundaries with an ESP-FLUS Model: A Case Study of the Min Delta Region, China. *Ecol. Indic.* **2022**, *135*, 108538. [CrossRef]
16. Zheng, X.; Lv, L. A WOE Method for Urban Growth Boundary Delineation and Its Applications to Land Use Planning. *Int. J. Geogr. Inf. Sci.* **2016**, *30*, 691–707. [CrossRef]
17. Long, Y.; Han, H.; Mao, Q. Establishing Urban Growth Boundaries Using Constrained CA. *Acta Geogr. Sin.* **2009**, *64*, 999–1008.
18. Wang, W.; Jiao, L.; Jia, Q.; Liu, J.; Mao, W.; Xu, Z.; Li, W. Land Use Optimization Modelling with Ecological Priority Perspective for Large-Scale Spatial Planning. *Sustain. Cities Soc.* **2021**, *65*, 102575. [CrossRef]
19. Yao, Z.; Jiang, C.; Shan-shan, F. Effects of Urban Growth Boundaries on Urban Spatial Structural and Ecological Functional Optimization in the Jining Metropolitan Area, China. *Land Use Policy* **2022**, *117*, 106113. [CrossRef]
20. Li, X.; Zhang, Y.; Liu, X.; Chen, Y. Assimilating Process Context Information of Cellular Automata into Change Detection for Monitoring Land Use Changes. *Int. J. Geogr. Inf. Sci.* **2012**, *26*, 1667–1687. [CrossRef]
21. Xia, C.; Zhang, A.; Wang, H.; Zhang, B. Modeling Urban Growth in a Metropolitan Area Based on Bidirectional Flows, an Improved Gravitational Field Model, and Partitioned Cellular Automata. *Int. J. Geogr. Inf. Sci.* **2019**, *33*, 877–899. [CrossRef]
22. Zheng, H.W.; Shen, G.Q.; Wang, H.; Hong, J. Simulating Land Use Change in Urban Renewal Areas: A Case Study in Hong Kong. *Habitat Int.* **2015**, *46*, 23–34. [CrossRef]
23. Wang, H.; Peng, P.; Kong, X.; Zhang, T.; Yi, G. Evaluating the Suitability of Urban Expansion Based on the Logic Minimum Cumulative Resistance Model: A Case Study from Leshan, China. *ISPRS Int. J. Geo-Inf.* **2019**, *8*, 291. [CrossRef]
24. Liu, X.; Wei, M.; Zeng, J. Simulating Urban Growth Scenarios Based on Ecological Security Pattern: A Case Study in Quanzhou, China. *Int. J. Environ. Res. Public Health* **2020**, *17*, 7282. [CrossRef]
25. Longato, D.; Cortinovis, C.; Balzan, M.; Geneletti, D. A Method to Prioritize and Allocate Nature-Based Solutions in Urban Areas Based on Ecosystem Service Demand. *Landsc. Urban Plan.* **2023**, *235*, 104743. [CrossRef]
26. Xu, L.; Huang, Q.; Ding, D.; Mei, M.; Qin, H. Modelling Urban Expansion Guided by Land Ecological Suitability: A Case Study of Changzhou City, China. *Habitat Int.* **2018**, *75*, 12–24. [CrossRef]

27. Huang, X.; Wang, H.; Xiao, F. Simulating Urban Growth Affected by National and Regional Land Use Policies: Case Study from Wuhan, China. *Land Use Policy* **2022**, *112*, 105850. [CrossRef]
28. Barredo, J.I.; Kasanko, M.; McCormick, N.; Lavalle, C. Modelling Dynamic Spatial Processes: Simulation of Urban Future Scenarios through Cellular Automata. *Landsc. Urban Plan.* **2003**, *64*, 145–160. [CrossRef]
29. Huang, Q.; Song, W. A Land-Use Spatial Optimum Allocation Model Coupling a Multi-Agent System with the Shuffled Frog Leaping Algorithm. *Comput. Environ. Urban Syst.* **2019**, *77*, 101360. [CrossRef]
30. Liu, G.; Jin, Q.; Li, J.; Li, L.; He, C.; Huang, Y.; Yao, Y. Policy Factors Impact Analysis Based on Remote Sensing Data and the CLUE-S Model in the Lijiang River Basin, China. *Catena* **2017**, *158*, 286–297. [CrossRef]
31. Clarke, K.; Hoppen, S.; Gaydos, L. A self-modifying cellular automaton model of historical. *Environ. Plan. B Plan. Des.* **1997**, *24*, 247–261. [CrossRef]
32. Tayyebi, A.; Pijanowski, B.C.; Tayyebi, A.H. An urban growth boundary model using neural networks, GIS and radial parameterization: An application to Tehran, Iran. *Landsc. Urban Plan.* **2011**, *100*, 35–44. [CrossRef]
33. Liang, X.; Liu, X.; Li, X.; Chen, Y.; Tian, H.; Yao, Y. Delineating Multi-Scenario Urban Growth Boundaries with a CA-Based FLUS Model and Morphological Method. *Landsc. Urban Plan.* **2018**, *177*, 47–63. [CrossRef]
34. Verburg, P.H.; de Koning, G.H.J.; Kok, K.; Veldkamp, A.; Bouma, J. A Spatial Explicit Allocation Procedure for Modelling the Pattern of Land Use Change Based upon Actual Land Use. *Ecol. Model.* **1999**, *116*, 45–61. [CrossRef]
35. Vermeiren, K.; Van Rompaey, A.; Loopmans, M.; Serwajja, E.; Mukwaya, P. Urban growth of Kampala, Uganda: Pattern analysis and scenario development. *Landsc. Urban Plan.* **2012**, *106*, 199–206. [CrossRef]
36. Xu, Q.; Wang, Q.; Liu, J.; Liang, H. Simulation of Land-Use Changes Using the Partitioned ANN-CA Model and Considering the Influence of Land-Use Change Frequency. *ISPRS Int. J. Geoinf.* **2021**, *10*, 346. [CrossRef]
37. Liu, X.; Liang, X.; Li, X.; Xu, X.; Ou, J.; Chen, Y.; Li, S.; Wang, S.; Pei, F. A Future Land Use Simulation Model (FLUS) for Simulating Multiple Land Use Scenarios by Coupling Human and Natural Effects. *Landsc. Urban Plan.* **2017**, *168*, 94–116. [CrossRef]
38. Liang, X.; Guan, Q.; Clarke, K.C.; Liu, S.; Wang, B.; Yao, Y. Understanding the Drivers of Sustainable Land Expansion Using a Patch-Generating Land Use Simulation (PLUS) Model: A Case Study in Wuhan, China. *Comput. Environ. Urban Syst.* **2021**, *85*, 101569. [CrossRef]
39. Lai, Z.; Chen, C.; Chen, J.; Wu, Z.; Wang, F.; Li, S. Multi-Scenario Simulation of Land-Use Change and Delineation of Urban Growth Boundaries in County Area: A Case Study of Xinxing County, Guangdong Province. *Land* **2022**, *11*, 1598. [CrossRef]
40. Xia, X.; Zhang, Y.; Shi, X.; Chen, J.; Rao, T. Simulation of Dynamic Urban Growth Boundary Combining Urban Vitality and Ecological Networks: A Case Study in Chengdu Metropolitan Area. *Land* **2022**, *11*, 179. [CrossRef]
41. Ridd, M.K. Exploring a V-I-S (vegetation-impervious surface-soil) model for urban ecosystem analysis through remote sensing: Comparative anatomy for cities. *Int. J. Remote Sens.* **1995**, *16*, 2165–2185. [CrossRef]

land

MDPI

Article

Long-Term Ecological and Environmental Quality Assessment Using an Improved Remote-Sensing Ecological Index (IRSEI): A Case Study of Hangzhou City, China

Cheng Cai [1], Jingye Li [2] and Zhanqi Wang [1,*]

[1] Department of Land Resource Management, School of Public Administration, China University of Geosciences (Wuhan), Wuhan 430074, China; yemaoshouling@cug.edu.cn
[2] Department of Land Resource Management, School of Public Administration, Hohai University, Nanjing 211100, China; jingye.li@hhu.edu.cn
* Correspondence: zhqwang@cug.edu.cn

Abstract: The integrity and resilience of our environment are confronted with unprecedented challenges, stemming from the escalating pressures of urban expansion and the need for ecological preservation. This study proposes an Improved Remote Sensing Ecological Index (IRSEI), which employs humidity (WET), the Normalized Difference Vegetation Index (NDVI), Land Surface Temperature (LST), a standardized Building–Bare Soil Index (NDBSI), aerosol optical depth (AOD), and the comprehensive salinity index (CSI). The IRSEI model was utilized to assess the ecological quality of Hangzhou over the period from 2003 to 2023. Additionally, the random forest model was employed to analyze the factors driving ecological quality. Furthermore, the gradient effect in the horizontal direction away from the urban center was examined using the buffer zone method. Our analysis reveals the following: (1) approximately 95% of the alterations in ecological quality observed from 2003 to 2023 exhibited marginal improvements, declines, or were negligible; (2) the transformations in IRSEI during this period, including variations in surface temperature and transportation networks, exhibited strong correlations (0.85) with human activities. Moreover, the influence of AOD and the comprehensive salinity index on IRSEI demonstrated distinct spatial disparities; (3) the IRSEI remained generally stable up to 30 km outside the city center, indicating a trend of agglomeration in the center and significant areas in the surroundings. The IRSEI serves as a robust framework for bolstering the assessment of regional ecological health, facilitating ecological preservation and rejuvenation efforts, and fostering coordinated sustainable regional development.

Keywords: ecological quality; spatiotemporal evolution; influencing factor; random forest; Hangzhou city

1. Introduction

Environmental protection and sustainable development constitute pivotal areas of contemporary research. In the context of economic growth and urban expansion, alterations in land use/cover and the repercussions of human activities have reshaped ecological landscapes and impacted their quality [1]. China has experienced remarkable strides in economic advancement and urbanization over recent decades; however, the consequences of prioritizing rapid development at the expense of ecological integrity and natural assets during initial expansion phases have become evident. Ecological challenges now emerge as pivotal factors influencing and constraining China's socioeconomic progress [2]. In 2006 and 2015, the Ministry of Environmental Protection issued trial and revised Technical Specifications for the Evaluation of the State of the Ecological Environment, which were presented as industry standards. These specifications proposed an ecological index (EI) encompassing factors such as biological richness, air pollution, water network density, vegetation coverage, land degradation, and related elements. Subsequently, in 2021, the

Ministry of Ecology and Environment introduced two trial measures for regional ecological quality evaluation, primarily targeting county and administrative divisions [3]. The evaluation outcomes are based on the overall ecological quality of the region. However, visualizing and comprehending the spatial distribution of ecological quality, as well as analyzing its spatial disparities and changes, remain challenging.

Remote sensing spatial information technology offers rapid, real-time, and large-scale monitoring capabilities, finding widespread application in studying ecological environments [4–7]. To better manage the urban environment and protect people's lives, a tool that can effectively evaluate and monitor the status of the urban ecological environment is needed. Therefore, the Remote Sensing Ecological Index (RSEI) was developed by Xu [8,9]. It integrates multiple intuitive indicators to reflect the ecological environment and can realize rapid monitoring and evaluation of the regional ecological environment [10]. The RSEI is based on remote-sensing information and established through principal component analysis by coupling greenness, humidity, dryness, and heat indices that reflect ecological environment status [8–11]. Through dynamic monitoring and analysis of ecological environment quality in these regions, it can provide a scientific basis for ecological environment protection and restoration. The popularity of RESI has been demonstrated in diverse natural ecosystems (such as forests, farmlands, deserts, and wetlands) [12] and man-made environments (mining areas, cities, and industrial areas) [13–15]; however, there are still some limitations when using RSEI. RSEI mainly selects indicators based on ecological environment characteristics, which may ignore the impact of social, economic, cultural, technological, and other factors on the ecological environment. Previous studies have reported a lack of homogeneity in application scenarios, randomness in models, and limited applicability in extreme ecological scenarios such as deserts and land degradation areas, but have ignored air quality in the atmosphere, among other factors [14,16]. Moreover, it fails to reveal what gradient effect RSEI has in the horizontal direction. In future studies, more indicators can be considered for the RSEI evaluation system.

Air pollution, particularly PM2.5, exerts a significant impact on the ecological quality within and surrounding major central cities. Aerosol optical depth (AOD) is extensively utilized to investigate the spatial distribution of PM2.5 and serves as a suitable proxy for an air quality index [17–19]. The comprehensive salinity index (CSI), with ecological factors such as air quality and vegetation cover, measures the soil fertility that impacts vegetation growth [20,21]. These effects may further influence local climate and air quality. While salinization may not be a widespread issue in Hangzhou overall, it can occur in specific areas or under certain conditions, significantly affecting soil quality and agricultural production. Remote-sensing technology enables a comprehensive evaluation of Hangzhou's ecological quality, providing a scientific foundation for land management and agricultural production. Furthermore, air quality is a vital component of ecological quality that directly affects human health and quality of life [22]. As a densely populated and economically developed city, Hangzhou faces considerable concerns regarding air quality [23]. Salinization and air quality are frequently neglected in regional ecological assessments, potentially leading to a distorted understanding of these interactions and the overall impacts on ecological quality. To address the deficiencies of remote-sensing ecological indices in characterizing ecosystems and air quality, this study introduces salinization and air quality indices based on RSEI and develops a comprehensive and Improved Remote-Sensing Ecological Index (IRSEI). Incorporating the salinization factor and air quality into Hangzhou's remote-sensing ecological quality assessment is crucial for comprehensively and accurately understanding Hangzhou's ecological quality.

By integrating remote-sensing technology with air quality monitoring data, we can gain a thorough understanding of Hangzhou's air pollution situation, encompassing the primary pollution sources as well as the distribution and transmission paths of pollutants. This information will facilitate the government in formulating targeted air pollution control measures aimed at enhancing Hangzhou's air quality. Furthermore, the incorporation of salinization factors and air quality into remote-sensing ecological quality assessments can reveal the interactions and influence mechanisms among various ecological quality elements. The primary objectives of this paper are threefold:

1. To develop an integrated IRESI assessment model that incorporates air quality and salinization indicators, providing a comprehensive assessment framework;
2. To assess and analyze the ecological environment quality of Hangzhou from 2003 to 2023 using the IRSEI model, with the aim of identifying trends and patterns and exploring the driving mechanisms through the application of the random forest algorithm;
3. To evaluate the urban–rural echelon effect of ecological quality in the horizontal direction of the urban central area by constructing buffer zones, in order to gain a deeper understanding of the spatial distribution and changes in ecological quality within the urban landscape.

2. Materials and Methods

2.1. Research Area

Hangzhou City was chosen as the research area for ecological monitoring and evaluation. Hangzhou serves as the capital of Zhejiang Province in southeastern China (refer to Figure 1). The urban area spans from approximately 29°11′ to 30°34′ N in latitude and 118°20′ to 120°44′ E in longitude. As of 2022, the city's population was 12.4 million. Renowned for its West Lake, Hangzhou's traditional urban core lies along the northeastern bank of this iconic water body. Over the past two decades, the Qianjiang CBD (also known as Qianjiang New Town) has undergone significant development east of the central city. The research area administers 10 districts, 2 counties, and 1 county-level city, which mainly include Shangcheng District, Gongshu District, Xihu District, Binjiang District, Xiaoshan District, Yuhang District, Linping District, Qiantang District, Fuyang District, Lin'an District, Tonglu County, Chun'an County, and Jiande City. Within the urban land area, mountains and hills, plains, and various water bodies occupy 65.6%, 26.4%, and 8% of the total area, respectively. Situated south of the Qiantang River, the study area is characterized by highly developed regions, cultivated land, and small-scale forested areas. Over the past decade, extensive land-use transformations have occurred, converting farmlands and wetlands into highly developed areas. Furthermore, the region has experienced rapid urbanization, leading to ecological challenges and a decline in ecological quality.

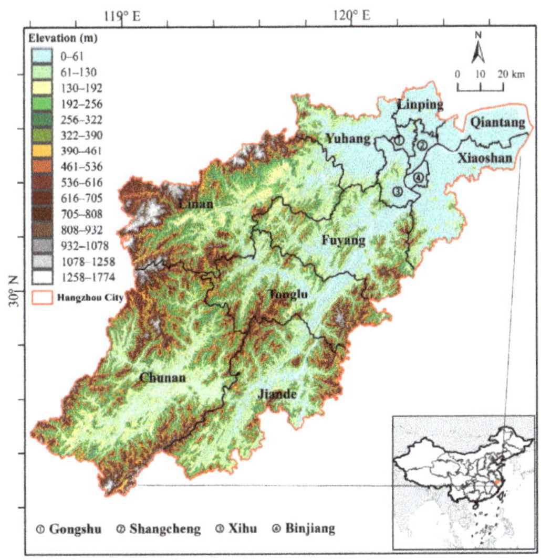

Figure 1. Study area location map.

2.2. Data Sources

This study primarily utilized remote-sensing images, Land Use and Land Cover (LULC), population, and economic data. Landsat Thematic Mapper (TM) and Operational Land Imager (OLI) remote-sensing images, along with MCD19A2 data, were predominantly employed to conduct the IRSEI evaluation, while other datasets were utilized for the IRSEI influencing factor analysis. Furthermore, Landsat TM/OLI images with high quality, featuring less than 10% cloud cover, were obtained between 30 June and 30 December of the respective years spanning from 2003 to 2023. In the Northern Hemisphere, the summer period, specifically spanning from June to September, is a crucial time when plant growth reaches its peak. During this phase, vegetation cover exhibits its most luxuriant state, enabling satellite remote-sensing data to capture and reflect the growth status of vegetation with greater precision. The decision to utilize satellite datasets spanning from 30 June to 30 December rather than annual data is primarily based on considerations of seasonal vegetation changes, phenological characteristics, and specific research objectives. Additionally, this selection aids in reducing the complexity of data processing and analysis. Following the acquisition of MCD19A2 data for the entire year, the annual mean AOD from 2003 to 2023 was computed to ensure the comparability of the research findings. The MCD19A2 data product is a gridded Level-2 product derived from the Moderate Resolution Imaging Spectroradiometer (MODIS) on both the Terra and Aqua satellites, utilizing the Multi-angle Implementation of Atmospheric Correction (MAIAC) algorithm for land AOD. This product is generated daily with a resolution of 1 kilometer, providing crucial aerosol information for Earth science research. The MCD19A2 product is widely used in various research fields such as aerosol science, atmospheric environment, and climate change.

The dataset preprocessing involved calibration and preprocessing by Google Earth Engine (GEE), utilizing a multi-spectral band grey value or digital number (DN), sensor reflection value conversion, and Spectral Hypercube Rapid Line-of-Sight Atmospheric Analysis (FLAASH). The nearest pixel method was employed in this study to rectify images from different periods by 0.5 pixels. Two polynomials and root-mean-square error analysis were utilized, and the cropped portion of the remote-sensing image was ultimately included within the research scope.

2.3. Research Methods

2.3.1. Research Framework

Humidity (WET), NDVI, Land Surface Temperature (LST), a standardized Building–Bare Soil Index (NDBSI), AOD, and the comprehensive salinity index (CSI) were employed to represent the regional ecological quality. The overall framework for calculating and analyzing the IRSEI consisted of three main steps. Firstly, the IRSEI index of Hangzhou was determined using principal component analysis. Secondly, the ecological indicators of Hangzhou were analyzed considering their spatiotemporal differences and changes. Finally, the factors influencing IRSEI and horizontal changes in ecological quality were studied.

2.3.2. IRSEI Model

The NDVI is typically utilized for monitoring vegetation growth, directly reflecting the quality of the regional ecological environment. LST is closely associated with vegetation growth, crop yield, the surface water cycle, urbanization, other natural phenomena and processes, and human activities. It serves as a heat index, reflecting the surface ecological environment. The Kauth–Thomas transformation method, a linear transformation based on multi-spectral imaging, derives moisture components reflecting soil and vegetation moisture information (WET). The dryness index, indicating soil dryness, can detrimentally impact the ecological environment. Given that urban construction land predominates in our study area, the dryness index was represented by combining the Bare Soil Index (SI) and Construction Index (IBI) into a standardized Building–Bare Soil Index (NDBSI). WET,

NDVI, LST, NDBSI, CSI, and AOD indices were utilized to represent the ecological quality of Hangzhou (Table 1).

Table 1. IRSEI parameter and formula.

IRSEI Indices			Equations	Reference
Humidity index	WET	/	$I\_wet(TM) = 0.0315\rho(Blue) + 0.2021\rho(Green) + 0.3102\rho(Red) + 0.1594\rho(NIR) - 0.6806\rho(SWIR1) - 0.6109\rho(SWIR2)$ $I\_wet(ETM+) = 0.2626\rho(Blue) + 0.2141\rho(Green) + 0.0926\rho(Red) + 0.0656\rho(NIR) - 0.7629\rho(SWIR1) - 0.5388\rho(SWIR2)$ $I\_wet(OLI) = 0.1511\rho(Blue) + 0.1973\rho(Green) + 0.3283\rho(Red) + 0.3407\rho(NIR) - 0.7117\rho(SWIR1) - 0.4559\rho(SWIR2)$ $MNDWI = (\rho(Green) - \rho(SWIR1))/(\rho(Green) + \rho(SWIR1))$	[2]
Greenness index	NDVI	Normalized Difference Vegetation Index	$NDVI = (\rho(NIR) - \rho(Red))/(\rho(NIR)+\rho(Red))$	[2]
Heat index	LST	Land Surface Temperature	$FVC=(NDVI-NDVI\_soil)/(NDVI\_veg-NDVI\_soil)$ $\varepsilon\_water = 0.995\ (NDVI \leq 0)$ $\varepsilon\_building = 0.9589 + 0.086 \times F(veg) - 0.0671 \times F^2veg\ (0 < NDVI < 0.7)$ $\varepsilon\_natural = 0.9625 + 0.0614 \times F(veg) - 0.0461 \times F^2veg\ (NDVI \geq 0.7)$ $L = g \times DN + b$ $Tb = K2/\ln(K1/L + 1)$ $LST = Tb/\{1 + [(\lambda Tb)/\rho] \ln \varepsilon\} - 273.15$	[2]
Dryness index	NDBSI	Normalized Difference Built-up and Soil Index	$NDBSI = (SI+IBI)/2$ $SI = [(\rho5+\rho(Red)) - (\rho(NIR)+\rho1)]/[(\rho5+\rho(Red)) + (\rho(NIR) + \rho(Blue))]$ $IBI = [2\rho(SWIR1)/(\rho(SWIR1) + \rho(NIR)) - (\rho(NIR)/(\rho(NIR) + \rho(Red)) + \rho(Green)/(\rho(Green) + \rho5))]/[2\rho(SWIR1)/(\rho(SWIR1) + \rho(NIR)) + (\rho(NIR)/(\rho(NIR)+\rho(Red)) + \rho(Green)/(\rho(Green) + \rho(SWIR1)))]$	[2]
Salinity index	CSI	Comprehensive Salinity Index	$CSI = (SI - T + NDSI + SI3)/3$ $SI - T = (\rho(Red)/\rho(NIR)) \times 100$ $NDSI = (\rho(Red) - \rho(NIR))/(\rho(Red) + \rho(NIR))$ $SI3 = Sqrt(\rho^2g + \rho^2r)$	[2,20]
Air index	AOD	Aerosol Optical Depth	The AOD data are collected by the MAIAC (Multi-angle Implementation of Atmospheric Correction) algorithm from MCD19, a new aerosol product of MODIS	[18]

1. Humidity index (I_{wet})

The humidity index is closely linked to the ecological environment's quality. Low humidity indicates severe soil degradation, low vegetation coverage, and a poor ecological environment, while high humidity suggests sufficient soil moisture, abundant vegetation cover on the surface, and a favorable ecological environment. In this study, the humidity index was denoted as the I_wet component. Due to the different sensors of the Landsat TM/ETM+ and Landsat OLI images, the extraction formulas for the humidity index varied (Table 1).

2. Greenness index (I_{ndvi})

The NDVI is closely related to vegetation coverage, biomass, and leaf area index, which are commonly used to monitor vegetative growth. The NDVI was selected to represent the green index using the formula, as shown in Table 1.

3. Heat index (I_{heat})

Fractional vegetation cover (FVC) refers to the percentage of the vertical projected area of vegetation in the soil relative to the total statistical area. Vegetation coverage was based on Landsat NDVI, and a mixed-pixel binary model was adopted where NDVI is the normalized vegetation index, NDVI_soil is the normalized vegetation index value of the bare land surface, and NDVI_veg is the normalized vegetation index value of complete vegetation cover. NDVI_soil and NDVI_veg selected NDVI_max and NDVI_min with a confidence level above 95%. The g and b represent the offset values of the thermal infrared band; DN is the grey value of the pixel affected by remote sensing; L denotes the radiation brightness; LST is the surface temperature; and K1 and K2 are the calibration parameters, and various sensors use different values; Tb indicates the brightness temperature of the sensor; ε is the specific emissivity; and λ is the central wavelength of the thermal infrared band; ρ is 1.438×10^{-2} mk.

4. Dryness index (I_{dry})

In this study, areas from bare soil and buildings were extracted by setting appropriate thresholds, and the NDBSI was then calculated using the area ratio as a weighted reference standard.

5. Salinity (I_{csi})

The CSI provides a more accurate reflection of the ecological impacts of salinization compared to other indices. Utilizing comprehensive learning, the CSI integrates the salinity index (SI-T), Normalized Difference Built-up and Soil Index (NDBSI), and salinity index 3 (SI3) to enhance the stability and reliability of the detection results. The CSI denotes the comprehensive salinity index. When calculating the CSI, it is necessary to normalize SI-T, NDSI, and SI3 to (0,1) to ensure that the CSI is obtained under the same standard and make the results representative.

6. Air index (I_{aod})

Aerosols are suspensions of liquid or solid particles dispersed in air or gases that circulate through numerous atmospheric chemical cycles and constitute an essential component of the atmospheric environment. Among the most fundamental optical properties of aerosols, AOD has emerged as a crucial parameter for studying atmospheric turbidity, providing insights into changes in aerosol distribution to a certain extent. The distribution of AOD is typically influenced by geographical factors, population density, and industrial distribution, making it an indicator of atmospheric turbidity and pollution. AOD accurately reflects the air quality concerning particulate matter within a specific area, exhibiting the highest degree of spatial coupling with pollutant emissions. Consequently, AOD was selected as a representative measure of air quality.

2.3.3. Random Forest Algorithm

The random forest algorithm is an ensemble learning method that enhances the generalization ability of a model by combining the predictions of multiple decision trees and incorporating randomness [24]. It is capable of addressing both classification and regression problems. One of its key features is its ability to rank the importance of different features, which aids in identifying the independent variables that contribute the most to model predictions and assessing the influence of each independent variable on the dependent variable. In this study, Python was utilized for machine learning purposes. The random forest regressor code was used to construct the regression model, obtain factor importance, and explain its effect on the dependent variable. The random forest algorithm was employed to analyze the driving forces behind IRSEI changes, with the selected independent variables being elevation (DEM), road conditions, humidity, green capital, heat, dryness, salinity, and air quality. The modeling process was performed using Python 3.12.

2.3.4. Equal Area Buffer Analysis Method

In the process of urbanization, the transformation from rural to urban areas leads to changes in economic, social, and environmental aspects. Therefore, the gradient effect of ecological quality in the horizontal direction between urban and rural areas is subject to the comprehensive influence of various factors such as urban–rural distance. By rationally utilizing the gradient effect, the dynamic balance and coordination of ecological quality between urban and rural areas can be effectively promoted, thereby enhancing the quality of the ecosystem. Taking the urban center of Hangzhou as the starting point of the circle, a circular buffer zone with a radius of 1km is established, and 500 concentric circles with gradually decreasing radii and equal areas are diverged outward, covering the main areas of the main urban area of Hangzhou. This study focuses on the relationship between ecological quality and the distance from the city center in this area from 2003 to 2023, analyzing the gradient effect of ecological quality in the horizontal direction between urban and rural areas.

3. Results and Discussion

3.1. Spatial Distribution Characteristics of Ecological Factors

By comparing the spatial distribution of the six ecological factors in Hangzhou from 2003 to 2023 (refer to Figure 2), we observed that high NDVI and WET values were primarily distributed around the main urban area. The spatial distribution characteristics of NDVI were consistent with land types, with high-value areas primarily consisting of vegetated areas such as forests and grasslands, while low-value areas mainly comprised central urban and residential areas. The spatial distribution of the NDBSI value was affected by the distribution of impervious ground and the degree of dryness of bare soil; hence, NDBSI values around water and forest areas were low, whereas those in cultivated land and human settlement areas were high. The spatial distribution of the LST value was influenced by human activities, the natural environment, and climate, with the highest values observed in the central urban area. AOD reflects the basic air quality situation, with variations in its spatial distribution caused by multiple factors such as human activities, natural climate, and the surrounding region. The results are consistent with other studies [17,18,25]. The CSI reflected changes in salinity and alkalinity, adversely affecting ecosystem structure and functions. The implementation of land management and ecological protection policies in Hangzhou in recent years, such as soil improvement and vegetation restoration, has also reduced the amount of saline–alkaline land. In arid areas, the adverse effects of soil salinization on the dynamic change in ecological quality cannot be ignored, and CSI also shows a correlation. For large areas, the salinity index reflects the regional ecological quality due to the diversity of soil texture and the topographic characteristics of landform [20].

In 2003, the LST values of Chun'an and Jiande were significantly higher than those of the central region, while the AOD values of the central and western regions were significantly higher than those of other regions. By 2008, the central city exhibited the highest NDBSI value, and the northwest of Lin'an District showed the highest AOD value. In 2013, the LST value of the main urban area was high, and the central part of Lin'an District exhibited a high AOD value. By 2018, the CSI in the central part of Fuyang was significantly lower than in 2003 and 2008, with the central part showing a high LST value. In 2023, most areas of Lin'an, Chun'an, and Tonglu exhibited high AOD values.

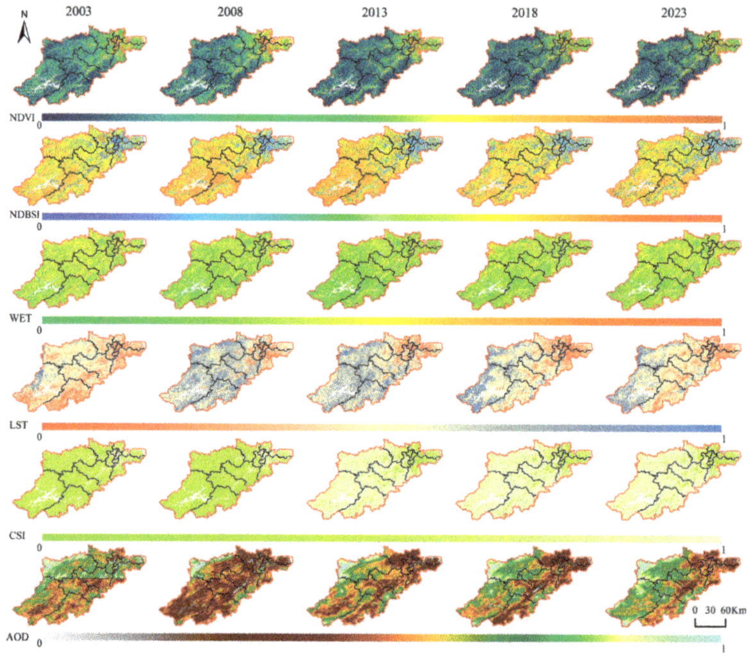

Figure 2. Spatial distribution map of ecological factors.

3.2. *Spatiotemporal Distribution of Ecological Quality in Hangzhou*

Combined with the spatial distribution characteristics of the ecological indicators (as discussed in Section 3.1) and the weights of each indicator (refer to Table 2), the spatial distribution characteristics and causes of the IRSEI in 2003, 2008, 2013, 2018, and 2023 were analyzed (refer to Figure 3). The comparatively low IRSEI values in the Gongshu, Shangcheng, West Lake, and Binjiang areas from 2003 to 2023 primarily resulted from the difference between urbanization and LST. In 2003, owing to the influence of AOD and NDBSI, the IRSEI in the central urban area was low. In 2013, the band area between Fuyang, Tonglu, and Jiande had a low IRSEI value, primarily due to the high AOD and poor air quality. In 2018, the IRSEI values in Gongshu and other central urban areas were low because of the high LST, whereas the spatial differences in WET, NDBSI, and AOD led to high IRSEI values in the central and northern parts of the region. Overall, the differences in the spatial distribution of AOD and air quality consistently had a significant impact on the IRSEI. The differences in the spatial distribution of WET and humidity also had a considerable impact on the IRSEI values.

Table 2. Principal component analysis.

Ecological Index	PC1				
	2003	**2008**	**2013**	**2018**	**2023**
NDVI	0.534	0.592	0.619	0.591	0.655
WET	0.182	0.146	0.155	0.104	0.116
LST	−0.130	−0.179	−0.282	−0.331	−0.251
NDBSI	−0.586	−0.529	−0.519	−0.484	−0.442
AOD	−0.565	−0.556	−0.485	−0.497	−0.528
CSI	−0.006	−0.068	−0.085	−0.215	−0.136
Eigenvalue	0.265	0.369	0.366	0.517	0.489
Contribution	76.81%	76.49%	83.65%	76.21%	84.33%
RSEI	4.14	4.23	4.25	4.25	4.32

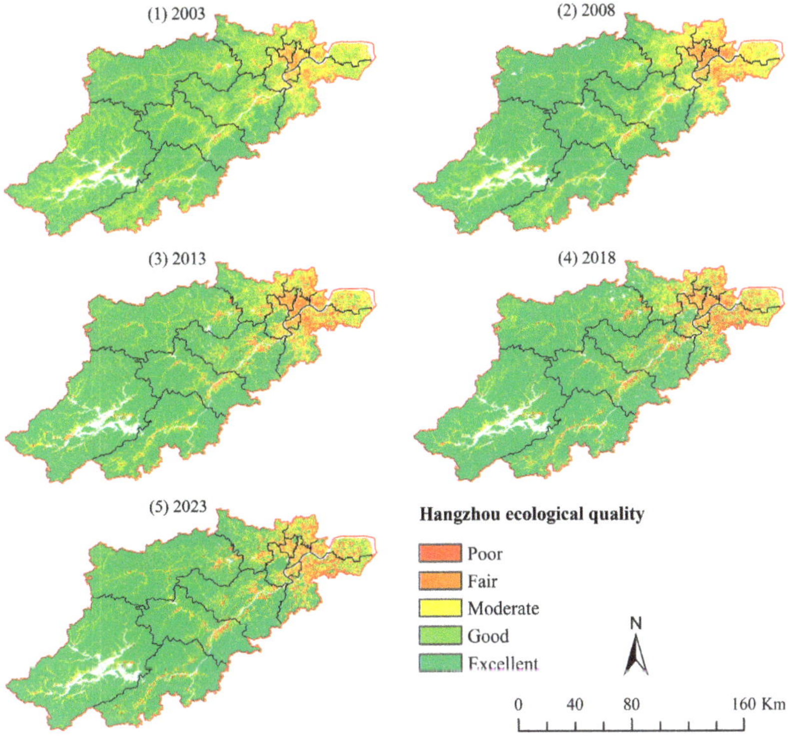

Figure 3. Spatial distribution map of ecological quality in Hangzhou.

Overall, the spatial distribution of the IRSEI in Hangzhou from 2003 to 2023 reveals that the six central urban areas exhibited significantly lower values than those of the remote urban areas, and the city center had lower values than the surrounding areas. Influenced by urban construction and expansion, the proportion of impervious water surfaces in the six central urban areas was high, resulting in overall low ecological quality (Table 3). From 2003 to 2023, the IRSEI values in Shangcheng, Xiacheng, Jianggan, Gongshu, Xihu, and Binjiang districts were notably lower than those in other districts, with most below 0.40. As a central urban area, Shangcheng District contains a large non-construction area, and its average ecological quality is higher than that of other central urban areas. Junan and Kende had the highest IRSEI values, ranging between 0.55 and 0.85.

Table 3. Area and proportion of ecological quality level in Hangzhou (unit: km^2/%).

Ecological Quality	2003		2008		2013		2018		2023	
Poor	61.67	0.33%	74.28	0.40%	291.23	1.58%	479.41	4.73%	339.82	3.21%
Fair	688.65	3.72%	939.83	5.11%	1186.34	6.42%	1079.32	10.65%	1068.11	10.08%
Moderate	1905.21	10.30%	2026.69	11.01%	1882.38	10.18%	1719.81	16.97%	1735.54	16.39%
Good	9718.29	52.55%	6576.60	35.73%	5358.51	28.99%	5046.68	49.80%	4831.47	45.61%
Excellent	6118.42	33.09%	8787.77	47.75%	9768.68	52.84%	10,134.23	100.00%	10,592.28	100.00%

From 2003 to 2023, as shown in Figure 3, the overall IRSEI in Hangzhou exhibited a steady changing trend. Hangzhou boasted a large proportion of forest area, with a forest coverage rate as high as 78.63%, indicating rich forest resources. Chun'an, particularly rich in forested wetlands, is home to the artificial Qiandao Lake. The ecological quality of the region's forestland remained high and stable, strongly influencing the ecological quality of the entire city.

Figure 4 illustrates the proportion of IRSEI grades and their conversions from 2003 to 2023. The grade transformation reveals that IRSEI values remained stable during this period, consistent with previous studies' conclusions. The areal proportion of grades 0–0.2 was the lowest, fluctuating between 0.33% and 2.55%; that of grades 0.2–0.4 was the second-lowest, fluctuating between 3.68% and 6.30%; the largest proportion belonged to grades 0.8–1.0, fluctuating between 33.18% and 57.05%. The proportions of grades 0.4–0.6 and 0.6–0.8 varied considerably, fluctuating from 9.25% to 52.82%. Between 2003 and 2008, there was a significant improvement in ecological quality, with approximately half of the area in grades 0.6–0.8 transitioning to grades 0.8–1. From 2008 to 2013, ecological quality continued to improve, with the area of grade 0.8–1.0 increasing from 47.84% to 53.05%. However, urban expansion led to significant land consumption, including farmland, forests, and even ecological protection areas for urban construction. This expansion resulted in severe air and water pollution, with emissions from traffic, industry, and residents negatively impacting the ecological environment. Consequently, the proportion of grade 0–0.2 increased from 0.40% to 1.56%. Ecological quality showed slight improvement from 2013 to 2018, remaining stable from 2018 to 2023.

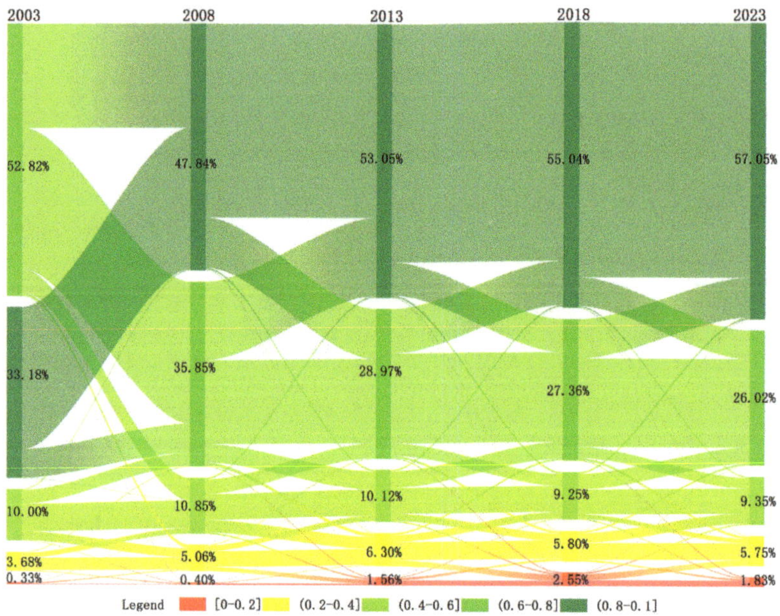

Figure 4. Spatial distribution map of ecological quality in Hangzhou.

3.3. *Influencing Factor Analysis of Ecological Quality Changes in Hangzhou*

Table 4 presents the degrees of influence of the AOD, CSI, WET, LST, NDVI, NDBSI, ROAD, and DEM variables on IRSEI value changes. The influence of each factor on the IRSEI exhibited significant spatial heterogeneity, with the coefficient of LST being large and the coefficients of AOD and WET being exceedingly small. This indicates that human activities, represented by LST, were the most significant and direct factors affecting ecological quality, while the direct influences of air quality and elevation changes on ecological quality were minor.

Table 4. Change in ecological classification in Hangzhou (unit: $km^2/\%$).

Factors	AOD	LST	WET	CSI	NDVI	NDBSI	ROAD	DEM
RMSE_noise	0.08	0.08	0.08	0.05	0.06	0.15	0.12	0.07
Contribution	0.16	0.41	0.15	0.07	0.09	0.06	0.04	0.02

For further analysis and understanding of the microscopic characteristics of ecological quality change in Hangzhou, a concentric ring with a radius of 1 km was established in the central urban area of Hangzhou to analyze the spatial horizontal distribution characteristics of areas with significant variations in Hangzhou (Figure 5). The area of these rings is the same, in order to truly reflect the true density of the change in ecological quality from different urban centers. From 2003 to 2008 and from 2008 to 2013, significant variations near the central urban area changed slightly in terms of their horizontal extent; however, after 2008 to 2013, these differences expanded, consistent with the areas of IRSEI variation. Between 2003 and 2023, areas experiencing significant changes in ecological quality conflicted in Hangzhou, distributed within ranges of 1–3 and 5–10 km. This trend suggests that the changes in IRSEI from 2003 to 2023 were closely related to urban expansion, population change, and economic development. Additionally, influenced by the construction of ecological civilization and adjustments in the development structure, the IRSEI remained generally stable up to 30 km outside the city center, indicating a trend of agglomeration in the center and significant areas in the surroundings. Overall, the area within the scope of the ecological quality changes increased as the increase in distance to the center of Nanjing City slowed, and near Nanjing City, land use change is more intense and the ecological quality of change is more obvious, illustrating that the change of land use will have great influence on ecological quality, which is consistent with other studies [2,15,21,26].

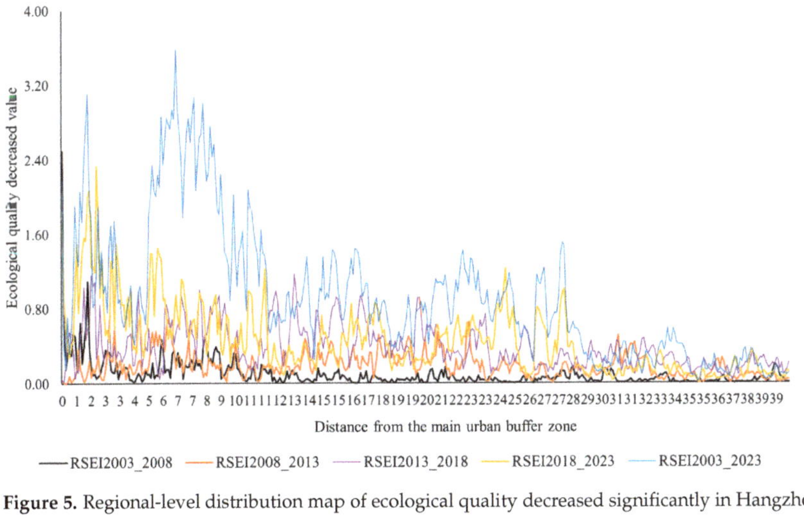

Figure 5. Regional-level distribution map of ecological quality decreased significantly in Hangzhou.

Hangzhou's overall ecological quality was high, demonstrating the positive impact of the municipal government's close attention to ecological environmental protection and sustainable development. In recent years, Hangzhou has implemented a series of measures to improve and enhance ecological quality, including strengthening pollution controls, promoting greening projects, and optimizing the industrial structure. Additionally, water quality in Hangzhou has significantly improved. The proportion of high-quality water in sections above the municipal control level remained high, with excellent assessment results for cross-administrative river-crossing sections, and centralized drinking water sources above the county level maintained a 100% water quality rating. These achievements are attributed to Hangzhou's strict supervision of sewage treatment and discharge as well as the implementation of ecological water replenishment and other measures. Remarkable progress has been made in air quality. Through the implementation of an air pollution prevention and control action plan and the strengthening of industrial pollution controls and vehicle exhaust emission controls, Hangzhou's air quality has significantly improved. The average concentration of $PM_{2.5}$ in urban areas has decreased annually, while the air quality rate has remained high. Hangzhou has also focused on ecological protection and restoration. Biodiversity in Hangzhou has been effectively protected through the implementation of ecological protection and restoration projects and the strengthening of the construction and management of ecological functional areas, such as nature reserves and forest parks. Concurrently, Hangzhou has actively promoted green development, optimized its industrial structure, developed a low-carbon and circular economy, and reduced pressure on the ecological environment. IRSEI takes into account various factors affecting the ecological environment more comprehensively by adding indicators such as AOD and CSI as well as utilizing principal component analysis for model construction. This improvement enables IRSEI to more accurately assess the quality of the ecological environment. IRSEI has been practically applied and validated in some areas, such as Hainan Island and the confluence area of the three rivers in Yibin City. These application cases demonstrate that IRSEI can more accurately reflect the quality of the ecological environment and its dynamic changes, providing a decision-making basis for ecological environmental protection and sustainable development.

These changes (Table 3) have primarily occurred because Hangzhou has focused closely on ecological protection and restoration in recent years. Through the implementation of ecological protection and restoration projects and the strengthening of the construction and management of ecological functional areas, such as nature reserves and forest parks, Hangzhou's biodiversity has been effectively protected. Concurrently, Hangzhou has also actively promoted green development, optimized its industrial structure, developed a low-carbon and circular economy, and reduced pressure on the ecological environment.

3.4. *Dynamic Monitoring of Ecological Quality in Hangzhou*

To further analyze their spatial differences, the IRSEI changes were divided into seven categories according to the changes in the IRSEI index.

Table 5 displays the shifts in the ecological quality of Hangzhou City from 2003 to 2023. Over the past two decades, changes in ecological quality have primarily fallen into three categories—slight improvement, no change, and slight deterioration—accounting for 85% of the total change. The proportion of slight improvement surpassed or closely approached 24% from 2003 to 2008, significantly outweighing the proportions of slight, moderate, and significant deteriorations. Between 2008 and 2013, the proportion of slight variations exceeded 16%, with evident improvement accounting for nearly 19.15%. From 2013 to 2018, the proportions of slight improvement and deterioration remained relatively stable. Furthermore, the proportion of evident improvement was 8.3% from 2000 to 2005, markedly higher than that of evident worsening (1.64%). Between 2018 and 2023, as well as from 2003 to 2023, the proportion of slight improvement (13.69% and 31.42%, respectively) notably exceeded that of less slight improvement (8.77% and 10.85%, respectively). This pattern underscores the continuous and steady enhancement of Hangzhou's regional ecological quality from 2018 to 2023.

An examination of the spatial distribution differences in IRSEI variations (Figure 6) from 2003 to 2023 revealed that IRSEI values generally improved across most regions. Between 2008 and 2013, while some parts of the central urban area experienced improvements in IRSEI values, others witnessed a decline. Moving on to the period between 2013 and 2018, the IRSEI in areas near the north of the central city displayed improvement, whereas those near the south witnessed a slight deterioration, attributable to the predominant direction of urban expansion towards the south during this timeframe. In the interval from

2018 to 2023, several regions exhibited no change in IRSEI, while others demonstrated a balance between improvements and deteriorations.

Table 5. Change in ecological classification in Hangzhou (unit: km^2/%).

Change	2003–2008		2008–2013		2013–2018		2018–2023		2003–2023	
Slight improvement	4396.89	24.02%	3500.90	19.15%	2487.21	13.58%	2518.89	13.69%	5779.39	31.42%
Generally better	66.26	0.36%	83.73	0.46%	135.72	0.74%	171.93	0.93%	194.95	1.06%
Significantly better	3.45	0.02%	2.41	0.01%	13.50	0.07%	22.66	0.12%	15.12	0.08%
No change	11,514.78	62.91%	11,414.47	62.43%	13,381.37	73.06%	13,863.75	75.33%	9905.65	53.85%
Slightly worse	2162.12	11.81%	3010.27	16.46%	2051.84	11.20%	1614.63	8.77%	1996.25	10.85%
Generally worse	147.87	0.81%	221.00	1.21%	192.35	1.05%	161.91	0.88%	399.58	2.17%
Significantly worse	12.65	0.07%	52.44	0.29%	54.83	0.30%	51.64	0.28%	103.64	0.56%

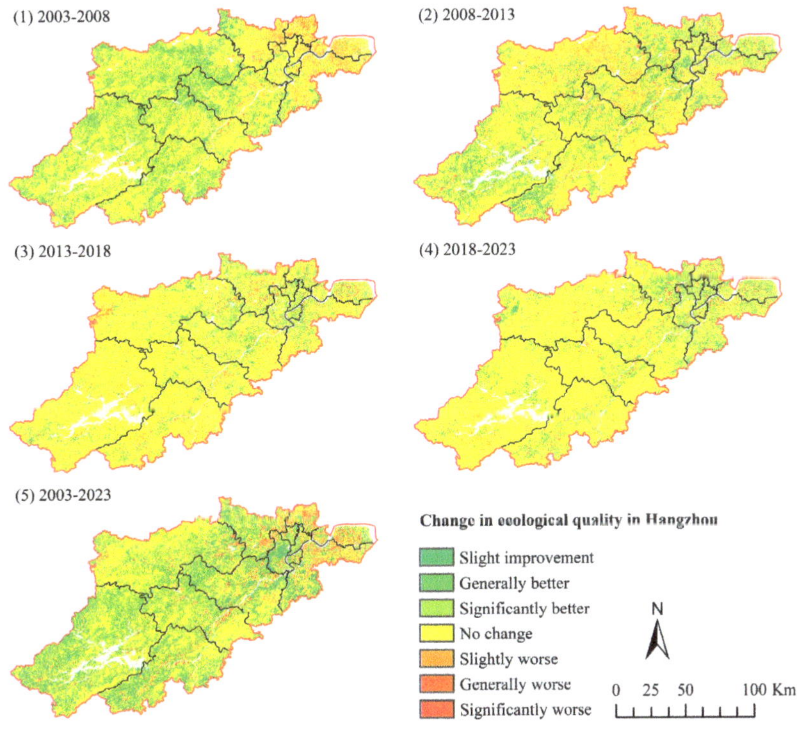

Figure 6. Spatial distribution map of ecological quality in Hangzhou, 2003–2023.

4. Conclusions

This study evaluated the ecological quality of Hangzhou City from 2003 to 2023. Over the past 20 years, ecological quality fluctuated significantly due to urban expansion. From 2003 to 2008, ecological quality improved, mainly due to enhancements in farmland, forests, and water areas. The average IRSEI in Hangzhou rose gently from 4.13 to 4.32. Changes in Hangzhou's ecological quality were linked to urban expansion, increased construction land, and reduced ecological land. However, intensified ecological protection, controlled urban expansion, and improved land intercrossing and fragmentation contributed to an overall enhancement in ecological quality. This study's main findings are as follows:

1. Overall, differences in the spatial distributions of AOD and CSI, representing air quality and salinization damage to the ecosystem, had a certain impact on the regional

IRSEI but were not dominant factors. The differences in the spatial distributions of LST and NDBSI also significantly impacted the regional IRSEI. Surface temperature, air quality (measured by aerosol optical depth, AOD), and humidity (as denoted by WET) substantially impact the ecological status as indicated by the IRSEI;
2. During 2003–2023, the IRSEI in Hangzhou exhibited a steady changing trend, with the area of grade 0–0.2 increasing from 33.18% to 57.05%, significantly improving ecological quality. The changes in the IRSEI from 2003 to 2023, such as surface temperature and traffic networks, were closely related to human activities, and the effects of AOD and CSI on IRSEI were spatially heterogeneous;
3. In Hangzhou, areas experiencing drastic changes in ecological quality were located within the ranges of 1–3 km and 5–10 km from the city center, while the IRSEI remained generally stable at distances of at least 30 km from the city center.

Author Contributions: Conceptualisation, C.C. and Z.W.; methodology, J.L.; software, C.C.; validation, C.C. and Z.W.; formal analysis, C.C.; data curation, C.C.; writing—original draft preparation, C.C.; writing—review and editing, C.C.; visualisation, C.C.; supervision, Z.W.; project administration, Z.W.; funding acquisition, Z.W. All authors have read and agreed to the published version of the manuscript.

Funding: This research was funded by a Key Project from the National Social Science Foundation of China (Grant No. 23AZD058), and supported by the Fundamental Research Funds for the Central Universities, Hohai University (Grant No. B240207085).

Data Availability Statement: The authors can provide the data upon reasonable request. The data are not publicly available due to privacy restrictions.

Acknowledgments: The authors are particularly grateful to all researchers for providing data support for this study.

Conflicts of Interest: The authors declare no conflicts of interest.

References

1. Ahani, S.; Dadashpoor, H. Urban growth containment policies for the guidance and control of peri-urbanization: A review and proposed framework. *Environ. Dev. Sustain.* **2021**, *23*, 1–30. [CrossRef]
2. Li, J.; Gong, J.; Guldmann, J.M.; Yang, J. Assessment of Urban Ecological Quality and Spatial Heterogeneity Based on Remote Sensing: A Case Study of the Rapid Urbanization of Wuhan City. *Remote Sens.* **2021**, *13*, 4440. [CrossRef]
3. Liao, W.; Jiang, W. Evaluation of the Spatiotemporal Variations in the Eco-environmental Quality in China Based on the Remote Sensing Ecological Index. *Remote Sens.* **2020**, *12*, 2462. [CrossRef]
4. Ermida, S.L.; Soares, P.; Mantas, V.; Göttsche, F.M.; Trigo, I.F. Google earth engine open-source code for land surface temperature estimation from the landsat series. *Remote Sens.* **2020**, *12*, 1471. [CrossRef]
5. Estoque, R.C. A Review of the Sustainability Concept and the State of SDG Monitoring Using Remote Sensing. *Remote Sens.* **2020**, *12*, 1770. [CrossRef]
6. Fei, W.; Zhao, S. Urban land expansion in China's six megacities from 1978 to 2015. *Sci. Total Environ.* **2019**, *664*, 60–71. [CrossRef]
7. Firozjaei, M.K.; Fathololoumi, S.; Weng, Q.; Kiavarz, M.; Alavipanah, S.K. Remotely sensed urban surface ecological index (RSUSEI): An analytical framework for assessing the surface ecological status in urban environments. *Remote Sens.* **2020**, *12*, 2029. [CrossRef]
8. Xu, H.; Wang, M.; Shi, T.; Guan, H.; Fang, C.; Lin, Z. Prediction of ecological effects of potential population and impervious surface increases using a remote sensing based ecological index (RSEI). *Ecol. Indic.* **2018**, *93*, 730–740. [CrossRef]
9. Xu, H.; Wang, Y.; Guan, H.; Shi, T.; Hu, X. Detecting ecological changes with a remote sensing based ecological index (RSEI) produced time series and change vector analysis. *Remote Sens.* **2019**, *11*, 2345. [CrossRef]
10. Hu, X.; Xu, H. A new remote sensing index for assessing the spatial heterogeneity in urban ecological quality: A case from Fuzhou City, China. *Ecol. Indic.* **2018**, *89*, 11–21. [CrossRef]
11. Yang, J.Y.; Wu, T.; Pan, X.Y.; Du, H.T.; Li, J.L.; Zhang, L.; Men, M.X.; Chen, Y. Ecological quality assessment of Xiongan New Area based on remote sensing ecological index. *Ying Yong Sheng Tai Xue Bao* **2019**, *30*, 277–284. [CrossRef] [PubMed]
12. Wang, J.; Liu, D.; Ma, J.; Cheng, Y.; Wang, L. Development of a large-scale remote sensing ecological index in arid areas and its application in the Aral Sea Basin. *J. Arid Land* **2021**, *13*, 40–55. [CrossRef]
13. Gong, C.; Lyu, F.; Wang, Y. Spatiotemporal change and drivers of ecosystem quality in the Loess Plateau based on RSEI: A case study of Shanxi, China. *Ecol. Indic.* **2023**, *155*, 111060. [CrossRef]

14. Hasanah, A.; Wu, J. Exploring dynamics relationship between carbon emissions and eco-environmental quality in Samarinda Metropolitan Area: A spatiotemporal approach. *Sci. Total Environ.* **2024**, *927*, 172188. [CrossRef] [PubMed]
15. Li, Y.; Tian, H.; Zhang, J.; Lu, S.; Xie, Z.; Shen, W.; Zheng, Z.; Li, M.; Rong, P.; Qin, Y. Detection of spatiotemporal changes in ecological quality in the Chinese mainland: Trends and attributes. *Sci. Total Environ.* **2023**, *884*, 163791. [CrossRef] [PubMed]
16. Airiken, M.; Zhang, F.; Chan, N.W.; Kung, H.T. Assessment of spatial and temporal ecological environment quality under land use change of urban agglomeration in the North Slope of Tianshan, China. *Environ. Sci. Pollut. Res.* **2022**, *29*, 12282–12299. [CrossRef] [PubMed]
17. Tang, Q.; Hua, L.; Tang, J.; Jiang, L.; Wang, Q.; Cao, Y.; Wang, T.; Cai, C. Advancing ecological quality assessment in China: Introducing the ARSEI and identifying key regional drivers. *Ecol. Indic.* **2024**, *163*, 112109. [CrossRef]
18. Zhang, H.; Liu, Y.; Li, X.; Feng, R.; Gong, Y.; Jiang, Y.; Guan, X.; Li, S. Combing remote sensing information entropy and machine learning for ecological environment assessment of Hefei-Nanjing-Hangzhou region, China. *J. Environ. Manag.* **2023**, *325*, 116533. [CrossRef]
19. Zhang, X.; Gao, Z.; Li, Y.; Sun, G.; Cen, Y.; Lou, Y.; Yao, Y.; Liu, W. Eco-Environment Quality Response to Climate Change and Human Activities on the Loess Plateau, China. *Land* **2023**, *12*, 1792. [CrossRef]
20. Chen, W.; Wang, J.; Ding, J.; Ge, X.; Han, L.; Qin, S. Detecting Long-Term Series Eco-Environmental Quality Changes and Driving Factors Using the Remote Sensing Ecological Index with Salinity Adaptability (RSEISI): A Case Study in the Tarim River Basin, China. *Land* **2023**, *12*, 1309. [CrossRef]
21. Li, W.; Samat, A.; Abuduwaili, J.; Wang, W. Spatiotemporal Pattern, Evolutionary Trend, and Driving Forces Analysis of Ecological Quality in the Irtysh River Basin (2000–2020). *Land* **2024**, *13*, 222. [CrossRef]
22. Kumar, P.; Singh, A.B.; Arora, T.; Singh, S.; Singh, R. Critical review on emerging health effects associated with the indoor air quality and its sustainable management. *Sci. Total Environ.* **2023**, *872*, 162163. [CrossRef] [PubMed]
23. Huang, J.; Wang, J.; Hu, Y.; Zhao, H. Impact of Air Emission Reduction Measures during the Asian Games on Air Quality and Health Risks in Hangzhou and Surrounding Cities. *Atmosphere* **2024**, *15*, 780. [CrossRef]
24. Kamusoko, C.; Gamba, J. Simulating urban growth using a Random Forest-Cellular Automata (RF-CA) model. *ISPRS Int. J. Geo-Inf.* **2015**, *4*, 447–470. [CrossRef]
25. Tang, L.; Liang, G.; Gu, G.; Xu, J.; Duan, L.; Zhang, X.; Yang, X.; Lu, R. Study on the spatial-temporal evolution characteristics, patterns, and driving mechanisms of ecological environment of the Ecological Security Barriers on China's Land Borders. *Environ. Impact Assess. Rev.* **2023**, *103*, 107267. [CrossRef]
26. Liu, Y.; Xu, W.; Hong, Z.; Wang, L.; Ou, G.; Lu, N.; Dai, Q. Integrating three-dimensional greenness into RSEI improved the scientificity of ecological environment quality assessment for forest. *Ecol. Indic.* **2023**, *156*, 111092. [CrossRef]

land

MDPI

Article

Exploring the Spatiotemporal Evolution Patterns and Determinants of Construction Land in Mianning County on the Eastern Edge of the Qinghai–Tibet Plateau

Yinbing Zhao [1,2,3], Zhongyun Ni [1,3,*], Yang Zhang [1,3], Peng Wan [1], Chuntao Geng [1], Wenhuan Yu [1], Yongjun Li [1] and Zhenrui Long [4]

1 College of Geography and Planning, Chengdu University of Technology, Chengdu 610059, China; zhaoyinbing06@cdut.cn (Y.Z.); zhangyang2020@cdut.edu.cn (Y.Z.); 2022020020@stu.cdut.edu.cn (P.W.); 2019050677@alu.cdut.edu.cn (C.G.); yuwenhuan@stu.cdut.edu.cn (W.Y.); liyongjun@stu.cdut.edu.cn (Y.L.)

2 State Key Laboratory of Urban and Regional Ecology, Research Center for Eco-Environmental Sciences, Chinese Academy of Sciences, Beijing 100085, China

3 Human Geography Research Center of Qinghai-Tibet Plateau and Its Eastern Edge, Chengdu University of Technology, Chengdu 610059, China

4 Sichuan Research Institute of Ecological Restoration of Land Space and Geohazard Prevention and Control, Chengdu 610063, China; longzrei@gmail.com

* Correspondence: nizhongyun2012@mail.cdut.edu.cn

Abstract: Studying the spatiotemporal evolution and driving forces behind construction land amidst the intricate ecological and geological setting on the eastern edge of the Qinghai–Tibet Plateau offers invaluable insights for local sustainable development in a landscape transition zone and ecologically fragile area. Using construction land data from four phases, spanning 1990 to 2020, in Mianning County, this study employs methodologies like the Landscape Expansion Index (LEI) and land use transfer matrix to delineate the spatiotemporal evolution characteristics of construction land. A comprehensive set of 12 influencing factors across five categories—geomorphology, geological activity, climate, river and vegetation environment, and social economy—were examined. The Geographically Weighted Regression (GWR) model was then employed to decipher the spatial distribution pattern of construction land in 1990 and 2020, shedding light on the driving mechanisms behind its changes over the three decades. The research reveals distinct patterns of construction land distribution and evolution in Mianning County, shaped by the ecological and geological landscape. Notably, the Anning River wide valley exhibits a concentrated and contiguous development mode, while the Yalong River deep valley showcases a decentralized development pattern, and the Dadu River basin manifests an aggregation development mode centered around high mountain lakes. Over the study period, all three river basins witnessed varying degrees of construction land expansion, transitioning from quantitative expansion to qualitative enhancement. Edge expansion predominantly characterizes the expansion mode, complemented by leapfrog and infilling modes, accompanied by conversions from cropland and forest land to construction land. An analysis of the spatial pattern and drivers of construction land change highlights human-induced factors dominating the Anning River Basin, contrasting with natural factors prevailing in the Yalong River Basin and the Dadu River Basin. Future efforts should prioritize climate change considerations and environmental capacity, aiming for an ecologically resilient spatial pattern of construction land.

Keywords: geomorphic gradient; geological environment; Landscape Expansion Index (LEI); spatiotemporal evolution; driving force analysis; Geographically Weighted Regression (GWR); eastern edge of the Qinghai–Tibet Plateau

1. Introduction

Construction land serves as a multifaceted indicator, offering insights into the spatial layout, expansion trends, land use efficiency, decision-making processes, and interactions with the natural environment within human activity domains. Through the vigilant monitoring and systematic study of construction land patterns and dynamics, we gain a deeper comprehension of human activities' impact on land resources and the environment, thus facilitating the formulation of scientifically informed land use planning and policies [1, 2]. Remote sensing technology [3–5] and land use surveys [6,7] provide essential tools for delineating the distribution patterns and dynamic changes of construction land. Several key indicators are instrumental in characterizing land use changes, including the speed of construction land expansion [8,9], alterations in construction land structure [10,11], land use intensity [12,13], land conversion rates [14], the nexus between construction land and population [15,16], and the economic construction land elasticity coefficient [17,18], among others.

The spatial distribution and dynamic changes of construction land are the culmination of both internal and external factors, which can be categorized into natural and human influences [19,20]. Over extended periods, natural driving forces predominantly shape the spatial patterns of construction land, characterized by gradual and stable transformations [21]. Conversely, regions endowed with favorable geographical conditions often experience human factors as the primary drivers of construction land change [22,23]. Particularly in mountainous regions with intricate terrain, the impact of natural factors is profound [24,25]. Natural factors encompass geological, topographical, climatic, and hydrological elements [26]. Geological factors include geological structure, lithological stratigraphy, seismic activity, and geological hazards [27,28]. Terrain features comprise altitude, slope, and aspect, among others [29,30]. Climate considerations encompass precipitation, temperature, and related variables [31,32]. Hydrological factors entail water resource distribution, accessibility, and submerged water levels [33,34]. Human factors encompass economic, social, technological, and policy dimensions. Economic factors span the level and stage of economic development, industrial structure, gross domestic product (GDP), investment, income levels, etc. [35–38]. Social elements involve population size and density, urbanization levels, and developmental stages [39–41]. Technical aspects encompass transportation, technology, construction practices, environmental protection measures, etc. [42–44]. Policy dimensions include land use planning, taxation and management policies, immigration, and relocation strategies [45–48].

Prior research has employed a diverse array of statistical and spatial analysis models to scrutinize the spatial patterns, dynamic shifts, and driving forces underlying construction land dynamics. These methodologies include linear regression analysis [49,50], multiple regression models [51–53], principal component analysis [54,55], logistic regression analysis [56–58], multi-index coupling models [59], system dynamics models [60–62], network/spatial lag models [63], random forest models [64], geographically weighted regression models [65,66], and Geodetector models [67,68]. Among these methodologies, those rooted in geographic spatial foundations hold particular promise, offering comprehensive insights into the multifaceted impacts of various factors across different spatial contexts. Consequently, the application of models such as Geodetector models and geographically weighted regression models is witnessing a notable surge in popularity.

In China, research on the drivers of land use change, including alterations in construction land, took root in the 1990s. Existing scholarship predominantly delves into the spatial characteristics and mechanisms governing construction land expansion, the driving forces propelling such expansion, and the ecological environment's response to this growth [54]. Research endeavors span diverse spatial scales, encompassing national [17,69], provincial [17,69], urban agglomerations [70], urban cores [71,72], peri-urban areas [73], and municipal and county levels [74,75]. However, scholarly attention dedicated to construction land and its transformations in China's western regions [76], particularly on the eastern periphery of the Qinghai–Tibet Plateau, remains relatively scant. Positioned within a transition zone of diverse topographical and climatic attributes, the geological and topographical complexity of the plateau's eastern edge, coupled with its variable climate, imparts unique regional characteristics to the spatial distribution and evolution of construction land. Although some studies have undertaken the geomorphological zoning [77] and city/autonomous prefecture-level analyses [24] of construction land distribution characteristics in this area, there remains a dearth of a thorough analysis elucidating the formation of its spatial patterns and dynamic changes. The inadequate spatial quantification and consideration of geological, climatic, and accessibility factors pose notable limitations. Addressing this gap, urgent attention is warranted to investigate the patterns, dynamics, and causal mechanisms underpinning construction land at the scale of typical cities, prefectures, and counties on the eastern fringes of the Qinghai–Tibet Plateau. Such endeavors are essential for a nuanced understanding of the driving forces stemming from natural and human factors, thereby furnishing a robust decision-making framework for optimal regional land resource allocation.

In addressing the limitations of prior research concerning regional coverage and indicator selection, this study undertakes a comprehensive examination of the intricate ecological and geological milieu within the transitional zone along the eastern fringe of the Qinghai–Tibet Plateau. Mianning County, traversing the geomorphic boundary of the plateau and encompassing the Yalong River Basin, Anning River Basin, and Dadu River Basin, serves as the focal area for this investigation. Leveraging Landsat satellite remote sensing imagery with a spatial resolution of 30 m, coupled with land survey data and field verification data, we obtained land use data spanning four distinct periods, 1990, 2000, 2010, and 2020, with a particular emphasis on construction land dynamics. Sequentially, the study achieves the following research objectives: (I) the identification of spatial characteristics pertaining to construction land; (II) the exploration of temporal and spatial variations in construction land; (III) the analysis of the factors contributing to the spatial pattern formation of construction land during the 1990 and 2020 phases; and (IV) the elucidation of the causes and mechanisms driving changes in construction land from 1990 to 2020.

2. Study Area and Data Processing

2.1. Study Area

Situated within the transition zone stretching from the eastern Qinghai–Tibet Plateau to the Sichuan Basin (Figure 1), Mianning County lies within the Liangshan Yi Autonomous Prefecture, Sichuan Province, China. The geographic coordinates of the research area span from 101°38′ E to 102°25′ E longitude and 28°05′ N to 29°02′ N latitude, encompassing a total land area of approximately 4420 km^2. The terrain within the research area exhibits an elevation gradient, characterized by higher elevations in the north and lower elevations in the south, with peaks reaching up to 5306 m and valleys descending to as low as 1255 m. The average elevation across the area is 2737 m, with an average topographic relief of 251 m.

From a tectonic perspective, the study area resides within the northern segment of the Sichuan–Yunnan structural belt, marked by significant geological structures such as the Xiaojin River Fault, the Jinhe–Chenghai Fault, and the Anning River Fault. The study area is divided into three basins: the Anning River Basin in the central and eastern parts, which includes Gaoyang Street, Yihai, Hui'an, Ruoshui, Fuxing, Hongmo, Shilong, Hebian, Lugu, Zeyuan, and Manshuiwan; the Yalong River Basin in the west, encompassing He'ai, Jinping, Mianshawan, Jianmei, Mofanggou, and Lizhuang; and the Dadu River Basin in the north, which includes Yele. The economic landscape of the county exhibits a notable geographical gradient, driven by a combination of natural and socio-economic factors. Notably, construction land is predominantly concentrated within the Anning River and Yalong River basins, mirroring the intricate interplay between geological features and human activities.

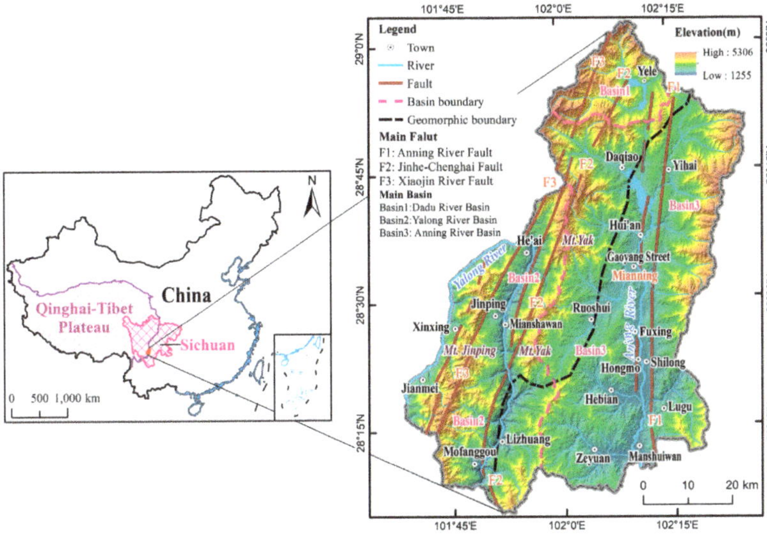

Figure 1. Location of the study area.

2.2. Data Sources

2.2.1. Construction Land Data

The study utilized Landsat-5 TM and Landsat-8 OLI images, comprising a total of four phases and eight scenes, obtained from the US Geological Survey website (http://glovis.usgs.gov, accessed on 26 March 2024). The temporal span averaged 10 years with a spatial resolution of 30 m (Figure 2), and there were no significant fluctuations in construction land at each selected time node. If the cloud cover exceeds 5% for a given year, data from neighboring years will be utilized for synthesis or replacement. By integrating data from the first, second, and third land use surveys of Mianning County, we derived land use data for the years 1990, 2000, 2010, and 2020 through interpretation, investigation, and correction processes (Figure 3). Routes and sampling verification across the three basins indicate that the accuracy of construction land identification exceeds 95%. Construction land encompasses six secondary categories: urban construction land, independent industrial and mining land, transportation land, rural residential land, water conservancy facility land, and special land [78]. The study area primarily emphasizes the first four secondary land categories. However, owing to the resolution constraints of Landsat satellite data, this study does not differentiate between these secondary categories within construction land; instead, it consolidates them into a unified category of construction land.

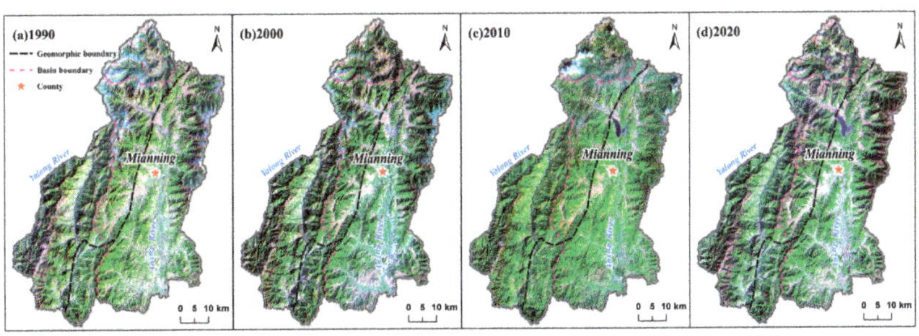

Figure 2. The fourth phase of Landsat remote sensing images.

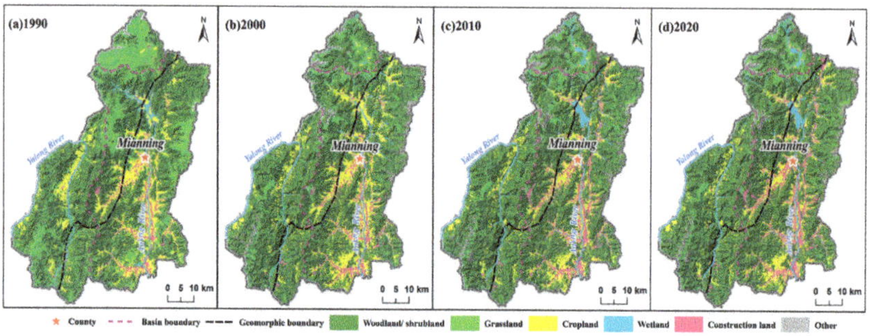

Figure 3. Land use types in the fourth phase.

2.2.2. Impact Factors

Drawing upon prior studies [79,80] and taking into account the geological and geographical conditions of Mianning County, we primarily consider factors closely related to the evolution of construction land based on their representativeness, comprehensiveness, and accessibility (Table 1). Natural factors that constrain or drive the evolution of construction land include geomorphology, geology, climate, rivers, and vegetation [81–83]. Conversely, human factors such as cropland, town accessibility, and road accessibility play a relatively active role in driving the spatiotemporal evolution of construction land [84–86].

Table 1. Data of influencing factors on the evolution of construction land.

Variable Category	Variable Name	Definition and Units	Data Sources	Spatial Resolution
Geomorphological	ELEVATION	Elevation represents macroscopic geomorphology (m)	Geospatial data cloud [a]	30 m
	SLOPE	Slope represents ground cutting condition (°)		
	ASPECT	Aspect represents ground orientation		
Geological activities	ED_FAULT	Euclidean distance of fault (m)	China Geological Survey [b]	Vector
	KD_GEOHAZARD *	Kernel density of geological hazard	Sichuan Provincial Institute of Land Space Ecological Restoration and Geological Disaster Prevention and Control	Vector
Climatic	TEM *	Annual precipitation (mm)	National Qinghai Tibet Plateau Scientific Data Center [c]	1 km
	PRE *	Annual mean temperature (°C)		1 km
Rivers and vegetation environment	ED_RIVER	Euclidean distance of river (m)	National Geomatics Center of China [d]	Vector
	NDVI *	Normalized Difference Vegetation Index	Remote sensing extraction from Landsat satellite data	30 m
Socio-economic	KD_CROPLAND *	Kernel density of cropland	Remote sensing interpretation from Landsat satellite data	30 m
	ED_TOWN	Euclidean distance of town (m)	National Geomatics Center of China [d]	Vector
	ED_ROAD *	Euclidean distance of road (m)	National Geomatics Center of China [d]	Vector

[a] http://www.gscloud.cn/ (accessed on 26 March 2024); [b] https://www.ngac.org.cn/ (accessed on 26 March 2024); [c] https://data.tpdc.ac.cn/ (accessed on 26 March 2024); [d] http://www.ngcc.cn(accessed on 26 March 2024). The symbol * indicates that the indicator has two periods of data from 1990 and 2020, and the change of the indicator from 1990 to 2020 needs to be calculated.

During the study period from 1990 to 2020, the geomorphology exhibited relatively stable characteristics; hence, geomorphological data from a single period were selected. Digital elevation data (ASTER GDEM) with a spatial resolution of 30 m were acquired from the Geospatial Data Cloud website (http://www.gscloud.cn, accessed on 26 March 2024). ASTER GDEM (Figure 4a) was utilized to derive slope (Figure 4b) and aspect (Figure 4c) data. These three factors, elevation, slope, and aspect, were employed to characterize the influence of geomorphology on the spatial pattern and temporal changes of construction land.

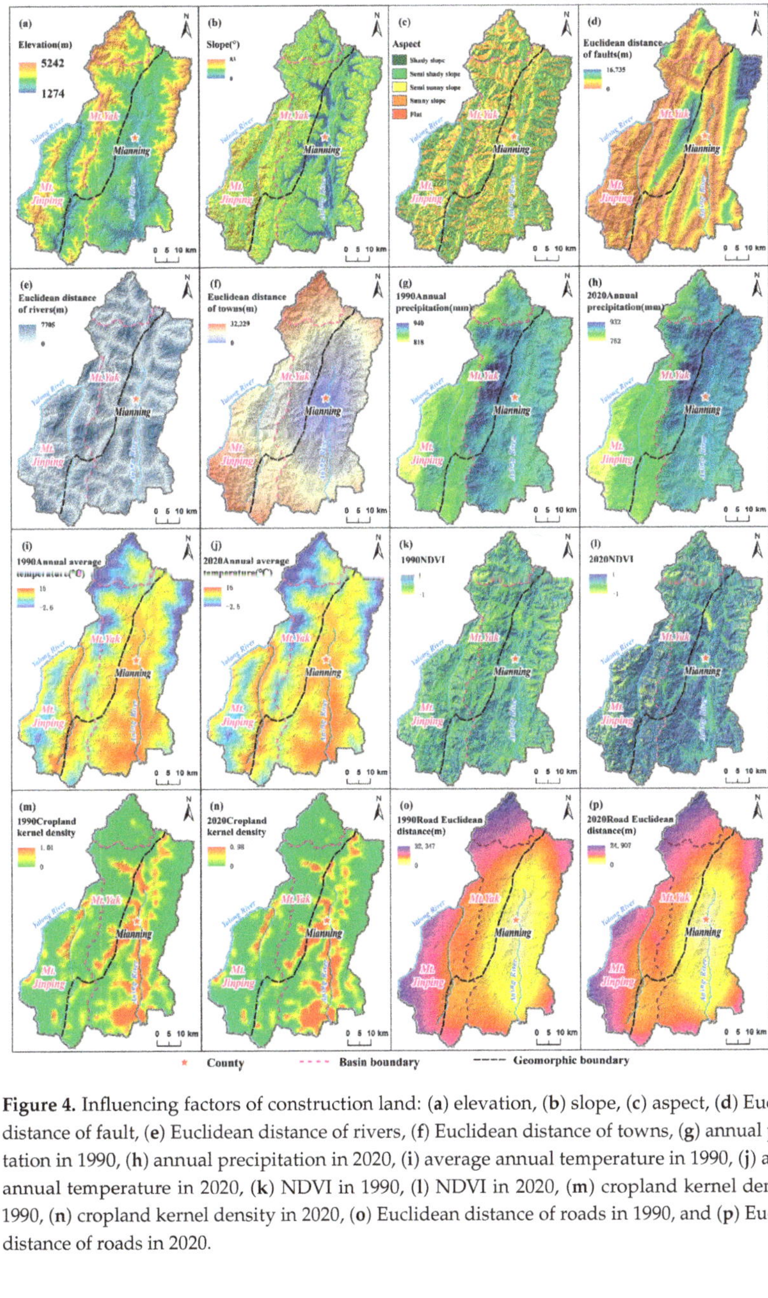

Figure 4. Influencing factors of construction land: (**a**) elevation, (**b**) slope, (**c**) aspect, (**d**) Euclidean distance of fault, (**e**) Euclidean distance of rivers, (**f**) Euclidean distance of towns, (**g**) annual precipitation in 1990, (**h**) annual precipitation in 2020, (**i**) average annual temperature in 1990, (**j**) average annual temperature in 2020, (**k**) NDVI in 1990, (**l**) NDVI in 2020, (**m**) cropland kernel density in 1990, (**n**) cropland kernel density in 2020, (**o**) Euclidean distance of roads in 1990, and (**p**) Euclidean distance of roads in 2020.

Geological considerations primarily involve faults and geological hazard factors. Throughout the study period, there was no significant activity observed along fault lines; therefore, a single period of geological data was utilized to extract fault information. Vector data regarding faults were obtained from the China Geological Survey (https://www.ngac.org.cn, accessed on 26 March 2024), and fault Euclidean distance data (Figure 4d) were generated to depict the influence of geological structures on construction land. Geological disaster occurrences exhibit notable spatiotemporal variations. We compiled geological disaster point data from 1990 to 2020, sourced from the Sichuan Provincial Institute of Land Space Ecological Restoration and Geological Disaster Prevention and Control. Using the kernel density analysis method, raster data for two time points, 1990 (Figure 5a) and 2020 (Figure 5b), were generated, and changes in geological disaster kernel density between these years were calculated (Figure 5c). The kernel density analysis is a spatial analysis method that can effectively display the concentration and spatial distribution trends of point data, particularly capturing the spatial distribution patterns of data without prior assumptions. These analyses were conducted to delineate the impact of geological disasters on the spatial distribution and dynamic changes of construction land.

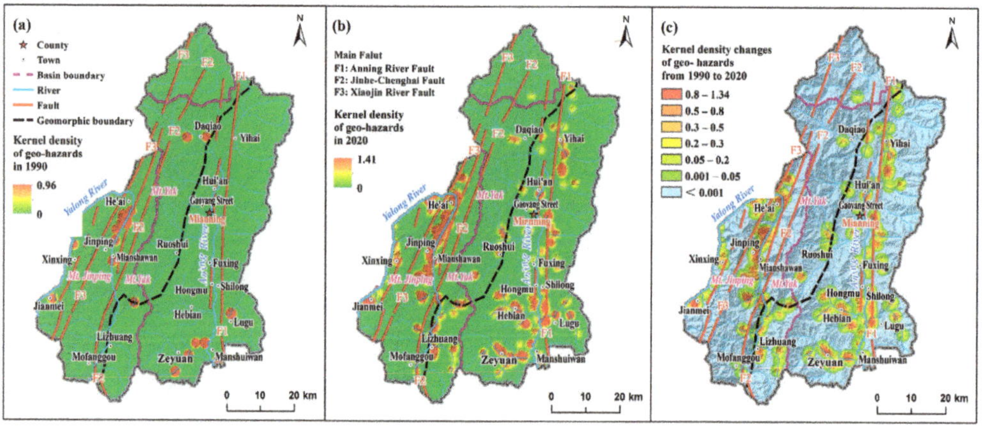

Figure 5. Geo-hazards factors affecting construction land: (**a**) kernel density of geo-hazards in 1990, (**b**) kernel density of geo-hazards in 2000, and (**c**) kernel density changes of geo-hazards from 1990 to 2020.

The influence of climate on construction land primarily revolves around two factors: annual precipitation and annual average temperature, sourced from the National Qinghai–Tibet Plateau Scientific Data Center (https://data.tpdc.ac.cn, accessed on 26 March 2024). The data span the period from 1990 to 2020. For the analysis of construction land patterns and evolution, data from 1990 (Figure 4g,i) and 2020 (Figure 4h,j) are predominantly utilized, along with change data spanning from 1990 to 2020.

Rivers and vegetation primarily influence the surrounding environment of construction land. Rivers typically undergo localized changes in width, and, for this study, single-period data sourced from the National Geomatics Center of China (http://www.ngcc.cn, accessed on 26 March 2024) were utilized. The Euclidean distance calculation method was applied to the river vector data to generate raster data (Figure 4e). Vegetation data were derived from Landsat remote sensing imagery for the years 1990 (Figure 4k) and 2020 (Figure 4l). The Normalized Difference Vegetation Index (NDVI) was computed to characterize the impact of vegetation.

Social and economic factors primarily encompass the influence of cropland density, town accessibility, and road accessibility. Cropland kernel density is utilized to depict the impact of cropland, with data sourced from remote sensing interpretation of Landsat imagery. Changes in cropland kernel density from 1990 (Figure 4m) to 2020 (Figure 4n) were calculated. Town points within Mianning County and surrounding areas were extracted from the National Geomatics Center of China to assess town accessibility. Euclidean distance was employed to describe the spatial impact of each town point. As town locations remained constant, single-period data were used to illustrate their spatial impact (Figure 4f). Road data from 1990 and 2020 were collected and revised to analyze road accessibility. The Euclidean distance of roads in 1990 (Figure 4o) and 2020 (Figure 4p) was calculated, and changes in the road Euclidean distance from 1990 to 2020 were analyzed to understand the role of roads in the spatial pattern and changes in construction land.

2.3. Methods

2.3.1. Landscape Expansion Index

The Landscape Expansion Index (LEI) is employed to quantitatively characterize the spatial expansion patterns of construction land, delineated into three modes (Figure 6): leapfrog ($LEI = 0$), edge expansion ($0 < LEI \leq 50$), and infilling ($50 < LEI \leq 100$). The edge expansion mode entails the augmentation of construction land based on existing areas, with the surrounding topography, economic conditions, and other factors meeting the requirements for construction land development [87]. The leapfrog mode involves selecting suitable areas for construction due to the saturation of existing construction land or constraints imposed by terrain, economic factors, and other considerations. The infilling mode primarily involves intensifying the use of existing construction land by adding or renovating within its boundaries. The calculation formula is as follows [88]:

$$LEI = 100 \times \frac{A_0}{A_0 + A_V} \tag{1}$$

where A_0 is the intersection point between the newly expanded construction land patch buffer zone and the original construction land, and A_V is the intersection of the buffer zone and non-construction land.

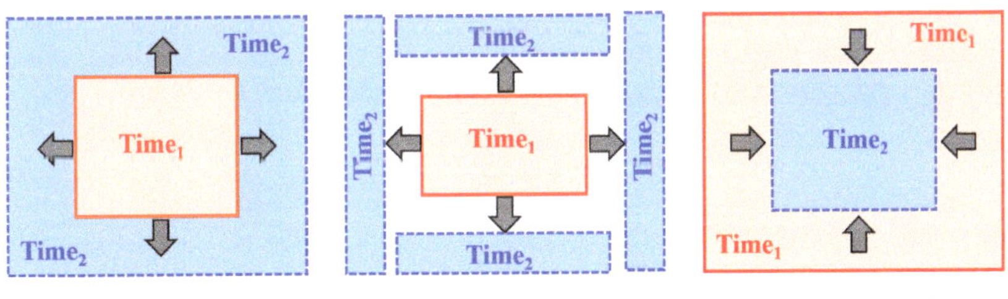

Figure 6. Three modes of landscape expansion.

2.3.2. Geographically Weighted Regression

The Geographically Weighted Regression (GWR) model serves as a spatial extension of conventional regression models, offering the capability to estimate local parameters [89]. In this model, parameters for each spatial point within the entire framework are independently quantified, typically utilized to assess the presence of spatial non-stationarity in the relationship between the dependent and independent variables [90]. The GWR model proves useful in discerning the influence of various natural and human factors on the spatial distribution pattern and evolution of construction land. Extending traditional global regression, the GWR model incorporates geographic location parameters. The calculation of Formula (2) is as follows:

$$y_i = \beta_o\,(\mu_i, v_i) + \sum_{k=1}^{p} \beta_k(\mu_i, v_i)x_{ik} + \varepsilon_i,\ i = 1, 2, \ldots n \tag{2}$$

where y_i is the dependent variable, x is the independent variable of the explanatory factor, $\beta_0(\mu_i, v_i)$ represents the intercept at position i, $\beta_k(\mu_i, v_i)$ represents the local parameter estimation of the explanatory variable x_{ik} at position i, and ε_i is the random error term at point i.

The estimation coefficients of GWR are weighted based on the observed values and the spatial proximity of a specific point i, and the rectangular equation can be used to estimate the parameters:

$$\hat{\beta}(\mu, v) = \left(X^T W(\mu_i, v_i) X\right)^{-1} X^T W(\mu_i, v_i) Y \tag{3}$$

where $\hat{\beta}(\mu, v)$ represents the unbiased estimate of the regression coefficient β, $W(\mu_i, v_i)$ is the weighting matrix, and X and Y are the matrices of independent and dependent variables. $W(\mu_i, v_i)$ ensures that observations close to a specific location have greater weight, expressed using a Gaussian weighted kernel function:

$$w_{ij} = exp(-\frac{d_{ij}}{b^2}) \tag{4}$$

where w_{ij} represents the weight of observation j at position i, d_{ij} represents the Euclidean distance between regression point i and adjacent observation j, and b represents the basic width of the kernel function.

Stationarity exists when the variable x_{ik} does not vary with position i, and the GWR-based stationarity index is used to estimate spatial stationarity [89]:

$$SI = \frac{\beta_{GWR\_iqr}}{2 \times GLM\_se} \tag{5}$$

where *SI* is the stationarity index, β_{GWR_iqr} is the standard error interquartile range of the GWR coefficient, and *GLM_se* is the standard error of the global regression analysis. When *SI* < 1, the explanatory variable y and the dependent variable x achieve spatial stationarity.

AIC can be used to determine the significance of the coefficients to compare relative measures of model performance [91]; the smaller the AIC is, the more reliable the model is, and *AICc* represents the limited sample size correction result of the AIC:

$$AICc = 2nIn(\hat{\sigma}) + nln(2\pi) + n(\frac{n + tr(S)}{n - 2 - tr(S)}) \tag{6}$$

where n is the number of samples, $\hat{\sigma}$ is the estimated value of the residual standard deviation, and *tr(S)* represents the trajectory of the hat matrix, and, when the *AICc* value is lower than three, the model performs better.

The GWR model is employed to elucidate the factors contributing to the spatial distribution of construction land in 1990 and 2020, as well as the drivers of its evolution from 1990 to 2020. For analyzing the spatial pattern of construction land in these years, the independent variables include ELEVATION, SLOPE, ASPECT, ED_FAULT, TEM, PRE, ED_RIVER, NDVI, KD_CROPLAND, ED_TOWN, and ED_ROAD corresponding to the respective years. The dependent variables are the kernel density of construction land in 1990 and 2020, respectively. To investigate the causes of the evolution of construction land from 1990 to 2020, six static factors (ELEVATION, SLOPE, ASPECT, ED_RIVER, ED_FAULT, and ED_TOWN) and six dynamic factors (KD_GEOHAZARD, TEM, PRE, NDVI, KD_CROPLAND, and ED_ROAD) from 1990 to 2020 are considered as independent variables. The dependent variable is the change in construction land kernel density from 1990 to 2020. To ensure the stability and explanatory power of the GWR model, we conducted a spatial autocorrelation analysis and checked for multicollinearity in the data. The results indicate that there is spatial correlation in the data, while there is no multicollinearity among the explanatory variables. During the calculation process, the raster data are transformed into point data to form a dataset, which is subsequently analyzed using GWR

4.0 software [92] developed at NCG (National Center for Geocomputation, National University of Ireland Maynooth) and the Department of Geography, Ritsumeikan University, Japan (https://gwr.maynoothuniversity.ie, accessed on 26 March 2024) to derive relevant results such as GWR estimation coefficients. Outliers are then removed from the GWR estimation coefficients. Finally, Kriging interpolation is applied to create a grid map of GWR estimation coefficients at the county scale.

3. Results

3.1. Spatiotemporal Characteristics of Construction Land

3.1.1. Temporal Changes in Construction Land

Within the study period, construction land exhibited sustained growth and displayed significant phased patterns (Table 2 and Figure 7). Figure 7 shows that high-density areas of construction land are primarily distributed along river valleys and relatively gentle slopes. Additionally, regions with significant changes in kernel density are also located in these areas. The area of construction land was 41.55 km^2 in 1990, 58.66 km^2 in 2000, 82.75 km^2 in 2010, and 95.26 km^2 in 2020, representing a 2.29-fold increase compared to 1990. Over the period from 1990 to 2020, construction land sustained a rapid and continuous growth trend, with a total increase of 53.71 km^2, corresponding to an average annual growth rate of approximately 2.80%, equivalent to 1.79 km^2 per year. During the growth phases of construction land, the period from 1990 to 2000 marked the initial stage of growth, while 2000 to 2010 witnessed the fastest growth phase. Subsequently, from 2010 to 2020, the growth rate stabilized, reflecting a stage of conservation and intensive growth. The number of construction land patches observed in each phase were 700 in 1990, 820 in 2000, 914 in 2010, and 1059 in 2020. From 1990 to 2010, the average patch area of construction land exhibited a generally increasing trend, indicating expansion into areas with favorable conditions. Conversely, from 2010 to 2020, the average patch area of construction land declined, suggesting that the scale of the newly expanded construction land was more constrained by land use conditions, resulting in a reduction in larger-scale expansion.

Table 2. Statistical characteristics of changes in construction land.

Statistical Indicators	1990	2000	2010	2020
Total area of construction land (km^2)	41.55	58.66	82.75	95.26
Number of patches on construction land	700	820	914	1059
Average patch area of construction land (m^2)	59,355	71,536	90,540	89,956
Area increase compared to the previous 10 years (km^2)	–	17.11	24.09	12.51
Percentage increase compared to 1990 (%)	–	41.18	99.16	129.26
Percentage increase compared to the previous 10 years (%)	–	41.18	41.07	15.11

The reciprocal conversion between construction land and other land use types remains relatively stable. Analyzing land use conversion from 1990 to 2020 reveals that the area converted to construction land far exceeds the area converted from construction land. Cropland, forestland, grassland, wetland, and other land types have all undergone varying degrees of conversion to construction land, with cropland and forestland accounting for 77.27% and 14.89% of the total converted area, respectively. Conversely, construction land has also been converted to other land types during different periods, primarily to cropland and forestland. Between 2000 and 2010, approximately 1.95 km^2 of construction land was transferred out, primarily occurring in localized relocation areas such as those designated for targeted poverty alleviation, ecological migration, and construction projects.

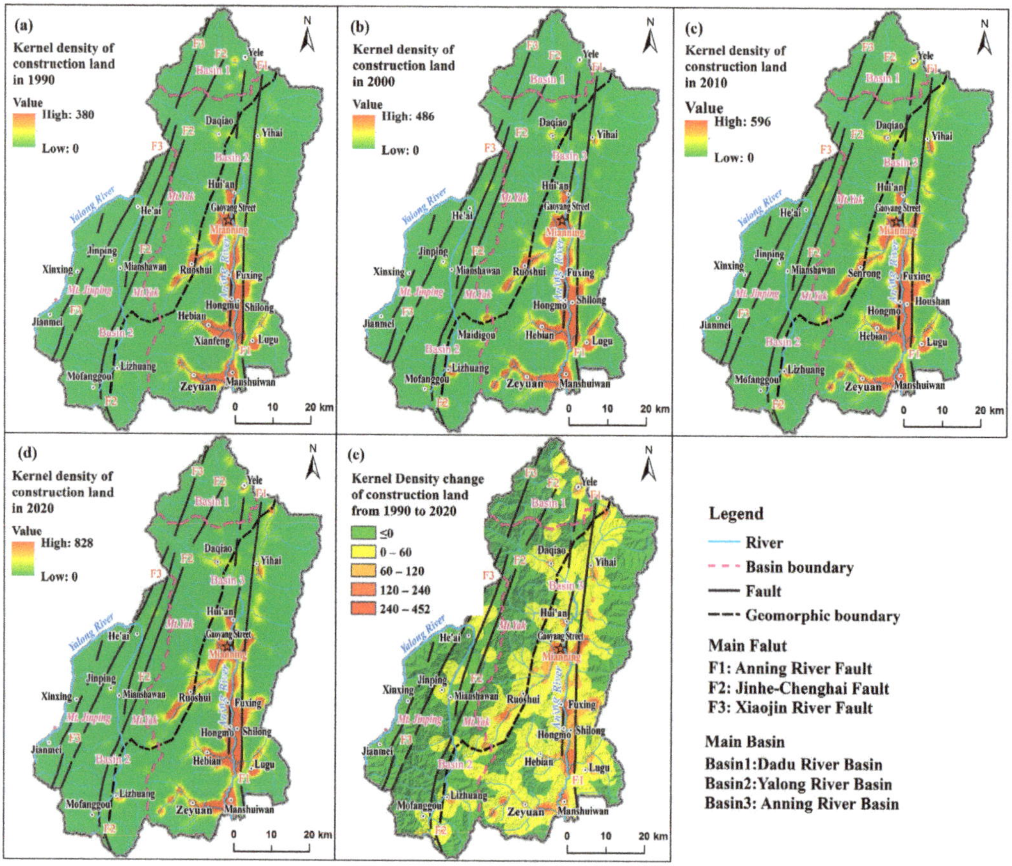

Figure 7. Kernel density of construction land in different periods: (**a**) 1990, (**b**) 2000, (**c**) 2010, (**d**) 2020, and (**e**) from 1990 to 2020.

3.1.2. Spatial Changes in Construction Land

According to the Landscape Expansion Index definition, three expansion modes emerge at different stages, with construction land in the study area predominantly characterized by the edge expansion mode, complemented by the leapfrog mode and infilling mode. In terms of the patch area proportion, the edge expansion mode represents the highest share of new construction land area, comprising approximately 64.83%, while the leapfrog mode accounts for 35.06%, and the infilling mode constitutes only 0.11%. Regarding the patch number proportion, patches exhibiting the edge expansion mode represent 71.30% of the total number of new construction lands, followed by the leapfrog mode at 28.69%, and the infilling mode at a mere 0.01%.

At the watershed scale, the distribution of the three expansion modes varies across different study periods. In the Anning River Basin, the expansion model of new construction land from 1990 to 2020 is primarily characterized by the edge expansion mode, with its proportion showing a trend of an initial decrease followed by an increase. The leapfrog mode exhibits an initial increase followed by a decrease, reaching 23.83% from 2000 to 2010. In the Yalong River Basin, the leapfrog mode and edge expansion mode proportions in the expansion modes of new construction land are roughly equivalent, with the leapfrog mode emerging as the primary expansion mode. Between 2000 and 2010, the number and area of new construction land in the Yalong River Basin were limited and mainly occurred in the leapfrog mode. In the Dadu River Basin, approximately 80% of the new construction land adopts the edge expansion mode, while approximately 19% adopts the leapfrog mode, with the infilling mode accounting for a negligible proportion.

3.1.3. Typical Spatial Patterns of Construction Land

Influenced by resource availability and environmental factors, the distribution and expansion of construction land in Mianning County have given rise to three distinctive spatial patterns: the Wide Valley Spatial Pattern, the Deep Valley Spatial Pattern, and the High Mountain Lake Basin Spatial Pattern.

The characteristics of the Wide Valley Spatial Pattern (Figure 8) are as follows: Construction land primarily spans the expansive valley of the Anning River, displaying a gradual decline in density as the distance from the river increases. The basin terrain is characterized by flat, broad areas with fertile soil, conducive agricultural conditions, well-established industrial infrastructure, a thriving economy, and a significant concentration of construction land. Rapid and edge expansion serve as the primary growth pattern for construction land within this mode.

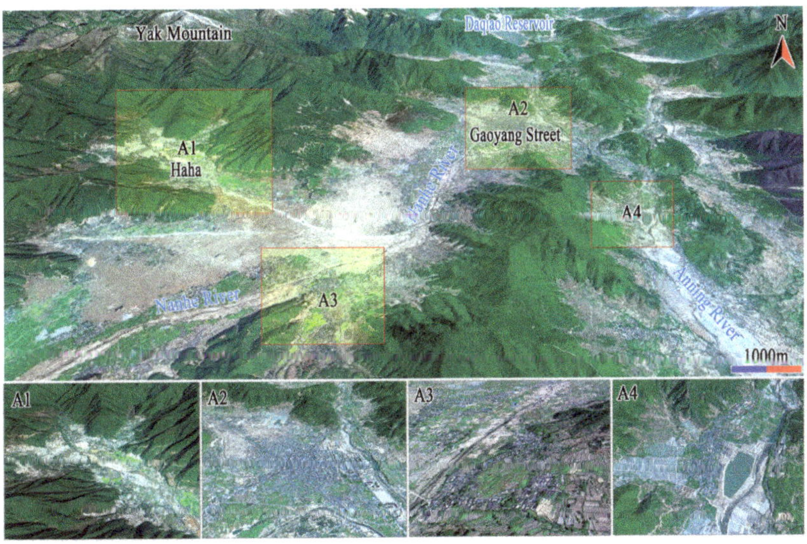

Figure 8. Wide Valley Spatial Pattern in the Anning River Basin (obtained on 25 April 2021, data from Century Space).

The Deep Valley Spatial Pattern (Figure 9) is characterized by the following features: Influenced by the rugged terrain, construction land is dispersed across numerous areas and exhibits localized concentrations. This mode can be further subdivided into three distinct cases. In the first scenario (B1), construction land is predominantly situated on flat slopes or valley bottoms near rivers, with landform types primarily comprising river terraces and alluvial fans. In the second scenario (B2), scattered and smaller-scale construction land is dispersed across small platforms within the middle of slopes. In the third scenario (B3), construction land is primarily located on upper slopes or even terraces near the summit of the slope. Changes in construction land predominantly occur in the form of the leapfrog mode and the edge expansion mode. These areas (B3) represent pivotal zones for construction land expansion within the Yalong River Basin.

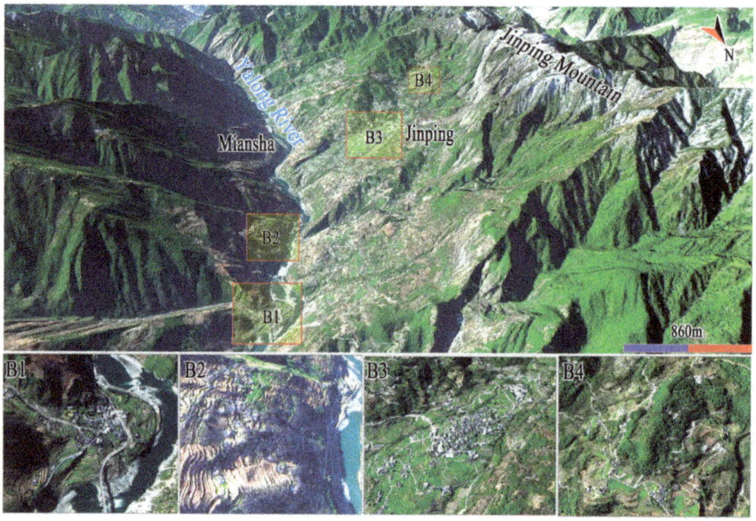

Figure 9. Deep Valley Spatial Pattern in the Yalong River Basin (obtained on 14 February 2022, data from Century Space).

The High Mountain Lake Basin Spatial Pattern (Figure 10) in the Dadu River Basin is distinguished by the following features: Construction land is concentrated around the Yele Basin, alternatively referred to as the Yele Reservoir. It primarily occupies the alluvial fan adjacent to the reservoir, forming a small cluster distribution centered around animal husbandry, agriculture, and tourism.

Figure 10. High Mountain Lake Basin Spatial Pattern in the Dadu River Basin (obtained on 30 March 2021, data from Century Space).

3.2. Characteristics of Multi-Factor Influences on Construction Land

3.2.1. Impact of Multiple Factors on Construction Land in 1990

In 1990, the spatial distribution pattern of construction land in the Yalong River Basin and Dadu River Basin was primarily influenced by natural factors, whereas both human and natural factors exerted similar control effects in the Anning River Basin (Figure 11). To effectively characterize the influence of other factors on construction land, the characteristics of GWR estimation coefficients were statistically analyzed using the 2020 construction land and its 1 km buffer zone (Table 3). The 1 km buffer zone was established based on the spatial expansion of construction land from 1990 to 2020, extending outward by two to three times the initial distance. This method was also applied to compute the statistics of the 2020 GWR estimation coefficient and the 1990–2020 GWR estimation coefficient. The Anning River Basin, boasting a larger scale of construction land, exhibited larger GWR estimation coefficients compared to the Yalong River Basin and Dadu River Basin. In the Anning River Basin, factors such as ED_TOWN, PRE, ED_ROAD, SLOPE, TEM, and KD_CROPLAND demonstrated significant impacts. Similarly, in the Yalong River Basin, significant impacts were observed for SLOPE, ELEVATION, ED_TOWN, PRE, and KD_CROPLAND. In contrast, in the Dadu River Basin, relatively significant impacts were attributed to KD_CROPLAND, SLOPE, ED_TOWN, and ELEVATION.

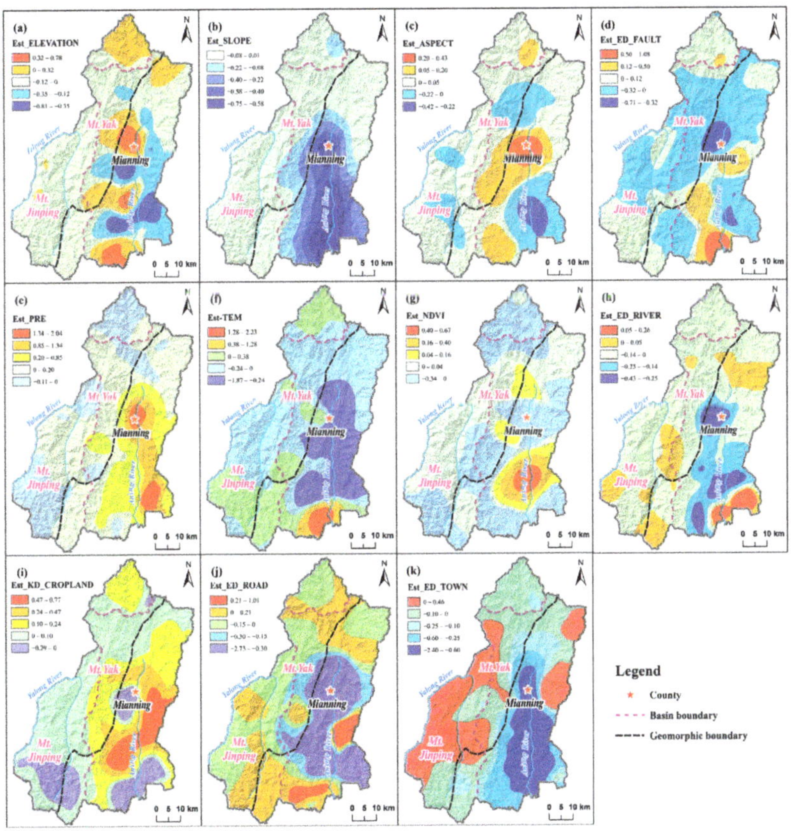

Figure 11. Estimated coefficient of GWR for the impact factors of construction land in 1990: (**a**) elevation, (**b**) slope, (**c**) aspect, (**d**) Euclidean distance of fault, (**e**) annual precipitation in 1990, (**f**) average annual temperature in 1990, (**g**) NDVI in 1990, (**h**) Euclidean distance of rivers, (**i**) cropland kernel density in 1990, (**j**) Euclidean distance of roads in 1990, and (**k**) Euclidean distance of towns.

Table 3. Average estimated coefficient of GWR for the impact factors of construction land in 1990.

Factor Categories	Factor	Anning River Basin	Yalong River Basin	Dadu River Basin
Geomorphological	ELEVATION	−0.0721	−0.0329	0.0606
	SLOPE	−0.3728	−0.0403	−0.0825
	ASPECT	0.0051	0.0089	0.0429
Geological activities	ED_FAULT	−0.0384	0.0001	0.0487
Climatic	PRE	0.5499	0.0179	−0.0180
	TEM	−0.2675	0.0021	0.0261
Rivers and vegetation environment	ED_RIVER	0.0537	−0.0017	−0.0130
	NDVI	−0.1253	−0.0023	−0.0572
Socio-economic	KD_CROPLAND	0.1811	0.0137	0.1129
	ED_TOWN	−0.6444	−0.0310	−0.0697
	ED_ROAD	−0.4343	−0.0045	0.0055

In terms of geomorphic factors, SLOPE exerts a greater influence compared to ELEVATION and ASPECT. ELEVATION demonstrates a negative correlation with the kernel density of construction land in both the Anning River Basin and the Yalong River Basin. In regions characterized by lower elevations, the conditions are more favorable for the establishment of construction land. However, in the Dadu River Basin, this is manifested by a broader distribution of construction land around higher-altitude waterlogging basins. SLOPE exhibits a negative correlation with the distribution of construction land across all three watersheds, indicating that steeper slopes tend to limit the placement of construction land. Conversely, ASPECT demonstrates a positive correlation with construction land, with sunlit and semi-sunlit slopes being more attractive for construction land development.

The geological activity factor solely assesses the influence of ED_FAULT. The GWR analysis results indicate its relatively minor correlation with the arrangement of construction land. In the Yalong River Basin and Dadu River Basin, construction land tends to be situated in regions farther from faults. Conversely, in the Anning River Basin, the presence of a wide valley landform attributed to the Anning River fault reduces its sensitivity to the impact of the construction land layout.

Within the realm of climatic factors, precipitation and temperature exert divergent effects across the three river basins. In the Anning River Basin, precipitation exhibits a primarily positive correlation with the kernel density of construction land, while temperature demonstrates a predominantly negative correlation, emerging as the two pivotal natural factors with significant influence. Conversely, in the Yalong River Basin, both precipitation and temperature showcase positive correlations with the kernel density of construction land, with precipitation wielding a greater impact. Meanwhile, in the Dadu River Basin, precipitation showcases a negative correlation with the kernel density of construction land, whereas temperature displays a positive correlation, indicative of the inclination towards warmer conditions within high-altitude basins, sought after for human settlement environments. This nuanced interplay of climatic factors underscores the intricate relationship between environmental dynamics and the spatial distribution of construction land across diverse geographical terrains.

In terms of environmental factors related to rivers and vegetation, both ED_RIVER and NDVI exhibit predominantly negative correlations with the kernel density of construction land across the three river basins, with the exception of ED_RIVER in the Anning River Basin, which displays a positive correlation with the construction land kernel density. Regions with construction land tend to feature lower NDVI values and the proximity to rivers. However, in the Anning River Basin, the influence of major rivers on water sources for construction land is relatively mitigated due to the support of the water network system, thus weakening their control effect.

Concerning socio-economic factors, regions characterized by concentrated cropland, proximity to urban areas, and accessibility to roads exert a more pronounced influence on the distribution of construction land. The Anning River Basin and Dadu River Basin, where cropland is more densely concentrated, exhibit a greater impact on the distribution of construction land compared to the Yalong River Basin, where cropland is more dispersed. Moreover, the Anning River Basin and Dadu River Basin, situated adjacent to more surrounding towns, demonstrate a higher kernel density of construction land distribution in areas closer to these towns, in contrast to the Yalong River Basin. Construction land in areas proximate to the road network within the Anning River Basin and Yalong River Basin exhibits a wider distribution. However, in 1990, the Dadu River Basin experienced relatively poor accessibility of construction land to main roads.

3.2.2. Impact of Multiple Factors on Construction Land in 2020

By 2020, the GWR analysis results of the spatial pattern of construction land remained generally consistent with those of 1990, albeit with changes in the intensity and spatial impact range of certain factors (Figure 12). To effectively analyze the impact of various factors on the formation of the spatial pattern of construction land, the 2020 construction land and its 1 km buffer zone were utilized as statistical areas to discern differences in the roles of various factors across the three watersheds (Table 4 and Figure 12). In the Anning River Basin, alongside the heightened influence of natural factors such as precipitation, the spatial pattern of construction land continues to be strongly influenced by human factors. Conversely, in the Yalong River Basin and the Dadu River Basin, the impact of human factors has further intensified, with urban and road accessibility exerting particularly pronounced effects in the Dadu River Basin.

The impact of geomorphic factors on the distribution of construction land remains consistent with the overall trends observed in 1990. Elevation continues to exhibit a negative correlation with the kernel density of construction land across all three watersheds, although the strength of this effect has diminished compared to 1990. While the slope demonstrates a negative correlation with the kernel density of construction land in the Anning River Basin and Yalong River Basin, it displays a positive correlation in the Dadu River Basin. This suggests the expansion of construction land towards areas characterized by steeper slopes. The aspect exhibits a positive correlation with the kernel density of construction land in all three watersheds, with the distribution of construction land in the Dadu River Basin showcasing a greater sensitivity to aspect variations.

The influence of geological factors on construction land primarily centers around the Euclidean distance of faults. In the Anning River Basin and the Yalong River Basin, as construction land expands, its spatial correlation with faults and the platforms or negative topography resulting from their effects has intensified. Conversely, in the Dadu River Basin, characterized by fewer faults, construction land tends to steer clear of fault-related effects.

The influence of precipitation and temperature, among climate factors, has intensified in the Anning River Basin and Dadu River Basin, but has relatively weakened in the Yalong River Basin. In the Anning River Basin, the positive correlation between construction land distribution and precipitation has notably increased, while the negative correlation with temperature has decreased. Conversely, in the Yalong River Basin, the positive correlation between temperature and the distribution of construction land outweighs the positive correlation between precipitation and construction land. In the Dadu River Basin, both the negative correlation between precipitation and the distribution of construction land and the positive correlation between temperature and the distribution of construction land have experienced significant increases.

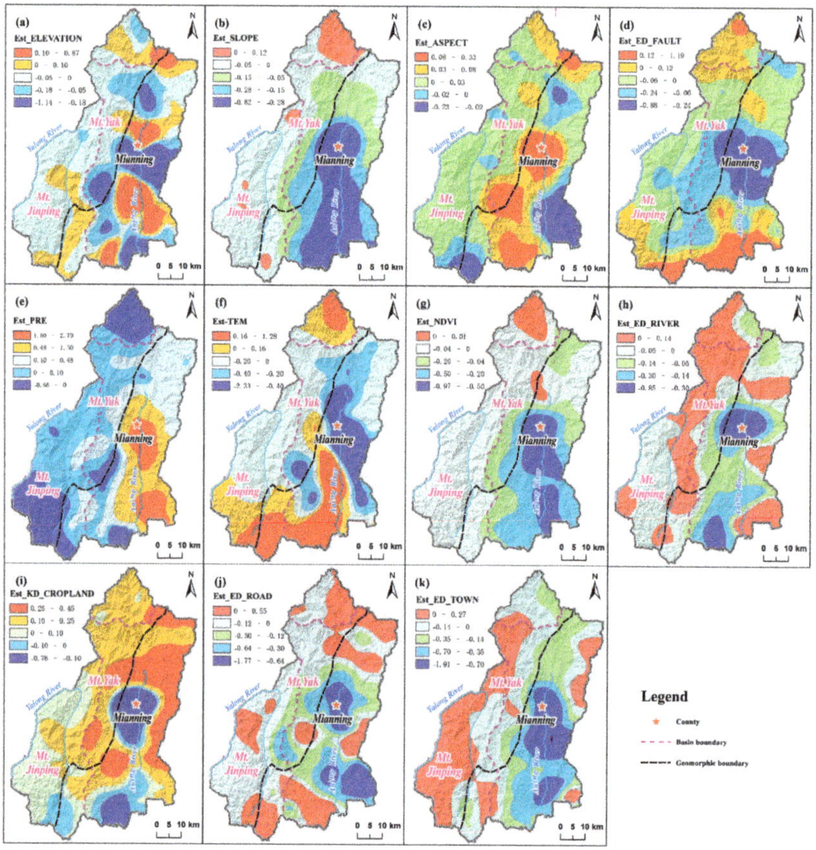

Figure 12. Estimated coefficient of GWR for the impact factors of construction land in 2020: (**a**) elevation, (**b**) slope, (**c**) aspect, (**d**) Euclidean distance of fault, (**e**) annual precipitation in 2020, (**f**) average annual temperature in 2020, (**g**) NDVI in 2020, (**h**) Euclidean distance of rivers, (**i**) cropland kernel density in 2020, (**j**)Euclidean distance of roads in 2020, and (**k**) Euclidean distance of towns.

Table 4. Average estimated coefficient of GWR for the impact factors of construction land in 2020.

Factor Categories	Factor	Anning River Basin	Yalong River Basin	Dadu River Basin
Geomorphological	ELEVATION	−0.0612	−0.0032	−0.0274
	SLOPE	−0.3451	−0.0310	0.0615
	ASPECT	0.0441	0.0053	0.0555
Geological activities	ED_FAULT	−0.1219	−0.0394	0.0304
Climatic	PRE	0.8412	0.0087	−0.2833
	TEM	−0.1833	0.0865	0.1639
Rivers and vegetation environment	ED_RIVER	−0.1535	0.0037	−0.0525
	NDVI	−0.3773	−0.0198	−0.0066
Socio-economic	KD_CROPLAND	0.0322	0.0385	0.1548
	ED_TOWN	−0.6054	−0.0477	−0.1697
	ED_ROAD	−0.3355	−0.0078	0.1079

Among the river and vegetation factors, the water system of the Anning River Basin exhibits a strong correlation with construction land, followed by the Dadu River Basin. Conversely, the Yalong River Basin, characterized by deep-cut terrain and a reliance on spring water sources, displays a relatively weak correlation with rivers. Across all three river basins, there exists a negative correlation between the distribution of construction land and NDVI, indicating a comparatively low vegetation coverage in construction land and its surrounding areas. The strength of this relationship follows the order of the Anning River Basin, the Yalong River Basin, and the Dadu River Basin.

Among socio-economic factors, the influence of cropland in the Anning River Basin has relatively diminished, whereas its impact has intensified in the Yalong River Basin and Dadu River Basin. The town accessibility's impact remains largely consistent in the Anning River Basin but has increased in the Yalong River Basin and Dadu River Basin, particularly in the latter. The road accessibility's impact has decreased relatively in the Anning River Basin but has grown in significance in the Yalong River Basin and Dadu River Basin. Notably, in the Dadu River Basin, the distribution of construction land has expanded into areas not directly linked to main roads.

3.2.3. Impact of Multiple Factors on Construction Land from 1990 and 2020

From 1990 to 2020, all three river basins underwent varying degrees of land use change. While the Anning River Basin was primarily influenced by human factors, the Yalong River Basin and Dadu River Basin were predominantly shaped by natural factors (Table 5 and Figure 13). In the Anning River Basin, factors such as ED_ROAD, ED_TOWN, and KD_CROPLAND, and geomorphological factors exerted a significant impact on the changes in construction land. In the Yalong River Basin, significant influences stemmed from KD_GEOHAZARD, PRE, KD_CROPLAND, and ED_ROAD, among others. Meanwhile, in the Dadu River Basin, significant factors included PRE, ED_TOWN, SLOPE, TEM, ED_ROAD, and ED_RIVER.

Table 5. Average estimated coefficient of GWR for the impact factors of construction land changes from 1990 to 2020.

Factor Categories	Factor	Anning River Basin	Yalong River Basin	Dadu River Basin
Geomorphological	ELEVATION	0.1013	0.0027	−0.0610
	SLOPE	−0.3099	−0.0100	0.1436
	ASPECT	−0.0246	0.0085	0.0529
Geological activities	ED_FAULT	−0.0303	−0.0187	−0.0318
	KD_GEOHAZARD	0.1437	0.1634	0.0666
Climatic	PRE	−0.0263	0.0795	−0.2337
	TEM	−0.0009	0.0225	0.1342
Rivers and vegetation environment	ED_RIVER	−0.1388	−0.0081	−0.0697
	NDVI	−0.2684	−0.0168	−0.0418
Socio-economic	KD_CROPLAND	−0.3261	0.0634	−0.0381
	ED_TOWN	−0.3404	0.0370	−0.1796
	ED_ROAD	0.3986	−0.0541	−0.1158

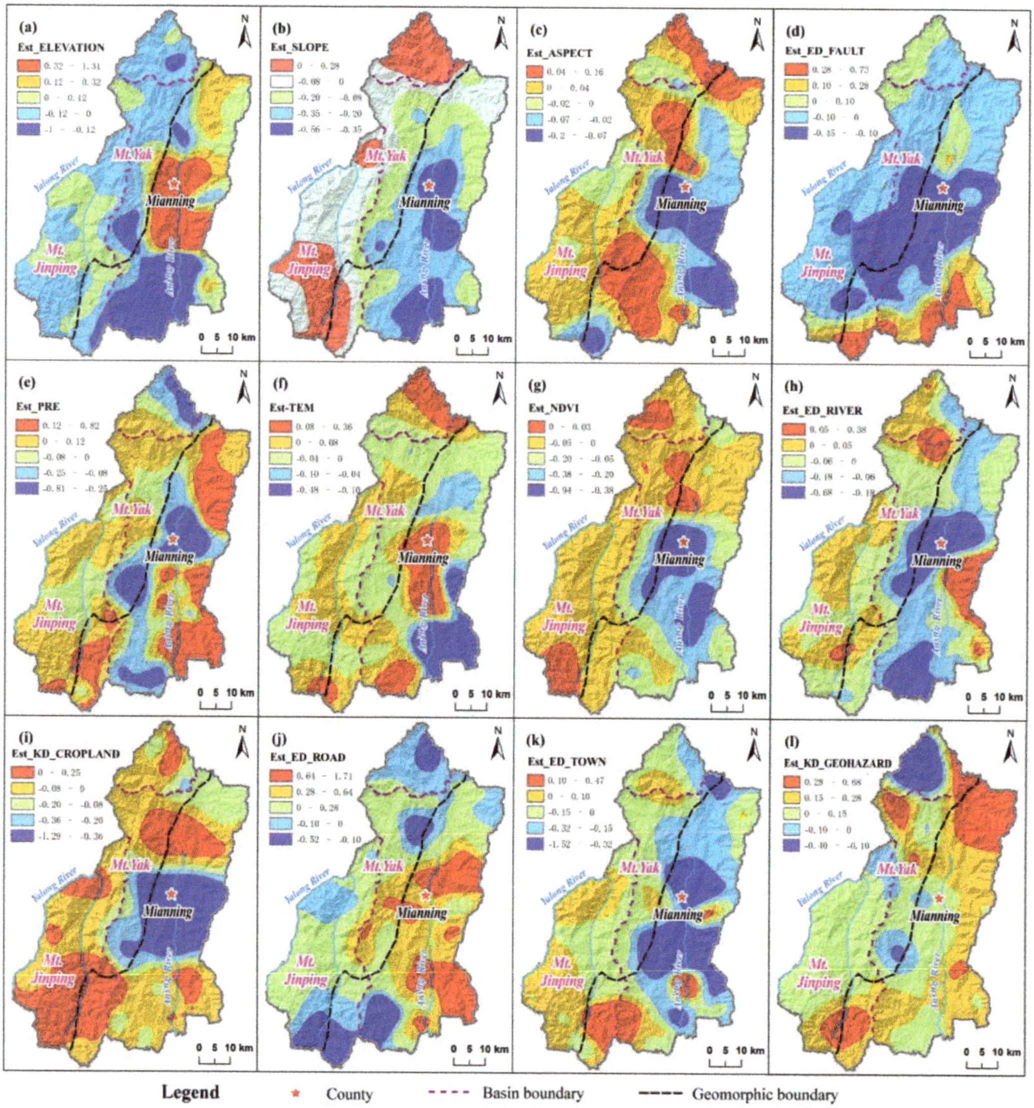

Figure 13. Estimated coefficient of GWR for the impact factors of construction land changes from 1990 to 2020: (**a**) elevation, (**b**) slope, (**c**) aspect, (**d**) Euclidean distance of fault, (**e**) annual precipitation changes from 1990 to 2020, (**f**) average annual temperature changes from 1990 to 2020, (**g**) NDVI changes from 1990 to 2020, (**h**) Euclidean distance of rivers, (**i**) cropland kernel density changes from 1990 to 2020, (**j**) Euclidean distance of changes from 1990 to 2020, (**k**) Euclidean distance of towns, and (**l**) changes in kernel density of geological hazards from 1990 to 2020.

Among the geomorphic factors, SLOPE has a more pronounced impact on the changes in construction land compared to ELEVATION and ASPECT, particularly evident in the Anning River Basin. Steeper slopes present a greater resistance to the expansion of construction land. In the Yalong River Basin, the slope exhibits a weak negative correlation with the expansion of construction land, whereas, in the Dadu River Basin, it shows a positive correlation, indicating land expansion in areas with slightly steeper slopes around the reservoir.

Among the geological activity factors, ED_FAULT shows no significant impact. However, changes in KD_GEOHAZARD are positively correlated with the expansion of construction land, indicating the influence of human construction activities on the geological environment, with the Yalong River Basin and Anning River Basin showing more pronounced effects.

In terms of climate factors, the changes in PRE and TEM from 1990 to 2020 exhibit a weak negative correlation with the expansion of construction land in the Anning River Basin, a positive correlation in the Yalong River Basin, and a negative correlation in the Dadu River Basin. This indirectly reflects the relative significance of climate change on construction land in high-altitude areas.

For the rivers and vegetation environmental factors, NDVI exhibits a negative correlation in all three watersheds, indicating a decrease in local vegetation coverage due to the expansion of construction land or the occurrence of construction land in areas with a reduced NDVI. ED_RIVER indicates that areas close to rivers provide more opportunities for construction land expansion, especially in the Anning River Basin. However, the overall impact of ED_RIVER is relatively weak.

Social economic factors exert a strong driving force on the expansion of construction land. The GWR analysis results of KD_CROPLAND indicate that, in the Anning River Basin and Dadu River Basin, construction land expands into areas where KD_CROPLAND has decreased, indicating a mutual conversion relationship between the two. Conversely, in the Yalong River Basin, there is a positive correlation between the density of construction land and changes in KD_CROPLAND. The GWR analysis results of ED_TOWN show that there is more expansion of construction land near towns in the Anning River Basin and Dadu River Basin, while the opposite is observed in the Yalong River Basin. Additionally, the GWR analysis results of ED_ROAD reveal that construction land expansion has occurred in areas relatively distant from roads in the Anning River Basin, whereas, in the Yalong River Basin and Dadu River Basin, the intensity of construction land expansion is higher in areas close to roads, particularly in the Dadu River Basin.

4. Discussion

4.1. Analysis of Spatiotemporal Change in Construction Land

4.1.1. Analysis of Temporal Changes in Construction Land

The development of construction land in Mianning County unfolds across three distinct stages, closely aligned with the county's broader economic and social trajectory. From 1990 to 2000, construction land witnessed a rapid expansion, spurred by vibrant pillar industries such as livestock, poultry, mulberry, pepper, fruit cultivation, and building materials. During this period, the county's population surged from 276,830 to 308,100, with a robust average annual GDP growth rate exceeding 20%. The subsequent period, spanning from 2000 to 2010, marked a sustained and accelerated growth in construction land. Mianning County diversified its economic base, fostering six key industries including rare earth mining, hydropower, building materials, tourism, livestock, poultry, and economic forestry. With an average annual population growth of 5500, the total population surged to 371,000, while the average annual growth rate of GDP value stood at 14%. Transitioning into the period from 2010 to 2020, the county's development landscape evolved, integrating new drivers such as new urbanization and infrastructure construction alongside the established six pillar industries. This phase heralded a comprehensive development pattern, underpinned by a multi-industry approach. Despite a slight decrease in population growth to an average of 3700 annually, the total population reached 408,000, with the annual average growth rate of GDP maintained at 5–7%. Over the entire 1990–2020 period, the expansion of construction land primarily relied on the conversion of cropland, forestland, and grassland to construction land, notably emphasizing the conversion of cropland. This trend underscores the intimate relationship between construction land, cropland, and human activities, along with the favorable natural and human environmental conditions associated with these land types.

4.1.2. Analysis of Spatial Changes in Construction Land

When examining the expansion scale of construction land, it becomes evident that larger watersheds tend to exhibit a greater spatial expansion intensity. The Anning River Basin emerges as the primary area for construction land expansion, followed by the Yalong River Basin, and, finally, the Dadu River Basin. In terms of the proportion of the area covered by the three types of construction land expansion models, a clear hierarchy emerges, with the edge expansion mode occupying the largest proportion, followed by the leapfrog mode, and, lastly, the infilling mode. As construction land resources become increasingly scarce, there is a notable shift towards conservation and intensive growth strategies. This trend underscores the evolving approach of the Mianning County government in construction land planning, development, and governance, reflecting a commitment to sustainable land management practices and efficient resource utilization.

The edge expansion mode stands out as the predominant method for construction land expansion across various stages. Its prominence can be attributed to its effectiveness in addressing key issues such as continuity, convenience, economy, and efficiency in expanding construction land [93]. This mode leverages the existing built environment, transportation networks, and infrastructure, maximizing their utilization. Throughout different stages of overall land use planning and national territory spatial planning in Mianning County, the government has underscored the importance of strengthening the marginal expansion of construction land through policy guidance and planning control. Consequently, this expansion model plays a pivotal role in optimizing the spatial structure of land use and enhancing the overall land use efficiency.

The leapfrog mode represents a unique approach heavily influenced by the geological, geographical, and socio-economic factors prevalent in Mianning County. Particularly notable in the deep-cut valleys of the Yalong River and the Yele Mountain Lake basin area, where space suitable for the continuous expansion of construction land is limited, this mode arises as a response to challenges such as mineral development, hydropower generation, and ecological poverty alleviation. Centralized relocation and resettlement efforts are often employed to address these issues, giving rise to a typical leapfrog pattern. Similarly, in the Anning River Basin, driven by infrastructure projects such as highway and high-speed rail construction, water conservancy hub development, urban expansion, and industrial park expansion, the leapfrog expansion of construction land of varying scales has occurred. Notably, the leapfrog expansion scale in the Anning River Basin surpasses that of the Yalong River Basin and the Dadu River Basin. Overall, the leapfrog mode emerges as a distinctive strategy, tailored to the specific conditions and development needs of Mianning County, showcasing its adaptability in navigating complex socio-economic and environmental landscapes [94].

Infill expansion represents a localized construction land expansion model observed in specific areas of Mianning County. This model emerges primarily due to the acute scarcity of construction land resources in certain regions, presenting a unique approach to land consolidation, renewal, and development within existing construction land boundaries [95]. For instance, in older urban areas like Gaoyang Street, Ruoshui Town, and Lugu Town, idle or underutilized land has undergone revitalization and repurposing, thereby enhancing the urban functionality of these locales. Similarly, in areas such as Jinping, Jianmei, and He'ai, the effective development of inefficient urban land has been achieved through the integration and optimization of internal transportation networks and infrastructure. Moreover, in smaller villages experiencing population growth, there have been instances of newly developed internal areas enclosed by existing construction land, facilitating the efficient utilization of available resources. Overall, infill expansion serves as a strategic response to address the pressing need for land optimization and utilization in Mianning County, fostering sustainable urban development and maximizing the efficiency of construction land usage.

4.1.3. Analysis of Spatial Patterns of Construction Land

The emergence of the three distribution patterns of construction land—namely, the Wide Valley Spatial Pattern, Deep Valley Spatial Pattern, and High Mountain Lake Basin Spatial Pattern—is predominantly influenced by natural factors, with human modification and adaptation to the natural environment playing a significant role in shaping these patterns.

The Anning River Wide Valley Spatial Pattern is primarily shaped by a confluence of factors, including topography, natural resources, environmental conditions, socio-economic development, and policy planning and guidance. Influenced by the Anning River and its tributaries, the wide plain area facilitates ample space for construction land. With abundant water resources and fertile soil, this region fosters favorable conditions for agricultural production and residential settlement. Moreover, its pristine ecological environment and scenic landscapes attract a population influx and industrial clustering, contributing to the expansion of construction land. The well-developed transportation network in the plain area facilitates efficient logistics and human mobility, further enhancing its appeal for industrialization and urbanization. The concerted efforts of the governments of Liangshan Prefecture, Xichang City, and Mianning County have led to the formulation of comprehensive land development and utilization plans, along with supportive policies tailored to the Anning River Valley. These initiatives have played a pivotal role in shaping the spatial distribution pattern of construction land in the Anning River wide valley area.

The Deep Valley Spatial Pattern is heavily influenced by geological structures, landforms, cropland availability, and transportation accessibility. In areas adjacent to the riverbed of the deeply incised Yalong River, the terrain of river terraces or alluvial fans tends to be relatively flat, facilitating the construction of land corridors along the riverbanks and promoting material exchange. Additionally, these areas are conducive to agricultural irrigation, domestic water supply, and industrial activities, contributing to the centralized distribution of construction land. Meanwhile, construction land in other areas of the deep valley, such as Jinping, is predominantly located on the upper and middle terraces of the mountains, influenced by local residential customs and traditions. These areas are chosen for their natural advantages, as they are less susceptible to floods and geological hazards, and are close to water sources such as mountain springs and streams, facilitating domestic water usage and agricultural irrigation. Benefiting from favorable sunshine exposure conditions, high humidity, and fertile soil, these regions are suitable for agricultural development, with relatively abundant animal and plant resources available for utilization. Consequently, a typical pattern emerges, characterized by the concentrated settlement at the bottom of the valley and a dispersed layout on the upper and middle slopes.

The High Mountain Lake Basin Spatial Pattern is primarily shaped by a combination of factors, including topography, climate, cropland availability, engineering construction, and transportation accessibility. Geological tectonic movements and river-lake erosion contribute to the formation of a relatively flat terrain around the lake, accompanied by high-quality cropland resources. Furthermore, the elevated location of these areas enhances the sensitivity of temperature and precipitation to the ecological environment, resulting in favorable water–heat combinations and oxygen conditions in the Yele Lake Basin and its surroundings, making it highly conducive to human habitation. The development of the Yele Reservoir, initiated in 2001, and subsequent hydropower projects, coupled with supporting infrastructure and tourism development efforts, have played a significant role in shaping the distribution of construction land in this region. Over time, construction land has gradually concentrated around the edges of the lake basin or in relatively flat mountain valleys, forming a distribution pattern centered around the Yele Reservoir.

4.2. Analysis of the Causes of Changes in Construction Land

4.2.1. Geomorphic Factors

The layout and evolution of construction land in mountainous regions are significantly influenced by geomorphic factors such as elevation, slope, and aspect [77]. Among these factors, the slope exerts the most pronounced impact on the distribution and dynamic evolution of construction land, followed by elevation and aspect. This hierarchy stems from the direct influence of the slope on the stability and load-carrying capacity of construction land foundations. In contrast, the elevation and aspect play crucial roles in determining the types of land use and the efficiency of land utilization.

Slope Factor

The slope directly influences construction conditions [96]. In regions characterized by relatively flat macroscopic landforms, the layout and expansion of construction land predominantly occur in areas with smaller slopes, while areas with steeper slopes tend to have less construction land. For instance, in the Anning River Basin, construction land is primarily concentrated in areas with slopes of less than 25°, with newly expanded construction land between 1990 and 2020 mainly situated in areas with slopes of less than 6°. Similarly, in the Yalong River Basin, construction land is primarily found in areas with slopes of less than 35°, with newly expanded construction land between 1990 and 2020 predominantly located in areas with slopes of less than 22°. In the Dadu River Basin, construction land is distributed within a range of slopes of less than 30°, with newly expanded construction land mostly situated in areas with slopes of less than 20°.

Elevation Factor

The elevation plays a pivotal role in determining the type, scale, and efficiency of construction land [97]. It is negatively correlated with the overall kernel density of construction land. Differences in climate, soil, and location conditions across different elevations contribute to distinct characteristics. The Anning River Basin, situated at lower elevations, boasts a relatively flat terrain, fertile soil, a mild climate, and favorable location, rendering it conducive to the construction land layout. In contrast, the Yalong River and Dadu River basins, characterized by higher elevations, feature a more complex terrain, variable climate, fragile ecology, and less convenient transportation, thereby increasing the cost of construction land. The main distribution range of construction land in the Anning River Basin typically spans between 1600 and 2200 m, while, in the Yalong River Basin, it ranges from 1500 to 2500 m, and, in the Dadu River Basin, it lies between 2600 and 3000 m. Notably, from 1990 to 2020, there was a trend of construction land expansion towards lower altitude areas in all three basins.

Aspect Factor

In mountainous regions, the aspect plays a role in the evolution process of mountain slope micro-topography, indirectly influencing the layout and evolution of construction land [98]. However, compared to the slope and elevation, the influence of the aspect in the research area is relatively minor. This is primarily due to the strong correlation between buildings or structures and slope direction. In mountainous areas, construction land with smaller spatial scales is distributed across all slope directions. More than half of the construction land in the Yalong River Basin and Dadu River Basin is situated on shady slopes, while, in the Anning River Basin, it predominantly occupies sunny slopes and flat lands. This distribution pattern reflects the limited availability of construction land resources and the complexity involved in selecting slope directions. From 1990 to 2020, the newly expanded construction land in the three watersheds predominantly favored sunnier areas. This trend underscores the ongoing competition for construction land resources and the strategic consideration of maximizing sunlight exposure for various purposes.

4.2.2. Geological Activities Factors

The influence of geological activities on the spatial distribution pattern and evolution of construction land is multifaceted, often exerting a comprehensive impact through geological structures, stratigraphic lithology, and geological disasters [99].

Fault Factor

Faults play a significant role in shaping the layout of construction land, affecting the foundation stability, earthquake risk, groundwater movement, building design and construction, and land use safety. Consequently, there is a tendency for construction land to be situated away from faults [100].

However, due to the limited availability of land resources in mountainous areas, construction land often needs to be located in relatively favorable positions near faults. The maximum impact distance of faults on construction land in the Anning River Basin is approximately 8 km, with an average impact distance of about 2 km. In the Yalong River Basin, the maximum impact distance is around 3 km, with an average impact distance of about 1 km. Similarly, in the Dadu River Basin, the maximum impact distance is about 7 km, with an average impact distance of approximately 2.4 km.

Initially, in 1990, the impact of faults on the layout of construction land was relatively small. However, by 2020, as the demand for construction land increased, construction land inevitably encroached closer to fault areas, leading to a gradual increase in the correlation between construction land and the Euclidean distance between faults. This trend is particularly pronounced in the Anning River Basin, where construction land expansion has been significant, followed by the Dadu River Basin, and, finally, the Yalong River Basin, where the expansion of construction land is greatly influenced by landforms.

Geological Disaster Factor

Geological disasters are among the most prevalent natural calamities in mountainous regions, often resulting in significant casualties and economic losses [101]. The overall spatial pattern of land use, particularly construction land, is highly susceptible to the adverse effects of geological disasters [102].

As of 2020, Mianning County has experienced over 180 geological disasters, encompassing every town, including debris flows, landslides, and collapses. These disasters are predominantly concentrated along fault lines such as the Xiaojin River Fault, the Jinhe–Chenghai Fault, the Anning River Fault, and the Yalong River, among others.

From 1990 to 2020, the kernel density of geological disasters increased by 0.27 in the Yalong River Basin, by 0.11 in the Anning River Basin, and by 0.01 in the Dadu River Basin. The kernel density of construction land in these basins exhibited a positive correlation with the kernel density of geological hazards, as evidenced by the GWR estimated coefficients: 0.1634 in the Yalong River Basin, 0.1437 in the Anning River Basin, and 0.0666 in the Dadu River Basin.

This strong spatial overlap between construction land and geological disasters underscores the disruption of the geological environment due to human activities, including construction endeavors. Geological disasters not only jeopardize the safety of construction land sites but also impede the road accessibility between construction lands, particularly in areas like Jinping, Mianshawan, and He'ai, where the Yalong River has carved deep valleys.

While efforts to prevent and control geological disasters have made considerable strides, continued monitoring and mitigation measures are imperative, especially in the face of changing climate conditions [103]. Optimizing the layout and intensity of construction land in the future will remain crucial for mitigating geological disaster risks and ensuring sustainable development.

4.2.3. Climatic Factors

Climatic factors primarily influence agricultural production by integrating precipitation and temperature, thereby shaping the layout of construction land. Additionally, climate impacts the human settlement environment, indirectly influencing residents' selection of construction land sites [104]. Over the period of 1990 to 2020, Mianning County has experienced two overarching trends—a decrease in precipitation and an increase in temperature—aligning with climate changes observed on the eastern edge of the Tibetan Plateau [105]. Notably, changes in precipitation have a more pronounced impact on the expansion of construction land compared to changes in temperature, likely due to precipitation's influence on agricultural production and the occurrence of geological disasters [106].

In the Anning River Basin, the precipitation was 897 mm with an average annual temperature of 12.06 °C in 1990. By 2020, precipitation decreased to 879 mm, while the average annual temperature rose to 12.77 °C. Over this period, precipitation decreased by 18 mm, while the temperature increased by 0.71 °C. Climatic factors exert a substantial influence on the construction land pattern in this basin, although the impact of climate change on the expansion of construction land is relatively mitigated by human activities.

Conversely, the Yalong River Basin experienced a decrease in precipitation from 865 mm and an average annual temperature of 11.50 °C in 1990 to 822 mm and 12.41 °C, respectively, in 2020. Here, precipitation decreased by 43 mm, and the temperature increased by 0.91 °C. While climatic factors have a lesser impact on the construction land pattern in this basin, the deep incision of the landform diminishes the influence of climate on construction land expansion.

In the Dadu River Basin, the precipitation decreased from 887 mm with an average annual temperature of 7.48 °C in 1990 to 881 mm and 8.49 °C, respectively, in 2020. During this period, precipitation decreased by 6 mm, while the temperature increased by 1.01 °C. Climatic factors wield a substantial influence on the construction land pattern in this basin. Furthermore, climate change significantly impacts the expansion of construction land due to the heightened climate sensitivity resulting from the high altitude, akin to the socio-economic environment's response to climate change in the high-altitude areas of the Himalayas [107].

4.2.4. River and Vegetation Environmental Factors

River Environmental Factor

As shapers of the landscape environment, rivers serve as vital suppliers of water resources [108] and contribute to the water environment [109]. Rivers play a crucial role in determining the location and development of construction land, with areas in close proximity to rivers offering greater development conveniences [110,111].

The expansion of construction land in Mianning County from 1990 to 2020 demonstrates a trend towards the closer proximity to rivers. In the Anning River Basin, the density of construction land was positively correlated with the Euclidean distance from the river in 1990, reflecting the risks of flooding in areas near the river, conflicts with cropland development, and challenges in infrastructure construction. The average distance between construction land and the river was approximately 1282 m.

However, with the completion of the Daqiao Reservoir in 1999 for power generation and the improved management of the Anning River, alongside enhancements in riverside infrastructure construction, the average distance between the newly expanded construction land and the river decreased to 870 m from 1990 to 2020. Consequently, the density of construction land and the Euclidean distance from the river exhibited a negative relationship.

In the Dadu River Basin, the layout of construction land and its changes were negatively correlated with the Euclidean distance from the river. In 1990, the average Euclidean distance between construction land and the river in this basin was 1221 m. However, since the construction of the Yele Reservoir in 1998, leading to rising water levels in reservoirs and rivers, the average distance between construction land and rivers or reservoirs has decreased by approximately 300–500 m by 2020.

In the Yalong River Basin, the average Euclidean distance between construction land and the river was 1363 m in 1990. The average distance between the expanded construction land and the river from 1990 to 2020 was reduced to 1181 m. The river's sensitivity to the layout and changes in construction land is relatively low in this area, primarily because construction land is predominantly distributed in limited spaces on both sides of the Yalong River Valley.

Vegetation Environmental Factor

The vegetation environment plays a crucial role in influencing construction land. Robust vegetation offers stable ecological services, providing a favorable ecological background for construction land [112]. Conversely, vegetation degradation can result in issues such as soil erosion, reduced biodiversity, and diminished agricultural productivity, thereby complicating the development and utilization of construction land. Through the implementation of key ecological projects, increased support for ecological initiatives, reinforced soil and water conservation efforts, promotion of comprehensive soil erosion management, and intensified ecological forestry construction, Mianning County has witnessed significant improvements in its vegetation environment [113,114].

Using NDVI as an indicator of the vegetation environment, Mianning County experienced varying degrees of improvement in vegetation cover in areas designated for construction land from 1990 to 2020 [115]. The NDVI increased by 0.12 in the Anning River Basin, 0.18 in the Yalong River Basin, and 0.01 in the Dadu River Basin, providing a relatively favorable ecological backdrop for construction land. However, due to construction activities leading to the conversion of other land types with relatively high NDVI values to construction land, NDVI is negatively correlated with construction land density in all three watersheds, indicating the relatively low vegetation coverage of construction land and its surrounding areas.

Furthermore, the expansion of construction land contributes to a decline in NDVI. This impact is most pronounced in the Anning River Basin, where the construction land density and the intensity of change are highest, while relatively less significant in the Yalong River Basin and Dadu River Basin.

4.2.5. Socio-Economic Factors

Socio-economic factors play a pivotal role in driving the dynamic evolution of land use and are closely intertwined with human factors in the distribution and evolution of construction land [24,116]. In the Anning River Basin, socio-economic factors contributed to 46% of the impact on land use spatial distribution in 1990. However, due to the heightened influence of climate factors, this overall impact decreased to 31% by 2020. Over the period spanning from 1990 to 2020, socio-economic factors accounted for approximately 50% of the overall impact on construction land expansion, a proportion roughly equivalent to that of natural factors.

Similarly, in the Yalong River Basin, socio-economic factors exerted an impact of approximately 32% on the land use spatial distribution in both 1990 and 2020, as well as on construction land expansion during the same period, reflecting a relatively stable pattern of human activity throughout this timeframe. Conversely, in the Dadu River Basin, socio-economic factors influenced the land use spatial distribution to a degree of around 35% in 1990. However, owing to factors such as engineering construction and infrastructure enhancement, this impact increased to 39% by 2020. In the expansion of construction land from 1990 to 2020, socio-economic factors accounted for about 29% of the overall impact, indicating a stronger influence of natural factors in this context.

Cropland Factor

Construction land and cropland share similar location and environmental requirements, characterized by a high spatial adjacency and a reliance on land type conversion [117]. In both 1990 and 2020, there existed a positive correlation between the density of construction land and cropland in Mianning County. However, the strength of this correlation weakened in the Anning River Basin, while increasing in the Yalong River Basin and Dadu River Basin from 1990 to 2020.

Over the same period, the expansion of construction land in the Anning River Basin resulted in a significant reduction in cropland, particularly in areas near rivers, towns, industrial parks, and transportation arteries. Similarly, in the Dadu River Basin, the expansion of construction land led to a slight decrease in cropland from 1990 to 2020, primarily occurring in the vicinity of the Yele Reservoir. Conversely, in the Yalong River Basin, there was a positive correlation between changes in construction land and cropland from 1990 to 2020, reflecting two distinct trends.

Firstly, the enhancement of local cropland in the Yalong River Basin contributed to an increase in cropland, while the expansion of construction land resulted from the conversion of non-cropland areas. Secondly, certain areas in the Yalong River Basin underwent a conversion from farmland to forests, accompanied by the withdrawal of construction land [118].

Town Accessibility Factor

The accessibility between administrative centers such as counties, towns, and townships can be utilized to gauge the clustering characteristics of human activities or population data [119]. The distance from the administrative center, to a certain extent, reflects the extent of the jurisdiction and socio-economic influence, directly influencing the distribution pattern of construction land. Moreover, this distance affects the speed and direction of the construction land expansion.

In Mianning County, the residences of town governments, along with the surrounding construction land, naturally emerge in areas with relatively favorable local transportation conditions. The average Euclidean distance between construction land and towns is approximately 21 km in the Anning River Basin, 25 km in the Yalong River Basin, and 27 km in the Dadu River Basin. The impact of accessibility to towns on the distribution and changes of construction land follows this order: Anning River Basin > Dadu River Basin > Yalong River Basin.

Road Accessibility Factor

Road accessibility significantly influences land use types, patterns, and values, and the economic and social environment derived from land use [120]. The road network plays a crucial role in shaping the distribution pattern and evolution of construction land, with the impact gradually diminishing as the distance from the road increases.

Mianning County has developed a comprehensive transportation system comprising highways, railways, and aviation, positioning itself to integrate into the Chengdu 2-h transportation circle in the future. Main roads in the research area, such as the G5 Beijing Kunming Expressway, National Highway G108, and National Highway G248, typically feature a concentrated construction land distribution. These areas boast convenient transportation, fostering the aggregation of economic activities and population flow. As influenced by land use planning and construction, the road network continues to enhance, infrastructure support strengthens, investment and population gather, construction land expands, and the structure and function of construction land optimize accordingly.

In the Anning River Basin, the impact of roads on the construction land layout decreased from 1990 to 2020, with the increase in road density supporting the construction land expansion. Conversely, in the Yalong River Basin, the impact of roads on the construction land spatial distribution increased during the same period, with new construction land concentrated in areas closer to roads. In the Dadu River Basin, the Euclidean distance of roads correlates positively with construction land density in 1990 and 2020, indicating construction land distributed at a certain distance from the main road, while the expansion from 1990 to 2020 occurred closer to roads.

4.3. Implications and Limitations

4.3.1. Implications

Construction land distribution in Mianning County is intricately influenced by natural and human factors. Geological and climatic conditions along the eastern Qinghai–Tibet Plateau drive variations in land use across the Anning, Yalong, and Dadu River basins. Considering the national territory spatial planning of Mianning County, customized development strategies are essential, taking into consideration the distinctive characteristics of each basin. Efficient land use in the Anning River Basin is paramount, while stricter controls are needed for the environmentally sensitive Yalong and Dadu River Basins. These strategies aim to promote sustainable development while preserving local ecosystems.

The Anning River Basin, pivotal for industry and agriculture, must address climate change and balance urban–rural development. Guided urban and rural planning is crucial for resource allocation and infrastructure development. The strategic expansion of construction land should target key areas like urban centers and industrial parks, ensuring economic growth while protecting the environment.

The Yalong River Basin, rich in hydropower and tourism resources, faces challenges due to its topography. Organic growth in favorable areas like Jinping is advisable, while stricter controls are needed in less favorable regions. Centralized construction zones and relocation initiatives can optimize resource use and mitigate environmental impacts.

The Dadu River Basin, prone to geological disasters, requires proactive measures for climate adaptation and disaster prevention. Sustainable development, focusing on ecotourism and hydropower, is recommended. The development around the Yele Reservoir scenic area aligns with ecotourism principles, enhancing resilience to climate change and promoting harmony with nature.

4.3.2. Limitations

This article employs Landsat data for extracting construction land information. However, due to the spatial resolution limitation of 30 m, finer details, such as roads in deep valleys and scattered residential areas, may not be adequately captured. Consequently, the spatial distribution and changes in construction land may not have reached a refined level. To enhance future research, it is suggested that we integrate high spatial resolution satellite remote sensing and UAV technology.

The spatial pattern and dynamic changes in construction land are influenced by various factors. While this article considers 12 factors, the selection and quantification of human factors could be refined. For instance, factors like construction land location customs and policies are challenging to quantify, and the regional and hierarchical nature of the economic impact may be difficult to accurately characterize. Future research could delve deeper into the influence of the regional geological background on the eastern edge of the Tibetan Plateau, comprehensively consider the response and adaptation of construction land changes driven by multiple factors such as climate change and the regional economy, and explore a synergistic model of the human–land coupling system in the landscape transition zone.

5. Conclusions

This study delves into the spatiotemporal evolution characteristics of construction land in Mianning County, situated on the eastern edge of the Qinghai–Tibet Plateau. Employing methodologies such as the landscape expansion index, geographic information system spatial analysis, and geographically weighted regression analysis, it delves into the causes behind the spatial distribution and changes in construction land. Here are the key findings:

The unique geological and climatic conditions on the eastern edge of the Qinghai–Tibet Plateau create a foundation for environmental gradients. These gradients play a crucial role in determining the distribution and evolution of construction land in the Anning River, Yalong River, and Dadu River basins at a macro level. Influenced by the region's resources and environment, the distribution and expansion of construction land in Mianning County have manifested into three typical spatial patterns: the Wide Valley Spatial Pattern, Deep Valley Spatial Pattern, and High Mountain Lake Basin Spatial Pattern. The development of construction land in Mianning County can be categorized into three stages: 1990–2000, 2000–2010, and 2010–2020. These stages align with the broader national economic and social development trends of the county, catering to the spatial growth requirements for both production and daily life through the allocation of construction land resources. The predominant expansion mode of construction land in the study area is edge expansion, supplemented by the leapfrog and infilling modes. Over the period of 1990 to 2020, all three river basins witnessed varying degrees of land use change. While human factors primarily drove the changes in the Anning River Basin, natural factors played a more significant role in the Yalong River Basin and Dadu River Basin.

These conclusions shed light on the intricate interplay between geological, climatic, and anthropogenic factors in shaping the spatiotemporal dynamics of construction land in Mianning County, offering valuable insights for sustainable land management practices in similar regions.

Author Contributions: Y.Z. (Yinbing Zhao) wrote the paper with contributions from all co-authors. Z.N. conceived and designed the research. Y.Z. (Yang Zhang) contributed to the methodology. P.W., C.G., W.Y. and Y.L. contributed to the data processing. Z.L. contributed to the geological hazard data. All authors have read and agreed to the published version of the manuscript.

Funding: This research was supported by the Open Foundation of the State Key Laboratory of Urban and Regional Ecology of China (SKLURE2022-2-6), the China Geological Survey Project (DD20190542), and the Key Laboratory of Supported by Open Foundation of the Research Center for Human Geography of Tibetan Plateau and Its Eastern Edge (Chengdu University of Technology) (RWDL2022-ZD002 and RWDL2021-ZD002).

Data Availability Statement: The data presented in this study are available upon request from the author. However, land use data and geological hazard data are excluded due to the lack of authorization from the cooperating party.

Conflicts of Interest: The authors declare no conflicts of interest.

References

1. Yang, H.; Xu, Y.; Zhou, K.; Wang, L.J.; Xu, L. Evaluation of suitability, adaptability, and reserve potential of construction land on the Qinghai-Tibet Plateau. *J. Geogr. Sci.* **2024**, *34*, 41–61. [CrossRef]
2. Cai, E.X.; Liu, Y.L.; Li, J.W.; Chen, W.Q. Spatiotemporal Characteristics of Urban-Rural Construction Land Transition and Rural-Urban Migrants in Rapid-Urbanization Areas of Central China. *J. Urban Plan. Dev.* **2020**, *146*, 05019023. [CrossRef]
3. Zhou, Y.; Liu, N.; Yu, R. Analysis of remote sensing monitoring methods for construction land at home and abroad. *IOP Conf. Ser. Earth Environ. Sci.* **2021**, *783*, 012127. [CrossRef]
4. Zhang, Z.; Liu, F.; Zhao, X.; Wang, X.; Shi, L.; Xu, J.; Yu, S.; Wen, Q.; Zuo, L.; Yi, L.; et al. Urban Expansion in China Based on Remote Sensing Technology: A Review. *Chin. Geogr. Sci.* **2018**, *28*, 727–743. [CrossRef]
5. Bera, D.; Chatterjee, N.D.; Ghosh, S.; Dinda, S.; Bera, S. Recent trends of land surface temperature in relation to the influencing factors using Google Earth Engine platform and time series products in megacities of India. *J. Clean. Prod.* **2022**, *379*, 134735. [CrossRef]
6. Wang, J.; Lin, Y.; Glendinning, A.; Xu, Y. Land-use changes and land policies evolution in China's urbanization processes. *Land Use Policy* **2018**, *75*, 375–387. [CrossRef]
7. Pissourios, I.A. Survey methodologies of urban land uses: An oddment of the past, or a gap in contemporary planning theory? *Land Use Policy* **2019**, *83*, 403–411. [CrossRef]
8. Liao, K.H.; Huang, W.Y.; Wang, C.J.; Wu, R.; Hu, Y. Spatio-Temporal Evolution Features and Impact Factors of Urban Expansion in Underdeveloped Cities: A Case Study of Nanchang, China. *Land* **2022**, *11*, 1799. [CrossRef]
9. Blakime, T.-H.; Adjonou, K.; Komi, K.; Hlovor, A.K.D.; Gbafa, K.S.; Zoungrana, J.-B.B.; Polorigni, B.; Kokou, K. Dynamics of Built-Up Areas and Challenges of Planning and Development of Urban Zone of Greater Lomé in Togo, West Africa. *Land* **2024**, *13*, 84. [CrossRef]
10. Huang, D.Q.; Lang, Y.; Liu, T. The Evolving Structure of Rural Construction Land in Urbanizing China: Case Study of Tai'an Prefecture. *Land.* **2021**, *10*, 65. [CrossRef]
11. Dong, L.; Li, X. Evolution of Urban Construction Land Structure Based on Information Entropy and Shift-Share Model: An Empirical Study on Beijing-Tianjin-Hebei Urban Agglomeration in China. *Sustainability* **2022**, *14*, 1244. [CrossRef]

12. Kuemmerle, T.; Erb, K.; Meyfroidt, P.; Müller, D.; Verburg, P.H.; Estel, S.; Haberl, H.; Hostert, P.; Jepsen, M.R.; Kastner, T.; et al. Challenges and opportunities in mapping land use intensity globally. *Curr. Opin. Environ. Sustain.* **2013**, *5*, 484–493. [CrossRef] [PubMed]
13. Canelas, J.V.; Pereira, H.M. Impacts of land-use intensity on ecosystems stability. *Ecol. Model.* **2022**, *472*, 110093. [CrossRef]
14. Yu, M.; Chen, Z.; Long, Y.; Mansury, Y. Urbanization, land conversion, and arable land in Chinese cities: The ripple effects of high-speed rail. *Appl. Geogr.* **2022**, *146*, 102756. [CrossRef]
15. Li, Z.; Luan, W.; Zhang, Z.; Su, M. Relationship between urban construction land expansion and population/economic growth in Liaoning Province, China. *Land Use Policy* **2020**, *99*, 105022. [CrossRef]
16. Marantz, N.J.; Lewis, P.G. Jurisdictional Size and Residential Development: Are Large-Scale Local Governments More Receptive to Multifamily Housing? *Urban Aff. Rev.* **2021**, *58*, 732–766. [CrossRef]
17. Liu, Y.; Zhang, Z.; Zhou, Y. Efficiency of construction land allocation in China: An econometric analysis of panel data. *Land Use Policy* **2018**, *74*, 261–272. [CrossRef]
18. Xu, G.; Yin, X.; Wu, G.; Gao, N. Rethinking the Contribution of Land Element to Urban Economic Growth: Evidence from 30 Provinces in China. *Land* **2022**, *11*, 801. [CrossRef]
19. Allan, A.; Soltani, A.; Abdi, M.H.; Zarei, M. Driving Forces behind Land Use and Land Cover Change: A Systematic and Bibliometric Review. *Land* **2022**, *11*, 1222. [CrossRef]
20. Harik, G.; Alameddine, I.; Zurayk, R.; El-Fadel, M. Uncertainty in forecasting land cover land use at a watershed scale: Towards enhanced sustainable land management. *Ecol. Model.* **2023**, *486*, 110515. [CrossRef]
21. Mitsuda, Y.; Ito, S. A review of spatial-explicit factors determining spatial distribution of land use/land-use change. *Landsc. Ecol. Eng.* **2011**, *7*, 117–125. [CrossRef]
22. Yuan, X.; Wang, C.; Li, B.; Wang, W.; Chen, N. Review of the Driving Forces and Impacts of Land Use/Cover Change in the Yangtze River Basin. *Geomat. Inf. Sci. Wuhan Univ.* **2023**, *48*, 1241–1255.
23. Irwin, E.G.; Bockstael, N.E. The evolution of urban sprawl: Evidence of spatial heterogeneity and increasing land fragmentation. *Proc. Natl. Acad. Sci. USA* **2007**, *104*, 20672–20677. [CrossRef]
24. Wang, Y.M.; Deng, Q.C.; Yang, H.Q.; Liu, H.; Yang, F.; Zhao, Y.K. Spatial-temporal differentiation and influencing factors of rural settlements in mountainous areas: An example of Liangshan Yi Autonomous Prefecture, Southwestern China. *J. Mt. Sci.* **2024**, *21*, 218–235. [CrossRef]
25. Zhang, Z.; Gong, J.; Li, J.; Yang, J.; Chen, G.; Ren, Y.; Plaza, A. Comparative study of construction land development suitability assessment: Insights from multiple models and driving forces. *Environ. Impact Assess. Rev.* **2024**, *107*, 107574. [CrossRef]
26. Schneider, A.; Friedl, M.A.; Potere, D. A new map of global urban extent from MODIS satellite data. *Environ. Res. Lett.* **2009**, *4*, 044003. [CrossRef]
27. Wang, Z.B.; Fang, C.L.; Zhang, X.R. Spatial expansion and potential of construction land use in the Yangtze River Delta. *J. Geogr. Sci.* **2015**, *25*, 851–864. [CrossRef]
28. Lee, S.; Ryu, J.-H.; Kim, I.-S. Landslide susceptibility analysis and its verification using likelihood ratio, logistic regression, and artificial neural network models: Case study of Youngin, Korea. *Landslides* **2007**, *4*, 327–338. [CrossRef]
29. Zhang, H.X.; Zhao, X.J.; Ren, J.; Hai, W.J.; Guo, J.; Li, C.Y.; Gao, Y.P. Research on the Slope Gradient Effect and Driving Factors of Construction Land in Urban Agglomerations in the Upper Yellow River: A Case Study of the Lanzhou-Xining Urban Agglomerations. *Land* **2023**, *12*, 745. [CrossRef]
30. Antrop, M. Landscape change and the urbanization process in Europe. *Landsc. Urban Plan.* **2004**, *67*, 9–26. [CrossRef]
31. Xu, Y.; Zhao, S.; Fan, J. Urban planning construction land standard and its revision based on climate and topography in China. *J. Geogr. Sci.* **2021**, *31*, 603–620. [CrossRef]
32. Hussain, A.; Cao, J.; Hussain, I.; Begum, S.; Akhtar, M.; Wu, X.; Guan, Y.; Zhou, J. Observed Trends and Variability of Temperature and Precipitation and Their Global Teleconnections in the Upper Indus Basin, Hindukush-Karakoram-Himalaya. *Atmosphere* **2021**, *12*, 973. [CrossRef]
33. Feng, C.; Zhang, N.; Habiyakare, T.; Yan, Y.; Zhang, H. Prospects of eco-hydrological model for sponge city construction. *Ecosyst. Health Sustain.* **2021**, *7*, 1994885. [CrossRef]
34. Danso, G.K.; Takyi, S.A.; Amponsah, O.; Yeboah, A.S.; Owusu, R.O. Exploring the effects of rapid urbanization on wetlands: Insights from the Greater Accra Metropolitan Area, Ghana. *SN Soc. Sci.* **2021**, *1*, 212. [CrossRef]
35. Xia, C.Y.; Li, Y.; Ye, Y.M.; Shi, Z. An Integrated Approach to Explore the Relationship Among Economic, Construction Land Use, and Ecology Subsystems in Zhejiang Province, China. *Sustainability* **2016**, *8*, 498. [CrossRef]
36. Kim, Y.; Newman, G.; Güneralp, B. A Review of Driving Factors, Scenarios, and Topics in Urban Land Change Models. *Land* **2020**, *9*, 246. [CrossRef] [PubMed]
37. Galvao, R.F.P.; Urushima, A.Y.F.; Hara, S.; De Jong, W. Analysis of Land Transition Features and Mechanisms in Peripheral Areas of Kyoto (1950–1960). *Sustainability* **2020**, *12*, 4502. [CrossRef]
38. Mahtta, R.; Fragkias, M.; Güneralp, B.; Mahendra, A.; Reba, M.; Wentz, E.A.; Seto, K.C. Urban land expansion: The role of population and economic growth for 300+ cities. *npj Urban Sustain.* **2022**, *2*, 5. [CrossRef]
39. Zhao, S.X.; Yin, M.M. Change of urban and rural construction land and driving factors of arable land occupation. *PLoS ONE* **2023**, *18*, e0286248. [CrossRef]

40. Yussif, K.; Dompreh, E.B.; Gasparatos, A. Sustainability of urban expansion in Africa: A systematic literature review using the Drivers–Pressures–State–Impact–Responses (DPSIR) framework. *Sustain. Sci.* **2023**, *18*, 1459–1479. [CrossRef]
41. Boillat, S.; Scarpa, F.M.; Robson, J.P.; Gasparri, I.; Aide, T.M.; Aguiar, A.P.D.; Anderson, L.O.; Batistella, M.; Fonseca, M.G.; Futemma, C.; et al. Land system science in Latin America: Challenges and perspectives. *Curr. Opin. Environ. Sustain.* **2017**, *26–27*, 37–46. [CrossRef]
42. Dong, O.Y.; Zhu, X.G.; Liu, X.G.; He, R.F.; Wan, Q. Spatial Differentiation and Driving Factor Analysis of Urban Construction Land Change in County-Level City of Guangxi, China. *Land* **2021**, *10*, 691. [CrossRef]
43. Zheng, Q.; Wang, K.; Huang, L.Y.; Zheng, Q.M.; Abubakar, G.A. Monitoring the Different Types of Urban Construction Land Expansion (UCLE) in China's Port City: A Case Study of Ningbo's Central City. *Sustainability* **2017**, *9*, 2374. [CrossRef]
44. Waddell, P. UrbanSim: Modeling Urban Development for Land Use, Transportation, and Environmental Planning. *J. Am. Plan. Assoc.* **2002**, *68*, 297–314. [CrossRef]
45. Hong, W.Y.; Wang, W.X.; Guo, R.Z. Policies for optimizing land-use layouts in highly urbanized areas: An analysis framework based on construction land clearance. *Habitat. Int.* **2022**, *130*, 102697. [CrossRef]
46. Gao, J.B.; Zhang, X.Y.; Yu, C.; Ma, Z.F.; Sun, J.W.; Guan, Y.J. How to Rebalance the Land-Use Structure after Large Infrastructure Construction? From the Perspective of Government Attention Evolution. *Land* **2023**, *12*, 1632. [CrossRef]
47. Henríquez-Dole, L.; Usón, T.J.; Vicuña, S.; Henríquez, C.; Gironás, J.; Meza, F. Integrating strategic land use planning in the construction of future land use scenarios and its performance: The Maipo River Basin, Chile. *Land Use Policy* **2018**, *78*, 353–366. [CrossRef]
48. Bimonte, S.; Stabile, A. Local taxation and urban development. Testing for the side-effects of the Italian property tax. *Ecol. Econ.* **2015**, *120*, 100–107. [CrossRef]
49. Wang, X.; Xiao, F.; Zhang, Y.; Yin, L.; Lesi, M.; Guo, B.; Zhao, Y.; Li, R. Thirty-year expansion of construction land in Xi'an: Spatial pattern and potential driving factors. *Geol. J.* **2018**, *53*, 309–321. [CrossRef]
50. Munroe, D.K.; Müller, D. Issues in spatially explicit statistical land-use/cover change (LUCC) models: Examples from western Honduras and the Central Highlands of Vietnam. *Land Use Policy* **2007**, *24*, 521–530. [CrossRef]
51. Zhang, L.F.; Fang, C.L.; Zhao, R.D.; Zhu, C.; Guan, J.Y. Spatial-temporal evolution and driving force analysis of eco-quality in urban agglomerations in China. *Sci. Total Environ.* **2023**, *866*, 161465. [CrossRef]
52. Hoyos, L.E.; Cabido, M.R.; Cingolani, A.M. A Multivariate Approach to Study Drivers of Land-Cover Changes through Remote Sensing in the Dry Chaco of Argentina. *ISPRS Int. J. Geo-Inf.* **2018**, *7*, 170. [CrossRef]
53. Ustaoglu, E.; Aydinoglu, A.C. Regional Variations of Land-Use Development and Land-Use/Cover Change Dynamics: A Case Study of Turkey. *Remote Sens.* **2019**, *11*, 885. [CrossRef]
54. Cai, W.J.; Tu, F.Y. Spatiotemporal characteristics and driving forces of construction land expansion in Yangtze River economic belt, China. *PLoS ONE* **2020**, *15*, e0227299. [CrossRef]
55. Meneses, B.M.; Reis, E.; Pereira, S.; Vale, M.J.; Reis, R. Understanding Driving Forces and Implications Associated with the Land Use and Land Cover Changes in Portugal. *Sustainability* **2017**, *9*, 351. [CrossRef]
56. Aroengbinang, B.W.; Kaswanto. Driving Force Analysis of Landuse and Cover Changes in Cimandiri and Cibuni Watersheds. *Procedia Environ. Sci.* **2015**, *24*, 184–188. [CrossRef]
57. Ren, Y.; Li, Z.H.; Li, J.N.; Ding, Y.; Miao, X.R. Analysis of Land Use/Cover Change and Driving Forces in the Selenga River Basin. *Sensors* **2022**, *22*, 1041. [CrossRef]
58. Alqurashi, A.F.; Kumar, L.; Al-Ghamdi, K.A. Spatiotemporal Modeling of Urban Growth Predictions Based on Driving Force Factors in Five Saudi Arabian Cities. *ISPRS Int. J. Geo-Inf.* **2016**, *5*, 139. [CrossRef]
59. Cao, M.; Zhao, J.; Zhang, S. Research on the Suitability Evaluation of Construction Land in Southwest Mountainous Areas of China: A Case Study of Baoxing County, Sichuan Province, China. *Appl. Ecol. Environ. Res.* **2018**, *16*, 6567–6587. [CrossRef]
60. Zhao, J.S.; Yuan, L.; Zhang, M. A study of the system dynamics coupling model of the driving factors for multi-scale land use change. *Environ. Earth Sci.* **2016**, *75*, 529. [CrossRef]
61. Wu, M.; Ren, X.; Che, Y.; Yang, K. A Coupled SD and CLUE-S Model for Exploring the Impact of Land Use Change on Ecosystem Service Value: A Case Study in Baoshan District, Shanghai, China. *Environ. Manag.* **2015**, *56*, 402–419. [CrossRef]
62. Rasmussen, L.V.; Rasmussen, K.; Reenberg, A.; Proud, S. A system dynamics approach to land use changes in agro-pastoral systems on the desert margins of Sahel. *Agric. Syst.* **2012**, *107*, 56–64. [CrossRef]
63. Wang, Y.; Zuo, C.C.; Zhu, M.K. How Semi-Urbanisation Drives Expansion of Rural Construction Land in China: A Rural-Urban Interaction Perspective. *Land* **2024**, *13*, 117. [CrossRef]
64. Wu, H.; Lin, A.; Xing, X.; Song, D.; Li, Y. Identifying core driving factors of urban land use change from global land cover products and POI data using the random forest method. *Int. J. Appl. Earth Obs.* **2021**, *103*, 102475. [CrossRef]
65. Meng, B.; Wang, X.X.; Zhang, Z.F.; Huang, P. Spatio-Temporal Pattern and Driving Force Evolution of Cultivated Land Occupied by Urban Expansion in the Chengdu Metropolitan Area. *Land* **2022**, *11*, 1458. [CrossRef]
66. Wu, R.; Li, Z.; Wang, S. The varying driving forces of urban land expansion in China: Insights from a spatial-temporal analysis. *Sci. Total Environ.* **2021**, *766*, 142591. [CrossRef] [PubMed]
67. Xu, D.; Zhang, K.; Cao, L.; Guan, X.; Zhang, H. Driving forces and prediction of urban land use change based on the geodetector and CA-Markov model: A case study of Zhengzhou, China. *Int. J. Digit. Earth* **2022**, *15*, 2246–2267. [CrossRef]

68. Kang, L.; Ma, L.; Liu, Y. Comparing the driving mechanisms of different types of urban construction land expansion: A case study of the Beijing-Tianjin-Hebei region. *J. Geogr. Sci.* **2024**, *34*, 722–744. [CrossRef]
69. Liu, T.; Liu, H.; Qi, Y.J. Construction land expansion and cultivated land protection in urbanizing China: Insights from national land surveys, 1996–2006. *Habitat Int.* **2015**, *46*, 13–22. [CrossRef]
70. Fang, C. Important progress and future direction of studies on China's urban agglomerations. *J. Geogr. Sci.* **2015**, *25*, 1003–1024. [CrossRef]
71. Cao, Y.G. Forces Driving Changes in Urban Construction Land of Urban Agglomerations in China. *J. Urban Plan. Dev.* **2015**, *141*, 05014011. [CrossRef]
72. Cao, Y.; Bai, Z.; Zhou, W.; Ai, G. Gradient Analysis of Urban Construction Land Expansion in the Chongqing Urban Area of China. *J. Urban Plan. Dev.* **2015**, *141*, 05014009. [CrossRef]
73. Wang, A.Q.; Chan, E.H.W.; Yeung, S.C.W.; Han, J.B. Urban Fringe Land Use Transitions in Hong Kong: From New Towns to New Development Areas. *Procedia Eng.* **2017**, *198*, 707–719. [CrossRef]
74. Wei, Y.; Zhang, Z. Assessing the fragmentation of construction land in urban areas: An index method and case study in Shunde, China. *Land Use Policy* **2012**, *29*, 417–428. [CrossRef]
75. Li, X.; Fu, J.; Jiang, D.; Lin, G.; Cao, C. Land use optimization in Ningbo City with a coupled GA and PLUS model. *J. Clean. Prod.* **2022**, *375*, 134004. [CrossRef]
76. Huang, X.; Huang, X.J.; Liu, M.M.; Wang, B.; Zhao, Y.H. Spatial-temporal Dynamics and Driving Forces of Land Development Intensity in the Western China from 2000 to 2015. *Chin. Geogr Sci* **2020**, *30*, 16–29. [CrossRef]
77. Shi, Z.Q.; Deng, W.; Zhang, S.Y. Spatio-temporal pattern changes of land space in Hengduan Mountains during 1990–2015. *J. Geogr. Sci.* **2018**, *28*, 529–542. [CrossRef]
78. Xu, Y.; Tang, Q.; Fan, J.; Bennett, S.J.; Li, Y. Assessing construction land potential and its spatial pattern in China. *Landsc. Urban Plan.* **2011**, *103*, 207–216. [CrossRef]
79. Xiang, J.W.; Li, X.M.; Xiao, R.B.; Wang, Y. Effects of land use transition on ecological vulnerability in poverty-stricken mountainous areas of China: A complex network approach. *J. Environ. Manag.* **2021**, *297*, 113206. [CrossRef]
80. Liu, C.; Zhang, H.Y.; Gan, F.P.; Lu, Y.E.; Wang, H.; Zhang, J.H.; Ju, X. Identifying the spatio-temporal variability of human activity intensity and associated drivers: A case study on the Tibetan Plateau. *Front. Earth Sci.* **2022**, *16*, 744–756. [CrossRef]
81. Yang, Y.; Yang, X.; Li, F.; Huang, W. Transitions in land use and cover and their dynamic mechanisms in the Haihe River Basin, China. *Environ. Earth Sci.* **2021**, *80*, 50. [CrossRef]
82. Yang, Z.; Li, C. Spatial and Temporal Characteristics of Rurality in Urban Suburb Town and Its Driving Factors Based on Land Use Transformation. *Complexity* **2020**, *2020*, 2806127. [CrossRef]
83. Cao, Y.H.; Liu, M.Y.; Cao, Y.D.; Chen, C.; Zhang, D.P. Change pattern and driving mechanism of construction land in China's undertaking industrial transfer demonstration area: Taking the Wanjiang City Belt along the Yangtze River as an Example. *Earth Sci. Res. J.* **2020**, *24*, 215–223. [CrossRef]
84. Gong, Y.L.; Li, J.T.; Li, Y.X. Spatiotemporal characteristics and driving mechanisms of arable land in the Beijing-Tianjin-Hebei region during 1990–2015. *Socio-Econ. Plan. Sci.* **2020**, *70*, 100720. [CrossRef]
85. Gao, Y.P.; Chen, W.J. Evolution and Influencing Factors of Township Spatial Form: A Two-Dimensional Perspective. *Complexity* **2020**, *2020*, 5617545. [CrossRef]
86. Yang, R.; Xu, Q.; Xu, X.; Chen, Y. Rural settlement spatial patterns and effects: Road traffic accessibility and geographic factors in Guangdong Province, China. *J. Geogr. Sci.* **2019**, *29*, 213–230. [CrossRef]
87. Xu, C.; Liu, M.S.; Zhang, C.; An, S.Q.; Yu, W.; Chen, J.M. The spatiotemporal dynamics of rapid urban growth in the Nanjing metropolitan region of China. *Landsc. Ecol.* **2007**, *22*, 925–937. [CrossRef]
88. Liu, J.Y.; Zhang, Z.X.; Xu, X.L.; Kuang, W.H.; Zhou, W.C.; Zhang, S.W.; Li, R.D.; Yan, C.Z.; Yu, D.S.; Wu, S.X.; et al. Spatial patterns and driving forces of land use change in China during the early 21st century. *J. Geogr. Sci.* **2010**, *20*, 483–494. [CrossRef]
89. Brunsdon, C.; Fotheringham, A.S.; Charlton, M.E. Geographically weighted regression: A method for exploring spatial nonstationarity. *Geogr. Anal.* **1996**, *28*, 281–298. [CrossRef]
90. Zhao, Y.B.; Sun, R.H.; Ni, Z.Y. Identification of Natural and Anthropogenic Drivers of Vegetation Change in the Beijing-Tianjin-Hebei Megacity Region. *Remote Sens.* **2019**, *11*, 1224. [CrossRef]
91. Leung, Y.; Mei, C.-L.; Zhang, W.-X. Statistical test for local patterns of spatial association. *Environ. Plan. A* **2003**, *35*, 725–744. [CrossRef]
92. Nakaya, T.; Fotheringham, A.S.; Brunsdon, C.; Charlton, M. Geographically weighted Poisson regression for disease association mapping. *Stat. Med.* **2005**, *24*, 2695–2717. [CrossRef]
93. Liu, Z.; Zhang, J.; Golubchikov, O. Edge-urbanization: Land policy, development zones, and urban expansion in Tianjin. *Sustainability* **2019**, *11*, 2538. [CrossRef]
94. Fulman, N.; Grinblat, Y.; Benenson, I. A project-based view of urban dynamics: Analyzing 'leapfrogging' and fringe development in Israel. *Cities* **2024**, *148*, 104908. [CrossRef]
95. Farris, J.T. The barriers to using urban infill development to achieve smart growth. *Hous. Policy Debate* **2001**, *12*, 1–30. [CrossRef]
96. Li, C.; Zhao, J.; Xu, Y. Examining spatiotemporally varying effects of urban expansion and the underlying driving factors. *Sustain. Cities Soc.* **2017**, *28*, 307–320. [CrossRef]

97. Duan, J.H.; Peng, Q.Z.; Huang, P.Y. Slope characteristics of urban construction land and its correlation with ground slope in China. *Open Geosci.* **2022**, *14*, 1524–1537. [CrossRef]
98. Li, M.; Shi, X.; Shen, Z.; Yang, E.; Bao, H.; Ni, Y. Effect of hillslope aspect on landform characteristics and erosion rates. *Environ. Monit. Assess.* **2019**, *191*, 598. [CrossRef] [PubMed]
99. Arouq, M.K.; Esmaeilpour, M.; Sarvar, H. Vulnerability assessment of cities to earthquake based on the catastrophe theory: A case study of Tabriz city, Iran. *Environ. Earth Sci.* **2020**, *79*, 354. [CrossRef]
100. Wang, J.; Wang, Z.; Cheng, H.; Kang, J.; Liu, X. Land Cover Changing Pattern in Pre- and Post-Earthquake Affected Area from Remote Sensing Data: A Case of Lushan County, Sichuan Province. *Land* **2022**, *11*, 1205. [CrossRef]
101. Lin, J.; Chen, W.; Qi, X.; Hou, H. Risk assessment and its influencing factors analysis of geological hazards in typical mountain environment. *J. Clean. Prod.* **2021**, *309*, 127077. [CrossRef]
102. Liu, Y.; Deng, W.; Peng, L. The coupling mechanism between the suitable space and rural settlements considering the effect of mountain hazards in the upper Minjiang River basin. *J. Mt. Sci.* **2020**, *17*, 2774–2783. [CrossRef]
103. Zhao, J.; Zhang, Q.; Wang, D.; Wu, W.; Yuan, R. Machine Learning-Based Evaluation of Susceptibility to Geological Hazards in the Hengduan Mountains Region, China. *Int. J. Disaster Risk Sci.* **2022**, *13*, 305–316. [CrossRef]
104. Liang, L.; Deng, X.; Wang, P.; Wang, Z.; Wang, L. Assessment of the impact of climate change on cities livability in China. *Sci. Total Environ.* **2020**, *726*, 138339. [CrossRef] [PubMed]
105. Tian, L.; Fu, W.; Tao, Y.; Li, M.; Wang, L. Dynamics of the alpine timberline and its response to climate change in the Hengduan mountains over the period 1985–2015. *Ecol. Indic.* **2022**, *135*, 108589. [CrossRef]
106. Liang, X.; Zhang, L.; He, S.; Song, K.; Zheng, Z. Characteristics and Projection of Rainfall Erosivity Distribution in the Hengduan Mountains. *Land* **2023**, *12*, 1435. [CrossRef]
107. Gupta, A.K.; Negi, M.; Nandy, S.; Kumar, M.; Singh, V.; Valente, D.; Petrosillo, I.; Pandey, R. Mapping socio-environmental vulnerability to climate change in different altitude zones in the Indian Himalayas. *Ecol. Indic.* **2020**, *109*, 105787. [CrossRef]
108. Nilsson, C.; Reidy, C.A.; Dynesius, M.; Revenga, C. Fragmentation and Flow Regulation of the World's Large River Systems. *Science* **2005**, *308*, 405–408. [CrossRef]
109. Sear, D.A.; Newson, M.D. Environmental change in river channels: A neglected element. Towards geomorphological typologies, standards and monitoring. *Sci. Total Environ.* **2003**, *310*, 17–23. [CrossRef]
110. Gregory, K.J.; Davis, R.J.; Downs, P.W. Identification of river channel change to due to urbanization. *Appl. Geogr.* **1992**, *12*, 299–318. [CrossRef]
111. Wu, L.; Xu, Y.; Yuan, J.; Xu, Y.; Wang, Q.; Xu, X.; Wen, H. Impacts of Land Use Change on River Systems for a River Network Plain. *Water* **2018**, *10*, 609. [CrossRef]
112. Stroud, S.; Peacock, J.; Hassall, C. Vegetation-based ecosystem service delivery in urban landscapes: A systematic review. *Basic Appl. Ecol.* **2022**, *61*, 82–101. [CrossRef]
113. Wang, Y.H.; Dai, E.F.; Yin, L.; Ma, L. Land use/land cover change and the effects on ecosystem services in the Hengduan Mountain region, China. *Ecosyst. Serv.* **2018**, *34*, 55–67. [CrossRef]
114. Yin, L.; Dai, E.; Guan, M.; Zhang, B. A novel approach for the identification of conservation priority areas in mountainous regions based on balancing multiple ecosystem services—A case study in the Hengduan Mountain region. *Glob. Ecol. Conserv.* **2022**, *38*, e02195. [CrossRef]
115. Liu, Y.; Tian, J.; Liu, R.; Ding, L. Influences of Climate Change and Human Activities on NDVI Changes in China. *Remote Sens.* **2021**, *13*, 4326. [CrossRef]
116. Handavu, F.; Chirwa, P.W.C.; Syampungani, S. Socio-economic factors influencing land-use and land-cover changes in the miombo woodlands of the Copperbelt province in Zambia. *For. Policy Econ.* **2019**, *100*, 75–94. [CrossRef]
117. Wang, L.Y.; Zhang, S.Y.; Liu, Y.F.; Liu, Y.L. Interaction between Construction Land Expansion and Cropland Expansion and Its Socioeconomic Determinants: Evidence from Urban Agglomeration in the Middle Reaches of the Yangtze River, China. *Front. Environ. Sci.* **2022**, *10*, 882582. [CrossRef]
118. He, W.; Di, B.F.; Zeng, Y.J.; Duan, Y.N.; Li, J.H.; Qiu, L.K.; Balikuddembe, J.K.; Peng, Q.Q.; Zeng, W.; Stamatopoulos, C.A.; et al. Reconsidering the eco-economic benefits of Grain for Green Program in Sichuan Province, China. *Ecol. Indic.* **2023**, *15*, 111225. [CrossRef]
119. Thiede, R.N.; Fabris-Rotelli, I.N.; Debba, P.; Cleghorn, C.W. A Markov chain model for geographical accessibility. *Spat. Stat.* **2023**, *55*, 100748. [CrossRef]
120. Oruonye, E. An assessment of the impact of road construction on land use pattern in urban centres in Nigeria, A case study of Jalingo LGA, Taraba State Nigeria. *Mediterr. J. Soc. Sci.* **2014**, *5*, 82–88. [CrossRef]

land

Article

Potential and Influencing Factors of Urban Spatial Development under Natural Constraints: A Case Study of the Guangdong-Hong Kong-Macao Greater Bay Area

Yukui Zhang [1,2,3,4,5], Tao Lin [2,3,4,5,*], Junmao Zhang [2,3,4,5], Meixia Lin [2,4,5], Yuan Chen [2,4,5], Yicheng Zheng [2,4,5,6], Xiaotong Wang [2,4,5], Yuqin Liu [2,4,5], Hong Ye [2,3,4,5] and Guoqin Zhang [2,3,4,5]

1 Fujian Agriculture and Forestry University, Fuzhou 350002, China; ykzhang@iue.ac.cn
2 Key Laboratory of Urban Environment and Health, Institute of Urban Environment, Chinese Academy of Sciences, Xiamen 361021, China; jmzhang@iue.ac.cn (J.Z.); mxlin@iue.ac.cn (M.L.); yuanchen@iue.ac.cn (Y.C.); yczheng@iue.ac.cn (Y.Z.); xtwang@iue.ac.cn (X.W.); yqliu@iue.ac.cn (Y.L.); hye@iue.ac.cn (H.Y.); gqzhang@iue.ac.cn (G.Z.)
3 College of Resources and Environment, University of Chinese Academy of Sciences, Beijing 100049, China
4 Fujian Key Laboratory of Digital Technology for Territorial Space Analysis and Simulation, Fuzhou 350108, China
5 Xiamen Key Laboratory of Smart Management on the Urban Environment, Xiamen 361021, China
6 School of Geographical Sciences, Faculty of Science and Engineering, University of Nottingham, Ningbo 315100, China
* Correspondence: tlin@iue.ac.cn

Abstract: As urbanization in China progresses, urban spatial development is transitioning from rapid expansion to more intensive and compact growth. This study examined the role of physical geography and environmental factors in shaping the urban spatial development in the Guangdong-Hong Kong-Macao Greater Bay Area (GBA). Based on the current natural conditions, we selected evaluation indices from topography, hydrogeology, climatic conditions, and natural disasters. These indices were used to create a carrying capacity and suitability evaluation system for development land under natural constraints. Finally, the spatial development potential of the city was finalized by taking into account the current state of the built-up area of the city. Meanwhile, we employed the Optimal Parameters-based Geographical Detector and assessed the impact of 14 natural factors on the spatial development of urban built-up areas. In 2020, the GBA had 52,168.77 km^2 of land suitable for construction, of which 34,241.13 km^2 was highly suitable (61.29%) and 17,927.64 km^2 was moderately suitable (32.09%). At the Bay Area level, 90.15% of the development potential remains untapped; at the city level, Zhaoqing City has the highest potential at 99.56%, while Macao has the lowest at 26.83%. Key factors influencing urban development include silty sand content, annual average relative humidity, and cumulative temperature above 0 °C, with varying impacts across different urban scales. At the Bay Area level, the silty sand content, annual average relative humidity, and cumulative temperature above 0 °C are the main influencing factors on the spatial development of urban built-up areas; at the city level, the main factors are annual average relative humidity and cumulative active temperature above 0 °C. This study reveals the important influence of natural environmental factors on urban spatial development, which is conducive to promoting sustainable development of land resources in GBA.

Keywords: urban spatial development potential; construction land carrying capacity evaluation; suitability assessment for development land; geographic detector; Guangdong-Hong Kong-Macao Greater Bay Area

1. Introduction

Land is an essential non-renewable resource to human survival providing the basic materials for human production and livelihoods. Their sustainable utilization can also be defined as a critical factor in societal sustainable development [1]. Land carrying capacity refers to the ability of land resources to sustainably support socio-economic activities over a period of time, including the production, living, and ecological functions of the land [2]. Land suitability assessment is an evaluation of the appropriateness of land for specific uses based on its natural conditions and socio-economic demands [1]. Land carrying capacity and land suitability assessment are fundamental bases for assessing land use conditions and planning land utilization. Effective management of urban spatial development is a challenge faced by planners and managers, among which carrying capacity assessment remains one of the useful planning tools [1,3]. Utilizing the carrying capacity to assess the suitability levels of construction land is an effective measure. This approach guides urban development and construction, thereby achieving sustainable urban development. For urban areas with rapid socio-economic development, accelerated urbanization, and significant imbalances in urban–rural land use, there is an urgent need to conduct studies focusing on the carrying capacity assessment of construction land [4–6].

In recent years, most scholars have initiated research into land-carrying capacity, focusing on the interactions among population, land, and food. As land development progressed, issues such as spatial imbalances in land use became apparent, prompting more comprehensive studies on overall land carrying capacity. In recent years, various methods, techniques, and frameworks have been used to assess land-carrying capacity [7]. The evaluations of land-carrying capacity typically concentrate on urban [8], regional [9], and specialized areas [5]. Common methodologies for assessing land-carrying capacity were utilized in various ways including systems dynamics [10], ecological footprint models [11], and multi-criteria evaluation [12]. The primary indicators selected for these assessments often relate to environmental capacity [13], population density [14], and ecological capacity [15]. However, existing studies predominantly focus on the carrying capacity of pre-urban development, relying on natural constraints like topography, climate, and hydrology to gauge the development potential of undeveloped land. These studies frequently overlook the link between existing urban areas and the potential for urban expansion under natural constraints. Clarifying this relationship is significant for understanding the dynamics between urban development and the natural environment.

The concept of urban spatial development potential refers to the economic growth, social progress, and environmental improvement of inherent conditions and potentials over a specified period. This interdisciplinary concept includes the economy, society, and environment, aiming to evaluate and predict a city's future development opportunities. Researchers assess urban development potential across several domains. In the economic domain, scholars focus on factors such as a city's growth potential [16], the optimization of its industrial structure [17], and innovation capabilities [18]. In the social domain, studies emphasize the urban population structure [19], cultural development [20], and public health [21]. In the field of technological innovation, the focus is on the city's capacity for innovation [22], low-carbon development [23], and smart industries [24]. Additionally, natural environmental factors, such as geographical location, climate, and natural resources, not only influence the quality of the urban ecological environment but also the city's capacity for sustainable development [25,26]. Previous assessments of urban development potential have predominantly concentrated on social, economic, and technological factors, often neglecting the complex interactions between natural environmental elements and urban development.

Assessing land suitability based on land-carrying capacity facilitates the effective utilization and management of land resources, thereby promoting sustainable development. Although various scholars have explored land-carrying capacity and land suitability from diverse perspectives, in-depth investigations into the alignment between natural constraints and current urban development are still scarce. Exploring the impacts of physical geography and environmental factors on urban spatial development offers insights into current and future potentials, which can provide a scientific basis for national spatial planning and the evolution of urban forms. Currently, China's urban spatial development is transitioning from rapid expansion to a more intensive and compact form. This transition underscores the importance of analyzing the effects of physical geography and environmental factors on both present and future urban spatial potentials, providing a scientific basis for national spatial planning and the evolution of urban forms. This study focuses on the Guangdong-Hong Kong-Macao Greater Bay Area (GBA) to develop an evaluation index system for the carrying capacity of construction land under natural constraints. It assesses the connection between the current state of urban construction land and its potential for suitable development under these constraints, attempting to answer the following two scientific questions: (1) Is the developed urban area in GBA fully utilizing the "urban spatial development potential" under natural constraints? (2) How did the different natural constraints impact the urban development space?

2. Materials and Methods

2.1. Study Area

The GBA is recognized as one of the world's four major bay areas, situated in the southern coastal region of China, spanning from 21°26′ to 24°28′ N latitude and 111°14′ to 115°24′ E longitude (Figure 1). It encompasses nine cities in the Pearl River Delta (including Guangzhou, Shenzhen, Zhuhai, Foshan, Huizhou, Dongguan, Zhongshan, Jiangmen, and Zhaoqing) and two Special Administrative Regions, Hong Kong and Macao. Collectively, these areas cover approximately 56,000 km^2, which constitutes 31.2% of Guangdong Province and only 0.6% of China's national territory. By the end of 2022, the GBA housed roughly 86.3011 million people and generated an economic output surpassing CNY 13 trillion, making it one of the most open and economically dynamic regions in China. This area is pivotal to the Belt and Road Initiative and China's broader national development strategy.

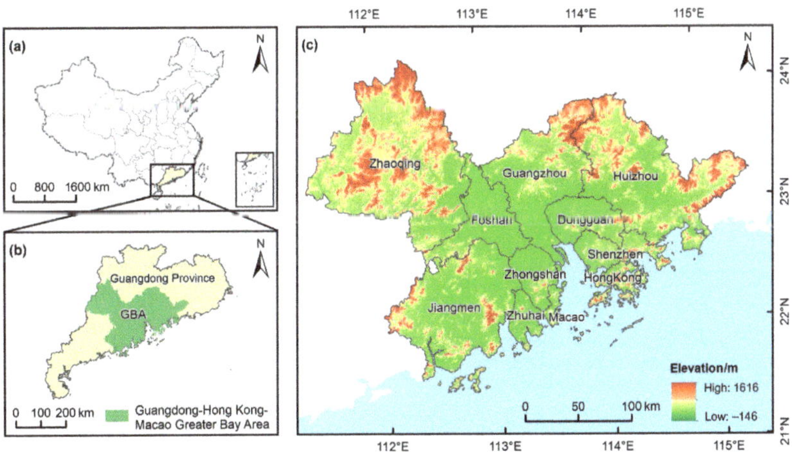

Figure 1. Location of the Guangdong-Hong Kong-Macao Greater Bay Area (GBA). (**a**) Location of Guangdong Province in China. (**b**) Location of GBA in Guangdong Province. (**c**) GBA administrative division and elevations.

The geographic landscape of the GBA features high terrain in the northwest and lower elevations in the southeast, with mountains predominantly in the northern sections near Zhaoqing, Guangzhou, and Huizhou, and plains chiefly in the central and coastal regions. The area benefits from a unique geographical setup, which is "surrounded by mountains on three sides and where three rivers converge", complemented by an extensive coastline, numerous ports, and a large sea area. It experiences a subtropical monsoon climate with an average annual temperature of 22 °C. The rainy season lasts from April to September, providing ample sunlight, warmth, and water resources which enrich the region.

The favorable natural geographical conditions have supported significant development; between 1990 and 2020, residential and construction land in the GBA expanded by 115.21%, reaching 9183.47 km^2. This land type has seen the greatest increase in the area and the fastest rate of change within the region [27]. Since 2010, the scale of construction land has tended towards stabilization, with improvements in the regularity and compactness of its form. Urban construction has thus entered a transitional phase from rapid expansion to intensive and compact development [28].

2.2. Data Sources

This study utilized elevation, soil, hydrology, climate, climatic zoning data, and built-up area data (Table 1) to develop a framework for assessing the carrying capacity and suitability of urban development land under natural constraints. This framework also facilitated the mapping and analysis of urban spatial development potential within the study area. All data were standardized to a uniform projection coordinate system, WGS_1984_Web_Mercator_Auxiliary_Sphere, and resampled to a 30 m resolution using the resampling tool in ArcGIS. Table 1 provides detailed information on the multi-source data utilized in this study.

Table 1. Data sources.

Dataset	Sources	Type of Dataset	Spatial Resolution	Time of Dataset	Note
Elevation data	Geospatial Data Cloud site, Computer Network Information Center, Chinese Academy of Sciences. (http://www.gscloud.cn, accessed on 15 November 2023)	tif	30 m	2009	Extraction of elevation, slope, and aspect data.
Soil data	Harmonized World Soil Database, HWSD (https://data.tpdc.ac.cn/en/data/, accessed on 15 November 2023)	tif	1:4 million	2009	Extraction of silty sand content
Drainage data	OpenStreetMap (https://www.openstreetmap.org, accessed on 15 November 2023)	shp	—	2020	—
Climatic zoning data	Resource and Environmental Science Data Platform of Chinese Academy of Sciences (https://www.resdc.cn/, accessed on 21 November 2023)	shp	—	1978	—
Active accumulated temperature	Swiss Federal Institute for Forest, Snow and Landscape Research (https://chelsa-climate.org/downloads/, accessed on 15 November 2023)	tif	1 km	1981–2010	For extracting data on active cumulative temperatures greater than 0 °C
Meteorology data	Resource and Environmental Science Data Platform of Chinese Academy of Sciences (https://www.resdc.cn/, accessed on 21 November 2023)	tif	1 km	1960–2010	Extraction of mean annual wind speed, mean annual air temperature, mean annual sunshine hours, mean annual relative humidity, and annual precipitation data
Fault data	National Earthquake Data Center (https://data.earthquake.cn/index.html, accessed on 15 November 2023)	shp	—	—	—
Built-up area data	Science Data Bank (https://www.scidb.cn/en, accessed on 15 November 2023)	shp	—	2020	—
Administrative division data	Resource and Environmental Science Data Platform of Chinese Academy of Sciences (https://www.resdc.cn/, accessed on 15 November 2023)	shp	—	2022	—

2.3. Methods

2.3.1. Construction of a Land-Carrying Capacity and Suitability Evaluation System under Natural Constraints

Evaluation Index Selection

The development of urban built-up areas is complex and influenced by multiple natural factors. Previous studies have highlighted several critical factors affecting urban development. (1) Topography: Plains and river valleys, characterized by flat terrain and fertile land, are generally more conducive to agriculture and construction, resulting in rapid urban expansion. Conversely, challenging topographic conditions such as mountains, hills, or deserts impose limitations on urban growth. (2) Water Resources: Essential for urban development, an adequate water supply supports residential life, industrial production, and agricultural irrigation. Consequently, cities near rivers, lakes, or other bodies of water tend to expand more rapidly. (3) Climate Conditions: Mild climates are favorable for human habitation and agriculture, attracting more population and investment, which in turn promotes urban expansion. In contrast, extreme climates—such as high temperatures, severe cold, or drought—can hinder urban development. (4) Natural Disasters: Earthquakes, floods, typhoons, and other natural disasters can severely damage cities and affect their expansion. In areas prone to disasters, urban planning and construction must be carried out more cautiously to mitigate risks.

Considering previous studies and the specific characteristics of the GBA, along with data availability and the quantifiability of indicators, we established a system comprising eight criteria to evaluate the carrying capacity of development land under natural constraints within the GBA. Combined with the existing research [29–31], we determined the evaluation system. The evaluation system includes (1) slope; (2) aspect; (3) silty sand content; (4) water resources; (5) climatic zone; (6) accumulated temperature above 0 °C; (8) wind effect index; (9) distance from fault lines.

Evaluation Metrics Quantification and Grading

Each indicator is classified into three to five levels, as outlined in Table 2. These levels correspond to five categories of carrying capacity: high, relatively high, moderate, medium, and relatively low. These categories are assigned scores of 9, 7, 5, 3, and 1, respectively. For instance, the "slope" indicator uses the following classifications: ≤3° (9 points), 3 to 8° (7 points), 8 to 15° (5 points), 15 to 25° (3 points), and >25° (1 point).

Table 2. Evaluation factors and indicators for grading the carrying capacity of development land.

Evaluation Factor	Indicators for Grading the Carrying Capacity of Development Land				
	High (9 Points)	**Middle-High (7 Points)**	**Middle (5 Points)**	**Middle-Low (3 Points)**	**Low (1 Point)**
Slope	≤3°	3 to 8°	8 to 15°	15 to 25°	>25°
			Corrected elevation and relief [1]		
Aspect	Southern slope	Southwest-facing slope, Southeast-facing slope	West slope, East slope	Northeast slope, Northwest slope	Northern slope
Silty sand content	<60%	——	60% to 80%	——	≥80%
Precipitation	>1400 mm	800 to 1400 mm	400 to 800 mm	200 to 400 mm	<200 mm
Distance to rivers and lakes [2]	≤1 km	1 to 2 km	2 to 5 km	——	>5 km

Table 2. *Cont.*

Evaluation Factor	Indicators for Grading the Carrying Capacity of Development Land				
	High (9 Points)	**Middle-High (7 Points)**	**Middle (5 Points)**	**Middle-Low (3 Points)**	**Low (1 Point)**
Climatic zone	Boreal zone, Temperate zone, Subtropical zone	Northern subtropical zone, Central subtropical zone, Southern subtropical zone	Plateau climate, Subtropical climate	——	——
Accumulated temperature above 0 °C	≥7600 °C	5800 to 7600 °C	4000 to 5800 °C	1500 to 4000 °C	≤1500 °C
Wind efficiency index	−299 to −100	——	−99 to −10, −400 to −300	——	>−10, <−400
Distance to fault lines	>36,000 m	8300–36,000 m	1600–8300 m	<1600 m	——

Note: [1] Elevation adjustment: Areas with elevation above 4900 m are degraded by two levels, while those between 3000 and 4900 m are degraded by one level. Terrain Undulation Correction: Locations with more than 200 m of terrain undulation have their construction land carrying capacity reduced by two grades, and those with undulations between 100 and 200 m are reduced by one grade. [2] Water resource adjustment based on precipitation and proximity: Sites rated as 3 for distance to rivers and lakes have their water resource rating degraded by one level. Sites rated as 1 for proximity see their precipitation rating degraded by two levels.

Assessment of the Indices

Each 30 m resolution pixel unit is used as an evaluation unit. To calculate the cumulative impact score for each evaluation unit, the quantification and grading results of the previously mentioned evaluation factors are combined. The formula used is as follows:

$$S_j = \sum_{j=1}^{n} X_{ij} W_j \tag{1}$$

In this formula, *Si* represents the composite carrying capacity score for the *i*-th cell. The score contributed by the *j*-th evaluation indicator in the *i*-th cell is weighted by *W*, which is the weight value of the *j*-th evaluation indicator. The total number of evaluation indicators is denoted by *n*.

In ArcGIS 10.8, the eight evaluation indicators for the GBA are analyzed using the raster calculator and an equal weight method to sum all indicators. The total scores of these indicators produce grid values ranging from 49 to 97. We further classified these scores into five categories by the natural breaks method. The intervals for these categories are detailed in Table 3. Additionally, the water system in the GBA was considered a no-development zone, received the lowest scores, and was automatically categorized as areas with low carrying capacity. Urban areas that achieve high or middle-high carrying capacity scores are identified as highly suitable for urban construction. Conversely, areas with middle or middle-low capacity scores are considered moderately suitable, while those with low carrying capacity are deemed unsuitable for urban development. The categories of high suitability and middle suitability are collectively referred to as "suitable areas".

Table 3. Carrying capacity and suitability classification criteria.

Value	Carrying Capacity Grade	Suitability Grade
49–70	Low	Unsuitable
70–76	Middle-low	Middle
77–82	Middle	Suitability
83–87	Middle-high	High
88–97	High	Suitability

2.3.2. Evaluation of Urban Spatial Development Potential

Urban spatial development potential quantifies the extent of land within the urban administrative boundaries that is suitable for development, excluding current built-up areas. This potential is directly related to the proportion of the remaining suitable areas within these boundaries compared to the total suitable areas in the city. The potential can be quantified using the following formula:

$$M = \frac{C - C_1}{C} \quad (2)$$

where M represents the urban development potential. C is the area of land suitable for construction within the administrative boundaries of the study area, and C_1 is the area of land suitable for construction currently within the built-up areas of the study area.

2.3.3. Optimal Parameters-Based Geographical Detector

The geographic detector is a novel spatial statistical method developed by Wang Jinfeng et al. (2020) [32] to examine the effects of various factors and their interrelationships across multiple spatial units. The traditional model of this detector necessitates manual adjustments for discretizing continuous data, a process that may introduce subjectivity and issues with inadequate discretization. To address these challenges, this study employed the Optimal Parameters-based Geographical Detector [29] to analyze the driving forces behind urban development in the GBA. This approach evaluates factors such as elevation, slope, temperature, and humidity to discern spatial differentiation in urban built-up areas and identify the principal driving forces. The formula used for the calculation is presented below:

$$q = 1 - \frac{\sum_{h=1}^{L} N_h \sigma_h^2}{N\sigma^2} \quad (3)$$

where q represents the detection value of factor X on the dependent variable Y; L is the number of categories for factor X; N_h and N are the numbers of units in category h and the study area, respectively; ${\sigma_h}^2$ and σ^2 are the variances of the y-values in category h and the entire study area, respectively. The range of q is [0, 1], where a higher q value indicates a greater influence of the selected factor X on the change in variable Y, and vice versa.

This study evaluated the influence of 14 natural factors on the spatial differentiation of urban built-up areas using geographic detectors. The factors analyzed were as follows: elevation (*X1*), slope (*X2*), aspect (*X3*), terrain undulation (*X4*), silty sand content (*X5*), distance from water bodies (*X6*), average annual temperature (*X7*), annual precipitation (*X8*), average annual wind speed (*X9*), annual sunshine duration (*X10*), average annual relative humidity (*X11*), accumulated temperature above 0 °C (*X12*), wind effect index (*X13*), and distance from fault lines (*X14*).

The calculation was performed using the "GD" package [29] in R version 4.3.1, where each continuous factor was discretized for geographic detector analysis. This study calculated the q values for each variable under different categorization methods: equal interval, natural breaks, quantile, geometric interval, and standard deviation. The method yielding the highest q value was deemed the optimal discretization for that variable, with categorizations ranging from 3 to 7 categories. For example, the optimal discretization of the accumulated temperature above 0 °C in Guangzhou in 2020 was achieved using the quantile classification method with six categories (Figure 2). Similarly, the accumulated temperature above 0 °C in Guangzhou in 2020 was classified into six categories using the same quantile interval method. This approach of selecting the optimal discretization method based on the highest q value is consistently applied across different continuous factors and across other years.

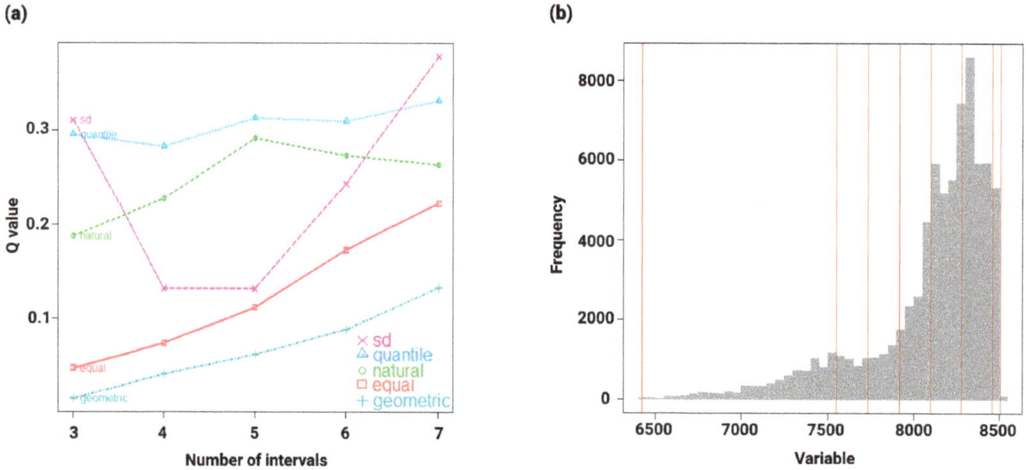

Figure 2. (**a**) Classification method of accumulated temperature above 0 °C in Guangzhou in 2020. (**b**) The optimal interval of accumulated temperature above 0 °C in Guangzhou in 2020.

3. Results

3.1. Distribution of Suitability and Carrying Capacity for Development Land under Natural Constraints

The results from the single-factor assessments were spatially overlaid. This process generated the final carrying capacity assessment for construction land in the GBA, as shown in Figure 3a. The results showed that the GBA has been categorized into five distinct zones based on the carrying capacity: low, middle-low, middle, middle-high, and high. Subsequently, these zones were further grouped into three categories reflecting their suitability for development: unsuitable, moderately suitable, and highly suitable areas, which are illustrated in Figure 3b. The areas and proportions of each carrying capacity and suitability zone are detailed in Table 4.

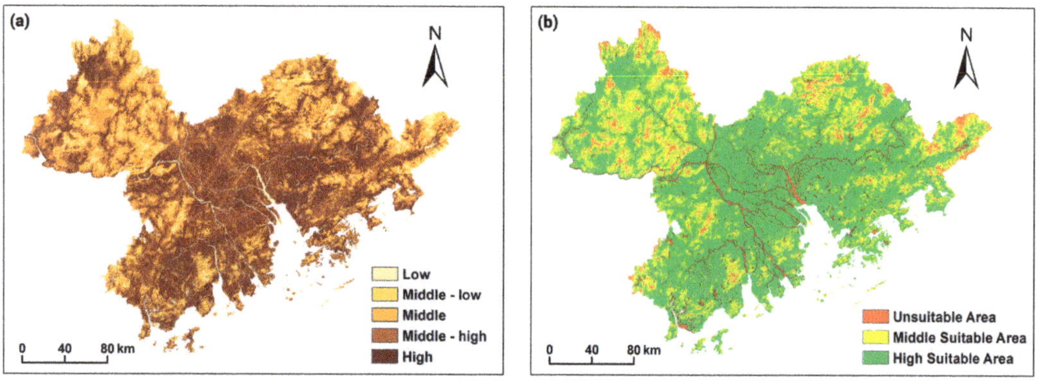

Figure 3. (**a**) Distribution of the carrying capacity grade in the GBA; (**b**) distribution of the carrying capacity and suitability grade in the GBA.

Table 4. Summary of carrying capacity and suitability grade area.

Carrying Capacity Grade	Area/km^2	Proportion/%	Suitability Grade	Area/km^2	Proportion/%
Low	3694.99	6.61	Unsuitable	3694.99	6.61
Middle-low	5515.54	9.87	Middle Suitable	17,927.64	32.09
Middle	12,412.10	22.22			
Middle-high	20,871.65	37.36	High Suitable	34,241.13	61.29
High	13,369.48	23.93			

The highly suitable areas are the most extensive, covering 34,241.13 km^2 and comprising 61.29% of the total area. The moderately suitable areas measure 17,927.64 km^2, comprising 32.09% of the total area. The unsuitable areas, being the smallest, occupy 3694.99 km^2, which represents 6.61% of the total area.

3.2. Analysis of Urban Spatial Development Potential under Natural Constraints

The overall urban spatial development potential of the GBA and its constituent cities is detailed in Figure 4 and Table 5. In 2020, the area of suitable zones already utilized in the urban built-up areas was 5137.15 km^2, which accounted for 9.85% of the total suitable area within the administrative boundaries of the GBA. The remaining suitable area within the administrative region amounted to 47,031.62 km^2, representing 90.15% of the total suitable area. This indicates that as of 2020, 90.15% of the areas suitable for urban development within the GBA remained undeveloped, highlighting the significant potential for further urban spatial development.

At the city level, Zhaoqing still has 99.56% of urban spatial development potential, which ranks first in the GBA, followed by Huizhou (98.99%), Jiangmen (98.04%), Zhuhai (88.96%), Hong Kong (86.48%), Zhongshan (86.41%), and Guangzhou (81.93%), which have urban spatial development potentials beyond 80%. Foshan has 76.72% of urban spatial development potential. Dongguan (49.09%), Shenzhen (45.28%), and Macao (26.83%) have urban spatial development potential of less than 50%. Macao (26.83%) has the lowest urban spatial development potential among all the cities in the GBA.

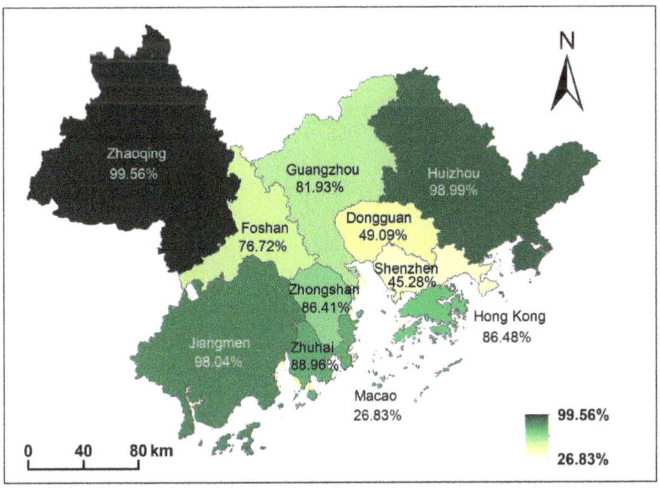

Figure 4. Urban spatial development potential of the GBA.

Table 5. Summary of suitable areas in the GBA.

Area	Built-Up Area		Administrative Division	
	Area of Suitable Area/km^2	% of Suitable Areas	Remaining Suitable Area/km^2	% of Suitable Areas
GBA	5137.15	9.85	47,031.62	90.15
Guangzhou	1235.37	18.07	5602.84	81.93
Shenzhen	1026.34	54.72	849.29	45.28
Zhuhai	162.91	11.04	1312.70	88.96
Foshan	826.95	23.28	2725.43	76.72
Huizhou	107.47	1.01	10,513.77	98.99
Dongguan	1159.19	50.91	1117.92	49.09
Zhongshan	222.07	13.59	1412.51	86.41
Jiangmen	171.05	1.96	8535.81	98.04
Zhaoqing	61.41	0.44	14,047.08	99.56
Hong Kong	141.57	13.52	905.89	86.48
Macao	22.82	73.17	8.37	26.83

3.3. Analysis of Urban Development Area Driving Forces under Natural Constraints

The results of the optimized parameter geographic detector from 2020 indicate that, within the GBA, all influencing factors, except aspect (*X3*), had q values above 0 and p values below 0.05. This demonstrates that, aside from aspect(*X3*), the remaining factors significantly explain the spatial heterogeneity of urban built-up areas, as detailed in Table 6 and Figure 5. The ranking of the explanatory power of each influencing factor is as follows: silty sand content (0.5669), average annual relative humidity (0.2208), growing degree days above 0 °C (0.1642), annual precipitation (0.1442), wind effect index (0.1137), average annual wind speed (0.1125), terrain undulation (0.1073), elevation (0.1043), average annual temperature (0.0858), slope (0.0621), annual sunshine duration (0.0523), distance to water systems (0.0485), distance to fault lines (0.0319), and aspect (0.0001).

Regional variations affect how these factors influence the spatial heterogeneity of urban built-up areas. For example, silty sand content (*X5*) exhibits strong explanatory power over 0.4 in the overall GBA, as well as in Foshan and Jiangmen, marking it as the most influential factor in these areas, but its explanatory power is notably weaker in Hong Kong, Macao, Zhaoqing, Zhongshan, and Zhuhai, where it falls below 0.1. Terrain undulation (*X4*), average annual relative humidity (*X11*), and accumulated temperature above 0 °C (*X12*) are significant explanatory factors in at least two regions each. In particular, terrain undulation (*X4*) is the most influential in Shenzhen and Dongguan, with values of 0.3838 and 0.2278, respectively. Average annual relative humidity (*X11*) holds significant influence in Zhuhai and Zhongshan, with values of 0.2282 and 0.2021, respectively. Accumulated temperature above 0 °C (*X12*) shows strong influence in Guangzhou and Hong Kong, with values of 0.3799 and 0.2036, respectively.

The q values of the influencing factors for each prefecture-level city, as well as the Special Administrative Regions of Hong Kong and Macao, were ranked from highest to lowest, and the top three factors, along with their frequency of occurrence, were tabulated (Table 7). The factors appearing in the top three are *X11*, *X12*, *X8*, *X5*, *X1*, *X4*, *X9*, *X13*, *X7*, and *X10*. Factors *X11* and *X12* appeared six times each; X8 appeared four times; *X4*, *X5*, and *X9* appeared three times each; *X7* and *X13* appeared twice; and *X10* appeared once. This indicates that at the city level, the development of urban built-up areas is predominantly influenced by the annual average relative humidity (*X11*) and the accumulated temperature above 0 °C (*X12*).

Table 6. Summary table of q values of each factor.

	X1	X2	X3	X4	X5	X6	X7	X8	X9	X10	X11	X12	X13	X14
GBA	0.1043 *	0.0621 *	0.0001	0.1073 *	0.5669 *	0.0485 *	0.0858 *	0.1442 *	0.1125 *	0.0523 *	0.2208 *	0.1642 *	0.1137 *	0.0319 *
Guangzhou	0.1992 *	0.0651 *	0.0005 *	0.1458 *	0.1206 *	0.0534 *	0.3453 *	0.3600 *	0.2256 *	0.0845 *	0.2053 *	0.3799 *	0.3253 *	0.0813 *
Shenzhen	0.2582 *	0.2239 *	0.0008	0.3838 *	0.1605 *	0.0180 *	0.1175 *	0.2684 *	0.2269 *	0.2152 *	0.1410 *	0.3135 *	0.2259 *	0.1294 *
Zhuhai	0.0547 *	0.0233 *	0.0010	0.0535 *	0.0213 *	0.0098 *	0.1381 *	0.0599 *	0.0943 *	0.1386 *	0.2282 *	0.1598 *	0.0893 *	0.0207 *
Foshan	0.0511 *	0.0303 *	0.0019 *	0.0622 *	0.6427 *	0.0643 *	0.1229 *	0.2020 *	0.3473 *	0.3125 *	0.4642 *	0.2857 *	0.2706 *	0.0274 *
Huizhou	0.0407 *	0.0208 *	0.0001 *	0.0350 *	0.4618 *	0.0111 *	0.1137 *	0.0758 *	0.0362 *	0.0293 *	0.2106 *	0.0755 *	0.0796 *	0.0036 *
Dongguan	0.2237 *	0.1454 *	0.0012 *	0.2278 *	0.1137 *	0.0428 *	0.0794 *	0.1637 *	0.2324 *	0.2185 *	0.1579 *	0.2519 *	0.1081 *	0.0453 *
Zhongshan	0.0573 *	0.0257 *	0.0018	0.0704 *	0.0654 *	0.0043 *	0.1869 *	0.0509 *	0.1867 *	0.1211 *	0.2021 *	0.0715 *	0.1482 *	0.0153 *
Jiangmen	0.0122 *	0.0063 *	0.0001	0.0123 *	0.5258 *	0.0042 *	0.0538 *	0.0234 *	0.0871 *	0.0622 *	0.1369 *	0.1143 *	0.0376 *	0.0143 *
Zhaoqing	0.0266 *	0.0068 *	0.0000	0.0171 *	0.0069 *	0.0071 *	0.0169 *	0.0443 *	0.0186 *	0.0177 *	0.0264 *	0.0265 *	0.0350 *	0.0183 *
Hong Kong	0.1512 *	0.0912 *	0.0009 *	0.1408 *	0.0304 *	0.0368 *	0.1315 *	0.1392 *	0.0483 *	0.0897 *	0.1720 *	0.2036 *	0.1620 *	0.0325 *
Macao	0.3710 *	0.1335 *	0.0363 *	0.2750 *	0.0958 *	0.0971 *	0.2406 *	0.1860 *	0.2687 *	0.2412 *	0.2704 *	0.2117 *	0.1145 *	0.2254 *

Note: "*" indicates $p < 0.05$.

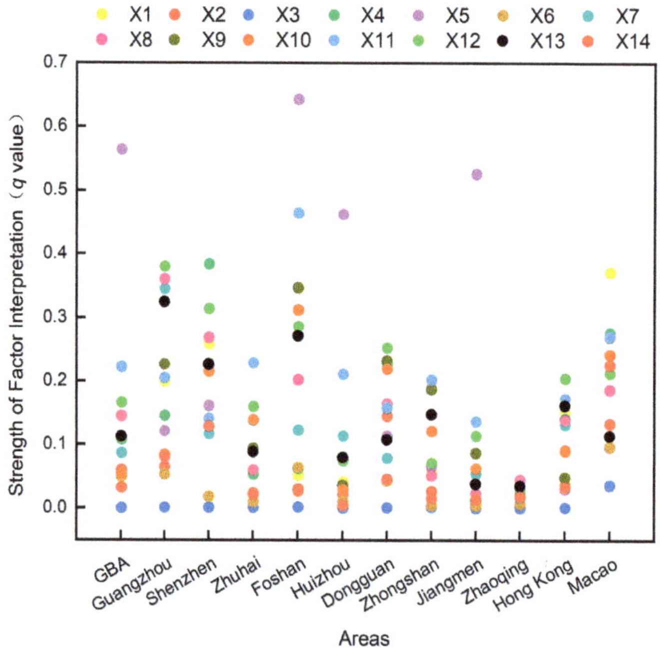

Figure 5. q value ranking map of the GBA.

Table 7. Summary of the frequencies of the main factors.

Factor	*X11*	*X12*	*X8*	*X1*	*X4*
Frequency	6	6	4	3	3
Factor	*X5*	*X9*	*X7*	*X13*	*X10*
Frequency	3	3	2	2	1

4. Discussion

From a regional perspective, the GBA has not yet fully tapped into its urban spatial development potential. Macao utilized 22.82 km^2 of suitable area in its urban built-up areas, which constituted 73.17% of all suitable areas within the Macao Special Administrative Region. This makes it the city with the highest proportion of utilized suitable areas in the GBA. Consequently, Macao has only 26.83% of its suitable construction area remaining, indicating the lowest urban spatial development potential at 26.83%. In contrast, Zhaoqing utilized only 61.41 km^2 of suitable area, a mere 0.44% of the total suitable area within administrative boundaries, marking the lowest proportion in the GBA. Accordingly, Zhaoqing possesses the highest urban spatial development potential among all cities in the GBA, at 99.56%. Meanwhile, the Hong Kong Special Administrative Region possesses a development potential of 86.48%, challenging prior studies that highlighted a scarcity of urban land [33–35]. Despite significant development in Hong Kong Island and in the Kowloon districts, substantial areas suitable for development still exist under natural constraints, particularly in the New Territories.

At the municipal level, average relative humidity and accumulated temperature above 0 °C significantly influence urban built-up areas. Previous studies have indicated that higher humidity exacerbates physiological stress from high temperatures, subsequently increasing mortality rates [36–38]. These findings suggest that humidity and temperature substantially affect human health and indirectly influence urban development. At the overall level of the GBA, silty sand content is the primary factor affecting spatial heterogeneity in urban areas. Similar to Zurui Ao et al. (2024)'s findings, the geological structure beneath cities plays a crucial role in urban subsidence, impacting surface stability and leading to structural damage [39]. This underscores the importance of soil conditions in urban development at the Bay Area level. Additionally, Juan Huang (2015) and other researchers partly attribute Hong Kong's limited built-up area to flawed land development policies [40,41], indicating that both natural conditions and urban policies significantly shape urban built-up areas [42], meriting further research.

This study's evaluation of urban spatial development potential under natural constraints presents several limitations. Firstly, there is significant room for improvement in the multi-indicator comprehensive evaluation approach. The selection, grading, and assignment of indicators introduce substantial uncertainty in the results. Despite efforts to objectify these processes, they inherently reflect subjective decisions by the evaluator, aiming to fulfill specific evaluation objectives but not necessarily capturing land suitability accurately. Secondly, the research data lacked refinement. First, the soil and elevation data are not synchronized in time, which, due to the absence of up-to-date officials, could impact the accuracy of the study. Furthermore, the spatial resolution of the data is insufficient for detailed analysis. Different sizes of spatial statistical grids affect the results of the model. Although the Optimal Parameters-based Geographical Detector is employed in this study, the selection of grid size for data extraction still directly influences data discretization and consequently impacts the analytical outcomes. Finally, the study does not consider the impact of other complex influences on urban areas. While it utilizes the differentiation and factor detection sub-models of the geographic detector model, which effectively identify key variables impacting urban development, it overlooks the complexity of urban systems. The urban environment is a complex and dynamic system, characterized by constant evolution and uncertainty. Its trajectory is influenced by a variety of factors, including climate change, technological innovation, and social unrest [43–45]. Urban development under natural constraints is typically influenced by multiple interacting factors, not just isolated variables. Moreover, in addition to external influences on cities, internal changes in cities also play a significant role in urban development [46].

To address the deficiencies identified in this study, future research should focus on several key areas: First, enhancing the evaluation indicator system is crucial. Further extensive research is necessary to refine the selection, grading, and assignment of land evaluation indicators, ensuring that they align with the evaluation's objectives and enhance both the comprehensiveness and accuracy of the indicators. Second, more refined data can be applied in future research. Employing finer grids would provide more detailed data, improving the model's accuracy in capturing local characteristics and dynamics in urban spatial development. Third, there is a need to consider the impact of different factors on urban built-up areas. For example, climate change would affect urban built-up areas, leading to changes within these areas themselves. This will enable a more comprehensive analysis and understanding of the complex interrelationships in the urban environment.

5. Conclusions

This study conducted a quantitative analysis of urban spatial development potential in the GBA, elucidating the relationship between the region's current urban space and its potential for future development under natural constraints. The findings indicate that Zhaoqing has the highest development potential at 99.56%, followed by Huizhou (98.99%), Jiangmen (98.04%), Zhuhai (88.96%), Hong Kong (86.48%), Zhongshan (86.41%), Guangzhou (81.93%), Foshan (76.72%), Dongguan (49.09%), and Shenzhen (45.28%), with Macao having the least at 26.83%. This ranking highlights significant variability in available urban development space across GBA cities under natural constraints. Further analysis reveals that different cities are affected by natural constraints to varying degrees. At the Bay Area level, urban development is predominantly influenced by silty sand content, average annual relative humidity, and accumulated temperature above 0 °C, contributing 0.5669,

0.2208, and 0.164 to spatial heterogeneity, respectively. At the city level, the same factors of relative humidity and temperature are the primary influences on urban built-up areas. The results of this study provide a scientific foundation for urban land space planning and the future development of urban forms in China during its transitional period. Future research should explore the effects of various influencing factors, including interactions between natural constraints and the impact of human policy factors, to promote sustainable and efficient urban spatial development.

Author Contributions: Conceptualization, Y.Z. (Yukui Zhang), T.L. and M.L.; methodology, Y.Z. (Yukui Zhang) and T.L.; software, Y.Z. (Yukui Zhang), Y.C. and J.Z.; validation, Y.Z. (Yukui Zhang); formal analysis, Y.Z. (Yukui Zhang), T.L. and J.Z.; investigation, Y.Z. (Yukui Zhang); resources, T.L. and Y.L.; data curation, Y.Z. (Yukui Zhang); writing—original draft preparation, Y.Z. (Yukui Zhang); writing—review and editing, Y.Z. (Yukui Zhang), T.L., J.Z., Y.Z. (Yicheng Zheng), X.W., M.L., Y.L., H.Y. and G.Z.; visualization, Y.Z. (Yukui Zhang), J.Z. and Y.C.; supervision, T.L.; project administration, T.L.; funding acquisition, T.L. and Y.L. All authors have read and agreed to the published version of the manuscript.

Funding: This research was funded by the National Natural Science Foundation of China (No. 42271299) and the National Key R&D Program of China (No. 2022YFC3800701).

Data Availability Statement: The data presented in this study are available upon request from the corresponding author. They are not publicly available because they are part of an ongoing study.

Conflicts of Interest: The authors declare no conflicts of interest.

References

1. Dong, G.L.; Ge, Y.B.; Jia, H.W.; Sun, C.Z.; Pan, S.Y. Land Use Multi-Suitability, Land Resource Scarcity and Diversity of Human Needs: A New Framework for Land Use Conflict Identification. *Land* **2021**, *10*, 1003. [CrossRef]
2. Xiong, J.X.; Wang, X.B.; Zhao, D.; Zhao, Y.Y. Spatiotemporal pattern and driving forces of ecological carrying capacity during urbanization process in the Dongting Lake area, China. *Ecol. Indic.* **2022**, *144*, 109486. [CrossRef]
3. Jin, X.X.; Wei, L.Y.; Wang, Y.; Lu, Y.Q. Construction of ecological security pattern based on the importance of ecosystem service functions and ecological sensitivity assessment: A case study in Fengxian County of Jiangsu Province, China. *Environ. Dev. Sustain.* **2021**, *23*, 563–590. [CrossRef]
4. Xiong, G.; Cao, X.; Hamm, N.A.S.; Lin, T.; Zhang, G.; Chen, B.J.S. Unbalanced Development Characteristics and Driving Mechanisms of Regional Urban Spatial form: A Case Study of Jiangsu Province, China. *Sustainability* **2021**, *13*, 3121. [CrossRef]
5. Cao, X.; Zhou, Y.J. Comprehensive Carrying Capacity of the Inland Node Cities along the Belt and Road. *Environ. Eng. Sci.* **2022**, *39*, 39–47. [CrossRef]
6. Shao, J.; Li, F.J.L. Multi-Function Tradeoffs of Land System in Urbanized Areas—A Case Study of Xi'an, China. *Land* **2021**, *10*, 640. [CrossRef]
7. Cao, X.; Shi, Y.; Zhou, L.; Tao, T.; Yang, Q.J.L. Analysis of Factors Influencing the Urban Carrying Capacity of the Shanghai Metropolis Based on a Multiscale Geographically Weighted Regression (MGWR) Model. *Land* **2021**, *10*, 578. [CrossRef]
8. Li, K.; Jin, X.; Ma, D.; Jiang, P.J.L. Evaluation of Resource and Environmental Carrying Capacity of China's Rapid-Urbanization Areas—A Case Study of Xinbei District, Changzhou. *Land* **2019**, *8*, 69. [CrossRef]
9. Wu, Y.; Zong, T.; Shuai, C.; Liao, S.; Jiao, L.; Shen, L. Does resource environment carrying capacity have a coercive effect on urbanization quality? Evidence from the Yangtze River Economic Belt, China. *J. Clean. Prod.* **2022**, *365*, 132612. [CrossRef]
10. Bao, K.; He, G.; Ruan, J.; Zhu, Y.; Hou, X. Analysis on the resource and environmental carrying capacity of coal city based on improved system dynamics model: A case study of Huainan, China. *Environ. Sci. Pollut. Res.* **2023**, *30*, 36728–36743. [CrossRef]
11. Świąder, M.; Lin, D.; Szewrański, S.; Kazak, J.K.; Iha, K.; Van Hoof, J.; Belčáková, I.; Altiok, S. The application of ecological footprint and biocapacity for environmental carrying capacity assessment: A new approach for European cities. *Environ. Sci. Policy* **2020**, *105*, 56–74. [CrossRef]
12. Bai, J.; Liu, S.; Wang, H.; Zhao, Y. Management, Evaluation of Urban Water Environment Carrying Capacity based on a Fuzzy Comprehensive Evaluation Model. *J. Environ. Account. Manag.* **2022**, *12*, 006.
13. Cui, Y. The coordinated relationship among industrialization, environmental carrying capacity and green infrastructure: A comparative research of Beijing-Tianjin-Hebei region, China. *Environ. Dev.* **2022**, *44*, 10075. [CrossRef]
14. Oezdemir, M.; Ayatac, H.; Ince, E.C. Proposed method for determining the population carrying capacity of cities: The city-mass index. *Proc. Inst. Civ. Eng.-Urban Des. Plan.* **2022**, *175*, 122–137.
15. Yang, Z.; Song, J.; Cheng, D.; Xia, J.; Li, Q.; Ahamad, M.I. Comprehensive evaluation and scenario simulation for the water resources carrying capacity in Xi'an city, China. *J. Environ. Manag.* **2019**, *230*, 221–233. [CrossRef] [PubMed]
16. De las Rivas Sanz, J.L.; Fernández-Maroto, M. Planning for Growth: Contradictions in the Framework of Economic and Urban Development from the "Spanish Miracle" (1959–1973). *J. Urban Hist.* **2023**, *49*, 41–59. [CrossRef]

17. Micheli, L.; Talavera, D.L.; Tina, G.M.; Almonacid, F.; Fernandez, E.F. Techno-economic potential and perspectives of floating photovoltaics in Europe. *Sol. Energy* **2022**, *243*, 203–214. [CrossRef]
18. Popescu, R.I. Study Regarding the Ways of Measuring Cities Competitiveness. *Econ. Ser. Manag.* **2011**, *14*, 288–303.
19. Kudelko, J.; Musial-Malago, M.J.C. The diversity of demographic potential and socioeconomic development of urban functional areas—Evidence from Poland. *Cities* **2022**, *123*, 103516. [CrossRef]
20. Cheng, H.-H.; Hsu, Y.-Y. Integrating spatial multi-criteria evaluation into the potential analysis of culture-led urban development €" A case study of Tainan. *Environ. Plan B-Urban Anal. City Sci.* **2021**, *49*, 335–357. [CrossRef]
21. Köstepen, A.; Öter, Z. Medical Tourism Potential in Turkey: The Case of Izmir City. In Proceedings of the 8th Silk Road International Conference "Development of Tourism in Black and Caspian Seas Regions", Tbilisi-Batumi, Georgia, 24–26 May 2013; Academic Publishers: Cambridge, MA, USA, 2016. Available online: https://www.researchgate.net/publication/309418800 (accessed on 10 March 2024).
22. Lin, B.; Ma, R. How does digital finance influence green technology innovation in China? Evidence from the financing constraints perspective. *J. Environ. Manag.* **2022**, *320*, 115833. [CrossRef]
23. Pan, D.; Bai, Y.; Chang, M.; Wang, X.; Wang, W.J.E. The technical and economic potential of urban rooftop photovoltaic systems for power generation in Guangzhou, China. *Energy Build.* **2022**, *277*, 112591. [CrossRef]
24. Koch, F.; Beyer, S.; Chen, C.Y. Monitoring the Sustainable Development Goals in cities: Potentials and pitfalls of using smart city data. *GAIA* **2023**, *32*, 47–53. [CrossRef]
25. Tan, F.; Gong, C.; Niu, Z.; Zhao, Y.J. An inquiry into urban carrying capacity of sustainable development demonstration belt of China: Multiscale evaluation and multidimensional interaction. *Sustain. Dev.* **2023**, *31*, 2892–2907. [CrossRef]
26. Lin, T.; Xue, X.Z.; Shi, L.Y.; Gao, L.J. Urban spatial expansion and its impacts on island ecosystem services and landscape pattern: A case study of the island city of Xiamen, Southeast China. *Ocean Coast. Manag.* **2013**, *81*, 90–96. [CrossRef]
27. Xu, J.; Zhao, X.; Liu, B.; Yi, L. Spatial-Temporal Characteristics and Driving Forces of Urban Sprawl for Major Cities of the Pearl River Delta Region in Recent 40 Year. *Acta Sci. Nat. Univ. Pekin.* **2015**, *51*, 1086–1131. (In Chinese)
28. Ju, H.; Zhang, S.; Yan, Y. Spatial pattern changes of urban expansion and multi-dimensional analysis of driving forces in the Guangdong-Hong Kong-Macao Greater Bay Area in 1980–2020. *Acta Geogr. Sin.* **2022**, *77*, 1086–1101. (In Chinese)
29. Department of Natural Resources of Guangdong Province, People's Republic of China (PRC). Technical Guidelines for Evaluation of Resource and Environmental Carrying Capacity and Suitability of Territorial Spatial Development of Guangdong Province (for Trial Implementation). Available online: https://nr.gd.gov.cn/zwgknew/tzgg/tz/content/post_3162101.html (accessed on 24 December 2020). (In Chinese)
30. Wang, A.G.; Ma, W.; Shi, Y.C. Study on the Potential Earthquake Deformation of Active Fault and the Safe Distance of Important Project Site to Active Fault. *J. Seismol. Res.* **2005**, *4*, 55–60+121. (In Chinese)
31. *GB/T 27963-2011*; Climatic Suitability Evaluating on Human Settlement. National Standard of the People's Republic of China: Beijing, China, 2011. (In Chinese)
32. Song, Y.; Wang, J.; Ge, Y.; Xu, C. An optimal parameters-based geographical detector model enhances geographic characteristics of explanatory variables for spatial heterogeneity analysis: Cases with different types of spatial data. *GISci. Remote Sens.* **2020**, *57*, 593–610. [CrossRef]
33. Hui, E.C.M.; Lam, M.C.M.; Ho, V.S. Development, Market Disequilibrium and Urban Land Shortages: Analysis of Policy and Patterns in Hong Kong. *J. Urban Plan. Dev.* **2006**, *132*, 80–88. [CrossRef]
34. Li, L.H.; Wong, S.K.; Cheung, K.S. Land supply and housing prices in Hong Kong: The political economy of urban land policy. *Environ. Plan. C-Gov. Policy* **2016**, *34*, 981–998. [CrossRef]
35. Ye, S. The land resource and landuse of Hong Kong. *Chin. Geogr. Sci.* **1998**, *8*, 12–24. [CrossRef]
36. Sobolewski, A.; Młynarczyk, M.; Konarska, M.; Bugajska, J. The influence of air humidity on the human heat stress in a hot environment. *Int. J. Occup. Saf. Ergon.* **2019**, *27*, 226–236. [CrossRef] [PubMed]
37. Curriero, F.C.; Heiner, K.S.; Samet, J.M.; Zeger, S.L.; Lisa, S.; Patz, J.A. Temperature and Mortality in 11 Cities of the Eastern United States. *Am. J. Epidemiol.* **2002**, *155*, 80–87. [CrossRef] [PubMed]
38. Guo, Y.J. The Role of Humidity in Associations of High Temperature with Mortality: A Multicountry, Multicity Study. *Environ. Health Perspect.* **2019**, *127*, 109001.
39. Ao, Z.; Hu, X.; Tao, S.; Hu, X.; Wang, G.; Li, M.; Wang, F.; Hu, L.; Liang, X.; Xiao, J.; et al. A national-scale assessment of land subsidence in China's major cities. *Science* **2024**, *384*, 301–306. [CrossRef] [PubMed]
40. Huang, J.; Shen, G.Q.; Zheng, H.W. Is insufficient land supply the root cause of housing shortage? Empirical evidence from Hong Kong. *Habitat Int.* **2015**, *49*, 538–546. [CrossRef]
41. Hui, E.C.M.; Ho, V.S.-M. Relationship between the land-use planning system, land supply and housing prices in Hong Kong. *Int. J. Strateg. Prop. Manag.* **2003**, *7*, 119–128.
42. Lin, T.; Li, X.H.; Zhang, G.Q.; Zhao, Q.J.; Cui, S.H. Dynamic analysis of island urban spatial expansionand its determinants: A case study of xiamen island. *Acta Geogr. Sin.* **2010**, *65*, 715–726.
43. De Roo, G. Being or becoming? That is the question! Confronting complexity with contemporary planning theory. In *A Planner's Encounter with Complexity*; De Roo, G., Silva, E.A., Eds.; Ashgate Publishing: Farnham, UK, 2010; pp. 19–40.
44. Mannucci, S.; Kwakkel, J.H.; Morganti, M.; Ferrero, M. Exploring potential futures: Evaluating the influence of deep uncertainties in urban planning through scenario planning: A case study in Rome, Italy. *Futures* **2023**, *154*, 103265. [CrossRef]

45. Rauws, W. Embracing Uncertainty without Abandoning Planning: Exploring an Adaptive Planning Approach for Guiding Urban Transformations. *DisP-Plan. Rev.* **2017**, *53*, 32–45. [CrossRef]
46. Carter, I.; Moroni, S. Adaptive and anti-adaptive neighbourhoods: Investigating the relationship between individual choice and systemic adaptability. *Environ. Plan. B Urban Anal. City Sci.* **2022**, *49*, 722–736. [CrossRef]

land

Article

Impact of Urban Expansion on Carbon Emissions in the Urban Agglomerations of Yellow River Basin, China

Zhenwei Wang [1], Yi Zeng [2], Xiaochun Wang [1,*], Tianci Gu [2] and Wanxu Chen [2]

1 School of Public Administration, Hubei University, Wuhan 430062, China
2 Hubei Key Laboratory of Regional Ecology and Environmental Change, School of Geography and Information Engineering, China University of Geosciences, Wuhan 430074, China
* Correspondence: wangxc@hubu.edu.cn

Abstract: Continued urban expansion (UE) has long been regarded as a huge challenge for climate change mitigation. However, much less is known about how UE affects carbon emissions (CEs), especially in the urban agglomerations of the Yellow River Basin (UAYRB), China. In this regard, this study introduced kernel density analysis, the Gini coefficient, and Markov chains to reveal the UE patterns and carbon emissions intensity (CEI) in the UAYRB at the county level, and explored the spatial heterogeneity of the impact of UE on CEI with the geographically and temporally weighted regression model. The results show that both CEI and UE in the UAYRB showed a steady growing trend during the study period. The kernel density of CEI and UE revealed that CEI in the UAYRB was weakening, while the UE rate continuously slowed down. The Gini coefficients of both CEI and UE in the UAYRB region were at high levels, indicating obvious spatial imbalance. The Markov transfer probability matrix for CEI with a time span of five years showed that CEI growth will still occur over the next five years, while that of UE was more obvious. Meanwhile, counties with a regression coefficient of UE on CEI higher than 0 covered the majority, and the distribution pattern remained quite stable. The regression coefficients of different urban landscape metrics on CEI in the UAYRB varied greatly; except for the landscape shape index, the regression coefficients of the aggregation index, interspersion and juxtaposition index, and patch density overall remained positive. These findings can advance the policy enlightenment of the high quality development of the Yellow River Basin.

Keywords: urban expansion; carbon emissions; landscape pattern index; geographically and temporally weighted regression; urban agglomerations; Yellow River Basin, China

1. Introduction

Global warming is a huge challenge facing humankind [1]. Greenhouse gases (GHGs) generated by human activities are the culprit [2]. It is reported that 2% of global urban areas generate approximately 75% of the global carbon emissions (CEs). The past century has witnessed a great change in urban population growth and urban expansion worldwide [3]. From 2000 to 2010, global urban areas grew by an average of 5694 km^2 per year, resulting in a net loss of 22.4 Tg carbon per year [4]. Urban expansion (UE) can also indirectly affect carbon stocks, which is difficult to fully quantify [5]. Research found that, from 1985 to 2015, approximately 12% and 9% of UE came at the expense of grassland and forest, respectively, while UE in China is expected to create a 1800 km corridor of coastal cities from Hangzhou to Shenyang [3]. China has been experiencing unprecedented urbanization since economic liberalization began in 1978 [6]. According to the National Bureau of Statistics, in 2015, China's urban built-up area was 1.6 times that in 2005, reaching 52,102 km^2. The non-negligible eco-environmental issues caused by continuing UE has concerned the global academics [7–9]. Urban agglomerations will be the main front of UE since it remains the primary form of urbanization until at least 2035 [10]. As the strategic core area of national economic development, urban agglomerations shoulder the historical responsibility of carrying the shift in the world's economic center of gravity [11,12]. Although agglomeration has promoted economies of scale and facilitated better infrastructure and services, the development of urban agglomerations has also been accompanied by unprecedented energy consumption, leading to growing challenges in eco-environmental issues related to climate change [13–16]. Owing to intensive human socioeconomic activities, urban agglomerations are inevitably becoming the major generators of CEs [10]. However, the spatial relationship between UE and CEs has not been comprehensively and thoroughly examined. Existing research has mostly focused on the impact of UE on CEs of city individuals, while studies are rare at the scale of urban agglomerations from the perspective of urban landscape metrics [17].

Currently, research is substantial concerning UE, and has mainly focused on the following aspects. (1) Characterization of UE. Quantitative indexes, such as expansion intensity and expansion direction, have been widely used to characterize the UE pattern [18,19]. Meanwhile, the extensive use of 3S has facilitated better understanding the spatiotemporal evolution of UE [20]. Jiao et al. (2018) proposed a new landscape metric to characterize the evolution process of UE and observed an increasingly decentralized spatial pattern [21]. (2) UE simulation. Since UE is a prevailing phenomenon worldwide, whether it exhibits some particular features arouses the attention of scholars. Seto et al. (2012) found that urban areas will increase by 1.2 million km^2 by 2030 under the current population density trend [3]. Guo et al. (2022) introduced the patch-based land use simulation (PLUS) model to simulated UE of the Harbin–Changchun urban agglomerations under ecological constraints and found that the PLUS model can better simulate UE at the scale of urban agglomerations [22]. Nevertheless, although exhibiting similar population or economic growth, countries will vary in the probabilities of UE. Seto et al. (2012) found that UE likelihood in individual countries tends to exhibit both high spatial variability and high spatial concentration, while some countries with low probabilities of UE show a high uncertainty of expansion pattern, such as Turkey [3]. (3) Drivers of UE. Natural, socioeconomic, and political factors are widely believed to be the traditional driving factors of land use change [23]. However, the unified analysis framework of drivers of urban land expansion has not reached a consensus [22]. Meanwhile, factors affecting UE vary greatly among regions. For example, the leading contributor to UE in China is GDP, while that in India and Africa is population [18]. (4) Impacts of UE on the eco-environment. The impacts of UE on the eco-environment can be divided into two perspectives of direct and indirect impacts. UE can directly affect natural habitats by converting them to urban use [24]. Mao et al. (2018) revealed that urbanization-induced wetland loss reached 2883 km^2 from 1990 to 2010 [25]. Among them, economically developed urban agglomerations are the hotspots of urbanization-induced wetland loss in China. Furthermore, UE from 1992 to 2016 has resulted in an average 0.8% loss of dryland habitat quality [26]. Liu et al. (2019) conducted research at the global scale to reveal the impacts of UE on terrestrial net primary productivity (NPP) and found that global terrestrial NPP loss equaled ~9% of the CEs from fossil fuel and cement emissions worldwide [4]. Although research is substantial on the direct impact of UE on the eco-environment, the indirect impact of UE on natural habitats is more severe than the direct one [9]. It is estimated that cropland expansion contributes the greatest to natural area losses globally, while the indirect impact of UE on natural area losses is significantly underestimated [9]. Ren et al. (2022) found that dryland UE has indirectly affected nearly 60% of threatened species [26]. We can conclude that the environmental impacts of UE have been extensively studied, while studies on the impacts of UE on CEs still leave much to be carried out, especially at the urban agglomerations level as they have become the major driver of CEs.

With the rapid development of urbanization in China, the impact of UE on CEs is one of the current research hotspots in the field of environment and economy [27–30], and understanding the impact of UE on CEs is crucial to the formulation of effective low-carbon development policies. However, little is known about how UE affects CEs, especially in the UAYRB. To this end, this study introduced kernel density, the Gini coefficient, and Markov chains to reveal the UE and CEI patterns in the UAYRB at the county level, and explored the spatial heterogeneity of the impact of UE on CEI with the geographically and temporally weighted regression (GTWR) model, and this provides a new perspective and methodological framework for urban CEs research. The impact of UE on CEs in the UAYRB in China has profound economic implications and practical significance. First, understanding the correlation between UE and CEs allows for more efficient resource allocation. Secondly, encouraging low-carbon sustainable urban development can promote the development of green technologies and industries, which, in turn, can create new jobs and promote economic growth. In terms of policymaking, identifying the impact of UE on CEs can provide decision makers with information that can help them to formulate effective urban planning and environmental policies to curb CEs while managing UE. This study theoretically explores and analyzes the impact of UE on CEs to provide new ideas for the theory of urban development and low-carbon transition. In the past, related studies lacked an exploration of the UAYRB as a specific region, and, by revealing the dynamic impact of UE on CEs, we provide experience and an important reference for CEs reduction studies in similar regions. And, methodologically, the kernel density analysis and Gini coefficient proposed for use in this study provide theoretical support and methodological exploration for further research in this area.

Studies on the impacts of UE on CEs are not scarce. Krayenhoff et al. (2018) found a nonlinear interaction between GHG-induced warming and corresponding UE in American cities [8]. Liu and Zhang (2022) found that the positive trade-offs between UE and ecological construction could mitigate CEs growth in China's urban agglomerations [10]. Cheng et al. (2022) revealed that cities with a larger population tend to have lower per capita CEs [31]. Actually, compared with low-density communities, high-density communities tend to have lower per capita energy use [32,33]. Urban population expansion is usually accompanied by urban area expansion, but an easily overlooked fact is that urban areas around the world are expanding twice as fast as their populations on average [18,34]. Much has been carried out to curb such expansion. However, contrary to what we believe—that land use planning is an effective way to curb UE [35]—it actually stimulates fragmented UE [36], though it is regarded as uneconomic, inefficient, and environmentally unfriendly [7,37]. The landscape pattern index can well characterize the UE pattern, which can facilitate better understanding the impacts of UE on CEs.

How urban area expansion affects CEs in the urban agglomerations area with the rapid development of urbanization remains an urgent issue that requires a prompt solution, most notably regarding urban agglomerations of the Yellow River Basin (UAYRB). As an important ecological barrier, food base, and economic zone in China, the Yellow River Basin concentrates a large amount of chemical, energy, and production industries, making it an ecologically fragile region with a high concentration of CEs and pollution [38]. The inherent problems of unbalanced development and unfriendly ecological environment in the Yellow River Basin require being solved by the development of urban agglomerations from point to area [39]. Within this context, this research intends to address the gaps mentioned above by proposing the following research objectives. (1) What are the UE patterns and CEs intensity (CEI) in the UAYRB? (2) What is the spatial heterogeneity of the impact of UE on CEI in the UAYRB?

2. Materials and Methods

2.1. Study Area

From west to east, the Yellow River flows through 9 provinces and regions, with a total length of 5464 km and an area of 795,000 km^2, accounting for 8.28% of the country's total. It is the second longest river in China and the fifth longest in the world. The Yellow River and its coastal basin are among the most important birthplaces of the Chinese nation, and they are also the main battlefield of national environmental civilization construction. In 2021, the State Council of the Central Committee of the Communist Party of China upgraded the ecological protection and high-quality development of the Yellow River Basin to a major national strategy, highlighting the strategic position of the Yellow River Basin in the overall situation of national development and socialist modernization [39]. The UAYRB consists of 7 urban agglomerations, Shandong Peninsula Urban Agglomerations (SPUA), Central Plains Urban Agglomerations (CPUA), Guanzhong Plain Urban Agglomerations (GPUA), Lanxi Urban Agglomerations (LXUA), Jinzhong Urban Agglomerations (JZUA), Hohhot–Baotou–Ordos–Yulin Urban Agglomerations (HBOYUA), and Ningxia Urban Agglomerations (NXUA) along the Yellow River (Figure 1). The seven urban agglomerations account for about 33.6% of the area of the Yellow River Basin, and the proportion of the population and main economic indicators in China's urban agglomerations is roughly between 20% and 25%. At the same time, that of the nine provinces and regions in the Yellow River Basin is as high as 60~70% [40]. It is a high-density population-gathering area in the Yellow River Basin, an important area for high-quality economic development, a heavy-loaded area for inheriting the Yellow River culture and promoting Chinese civilization, and a key area for comprehensive environmental pollution control and ecological protection. Therefore, it has a very important strategic position in the high-quality development of the Yellow River Basin.

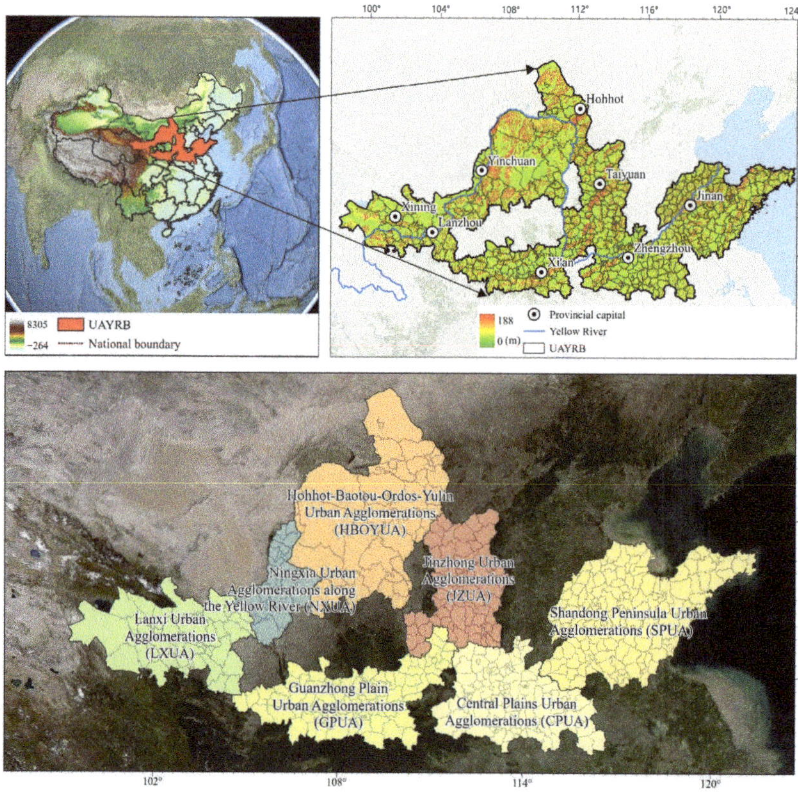

Figure 1. Study area.

2.2. Data Sources

The data in this study involved CEs and land use data. CEs data were sourced from the open-source data inventory (https://db.cger.nies.go.jp/dataset/ODIAC/, accessed on 7 May 2024) [41]; these CEs data successfully estimate the spatial distribution of fossil fuel CEs on a global scale by combining night-time lighting data and emission location profiles of individual power plants using an innovative emission modeling approach. The spatial resolution is 1000 m. The land use change data were downloaded from the Data Center for Resources and Environmental Sciences and the Chinese Academy of Sciences (https://www.resdc.cn/, accessed on 7 May 2024). The spatial resolution is 30 m. This study used the proportion of urban land and the landscape pattern index to characterize UE [42]. The proportion of urban land is the ratio of urban land area to the area of the study unit, and the landscape pattern indexes were calculated in Fragstats v4.2.1 (Oregon State University, Corvallis, OR, USA).

2.3. Methods

This study aims to analyze the impact of UE on CEs. We adopt a series of models to assess the association between them to achieve this goal. Through these models (Figure 2), we expect to reveal the impact of UE on CEs and provide a scientific basis for formulating low-carbon development policies. To demonstrate the impact of UE on CEs in the UAYRB, this study uses the urban land area share and the landscape pattern index to characterize the spatiotemporal changes of UE, and, at the same time, we use the CEI as an indicator to analyze the changes in CEs. By calculating these indicators, we use kernel density analysis to reveal the dynamic change patterns of UE and CEI and analyze the imbalance of UE and CEI based on the Gini coefficient. In addition, we apply Markov chains to predict future changes in the trends of UE and CEI and reveal the impact of UE on CEI through the GTWR model. The core analytical approach of this study focuses on showing the changing patterns of UE and CEs in the UAYRB and further exploring the mechanism of UE's impact on CEI.

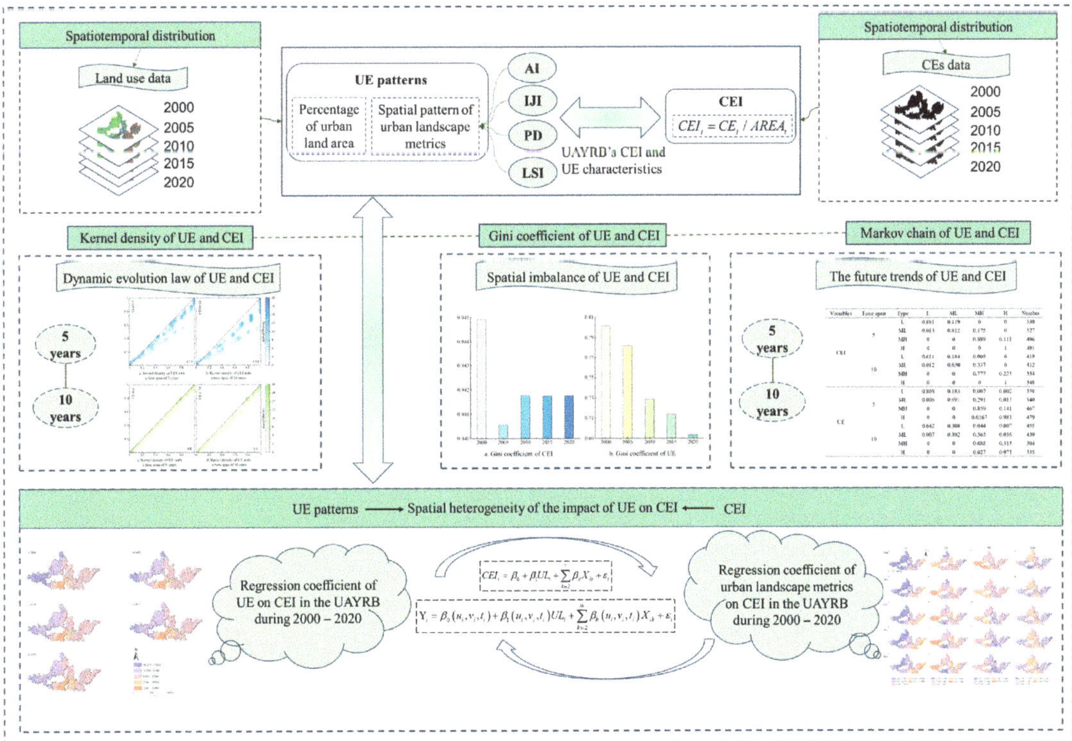

Figure 2. Basic flowchart of the method.

2.3.1. UE Measurement

This study uses the proportion of urban land to visualize the spatial pattern and changes in UE. We also introduced four landscape metrics—aggregation index (AI), interspersion and juxtaposition index (IJI), patch density (PD), and landscape shape index (LSI)—to describe the urban landscape in the UAYRB [43–45]. The AI examines the connectivity between patches of each landscape type and is used to measure the degree of aggregation of urban land. The IJI is used to assess the degree of interlocking and juxtaposition between different land types. PD expresses the density of urban land in the landscape, which can reflect the heterogeneity and fragmentation of urban land. The LSI, as a landscape shape index, reflects the changes in the shape of land use patches in the process of urbanization, and a high LSI value may indicate that UE has led to a more complex shape of land use patches. These indicators can visualize the changing characteristics of UE in the UAYRB.

2.3.2. CEI Assessment

The measurement of CEI in this study is expressed as the ratio of total CEs to area for each unit [46], and the unit is t CO_2/km^2. The calculation equation is as follows:

$$CEI_t = CE_t / AREA_t \tag{1}$$

where CEI_t is the CEs intensity in year t; CE_t is CEs in year t; and $AREA_t$ is the area of administrative units.

2.3.3. Kernel Density

Kernel density estimation is a nonparametric estimation, which has the advantage of not relying on sample data and has been widely used by scholars [47,48]. The dynamic evolutionary trend of the research elements is estimated by the density function to reveal its changing rules. Here, we introduced the method of kernel density estimation to explore the dynamic evolution law of UE and CEI in the UAYRB. We used the Gaussian function as the kernel function and referred to Yang et al. (2023a) to calculate the kernel density analysis [49].

2.3.4. Gini Coefficient

The Gini coefficient is often used to measure the degree of imbalance and insufficiency of regional economic development [50]. This study introduced the Gini coefficient to analyze the CEI and the spatial imbalance of the UE of the UAYRB. The Gini coefficient is applied to assess the uneven spatial distribution of UE and CEI, thus revealing the uneven spatial distribution of UE on CEI. It can provide us with a new perspective to understand the relationship between UE and CEI.

2.3.5. Markov Chain

Markov chains are characterized by non-aftereffects, analyzing the transfer patterns based on the current state of change in the study elements and thus predicting the future trend of change [49,51]. Markov chains are introduced to analyze the trend of interconversion between different orders of magnitude by constructing Markov transfer probability matrices. Here, we introduced Markov chains to analyze the future trends of UE and CEI in the UAYRB. Markov chain analysis can help us understand the spatiotemporal evolution of UE and thus predict the future trend of CEI. By analyzing the Markov chain of UE, we can reveal the long-term impact of UE on CEI and provide a scientific basis for future urban planning and CEI control.

2.3.6. GTWR Model

This study employed the global regression (ordinary least squares method, OLS) model without considering spatial factors to investigate the impact of UE on CEI in Chinese urban agglomerations. The model was constructed as follows:

$$CEI_i = \beta_0 + \beta_1 UL_i + \sum_{k=2}^{7} \beta_k X_{ki} + \varepsilon_i \tag{2}$$

where i represents a district or county, CEI_i represents the CEs intensity value of a district or county i, UL_i represents the UL of district or county i, X_{ki} represents the urban landscape pattern index affecting the CEI of district or county i, and ε_i is the residual term.

The changes in urban land and CEI are panel data with multiple time series, while changes in urban land do not immediately cause changes in CEI, and their effects may have some lag effect [52]. The GWR only considers the spatial relationship of the cross-sectional data at a single time, which is insufficient for studying time-series data. GTWR solves this problem. The equation is as follows:

$$Y_i = \beta_0(u_i, v_i, t_i) + \beta_1(u_i, v_i, t_i) UL_i + \sum_{k=2}^{m} \beta_k(u_i, v_i, t_i) X_{ik} + \varepsilon_i \tag{3}$$

where (u_i, v_i, t_i) is the sample point with spatial coordinates and timestamps, m is the number of samples, ε_i is the random error term, and β_k is the estimated local regression coefficients. To make the time-series variation more apparent, in this study, the regression results of every five years are averaged to compare the changes in driving mechanisms on a ten-year scale.

3. Results

3.1. CEI in the UAYRB

The CEIs of the UAYRB in 2000, 2005, 2010, 2015, and 2020 were 10.131, 18.191, 26.816, 30.991, and 32.712 t CO_2/km^2, respectively, showing a steady growing trend during the study period. The highest CEI in 2020 reached 1732.712 t CO_2/km^2, almost three times as much as that in 2000. However, the mean CEI of the UAYRB is no more than 40 t CO_2/km^2, with the highest value of only 32.712 t CO_2/km^2 in 2020. The standard deviation of CEI in the study period exhibited a rocketing upward trend, indicating an obvious spatial difference in the CEI of the UAYRB. Spatially, the high CEI was concentrated in the SPUA and CPUA and scattered in the GZUA and JZUA during the study period (Figure 3). As time went by, the CEI in these areas showed an increasing trend, while the proportion of high CEI increased accordingly. The SPUA and CPUA showed the most obvious change characteristics. The regions with low CEI west of the HBOYUA, west and the north of the LXUA, and south of the NXUA exhibited no evident changes during the study period.

3.2. UE Patterns in the UAYRB

The proportions of urban land in the UAYRB in 2000, 2005, 2010, 2015, and 2020 were 6.7%, 7.8%, 10.3%, 10.8%, and 11.4%, respectively, showing an obvious increasing trend during the study period (Figure 4). Similar to the spatial pattern of CEI, the regions with a high proportion of urban land were concentrated in the SPUA and CPUA. Meanwhile, some major cities of the other urban agglomerations also have a high proportion of urban land. Over time, the spatial pattern of the regions with a high proportion of urban land showed a westward expansion trend. The SPUA and CPUA were still urban agglomerations with the most significant changes. Compared to the study period of 2000–2005, 2005–2020 witnessed the proportion of urban land increasing more evidently.

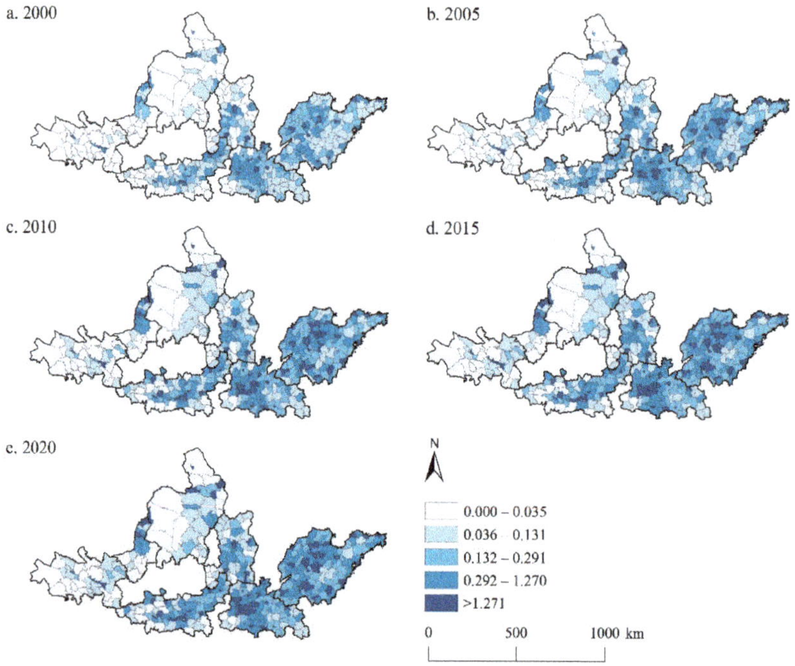

Figure 3. CEI in the UAYRB during 2000–2020. (**a**–**e**) are the CEI in the UAYRB in 2000, 2005, 2010, 2015, and 2020, respectively.

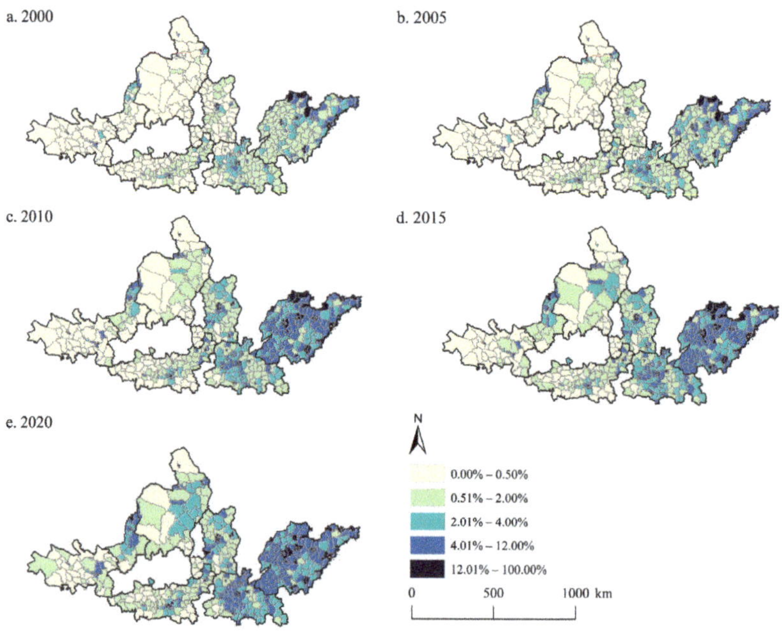

Figure 4. UE patterns in the UAYRB during 2000–2020. (**a**–**e**) are the UE pattern in the UAYRB in 2000, 2005, 2010, 2015, and 2020, respectively.

To better reveal the spatial pattern of UE, we introduced four landscape metrics of AI, IJI, PD, and LSI to describe the urban landscape in the UAYRB (Figure 5). Statistically, the AI in the UAYRB in 2000, 2005, 2010, 2015, and 2020 was 96.648, 96.655, 96.244, 96.284, and 95.982, respectively, exhibiting an overall declining trend but remaining relatively stable during the study period. Spatially, an AI greater than 90 covered the vast majority of regions, while an AI greater than 96 shrunk quite obviously over time, especially in the west of the UAYRB. The counties with a high IJI were most concentrated in the HBOYUA, LXUA, and NXUA, while the SPUA and CPUA clustered in low -IJI counties. Notably, the spatial pattern of a high IJI showed an eastward expansion trend over time. PD in the UAYRB in 2000, 2005, 2010, 2015, and 2020 was 0.026, 0.029, 0.052, 0.054, and 0.065, respectively, showing a steady growing trend. Spatially, SPUAs remained the regions with higher PD, while counties with high PD tended to move west over time. The LSI in the UAYRB in 2000, 2005, 2010, 2015, and 2020 was 2.112, 2.286, 3.854, 4.046, and 4.404, respectively, also showing a growing trend. However, the spatial distribution pattern of the LSI was quite different from the former three. In 2000 and 2005, an LSI lower than 15 covered almost the whole UAYRB. As time passed, counties with a high LSI mainly occurred in the HBOYUAH and the JZUA.

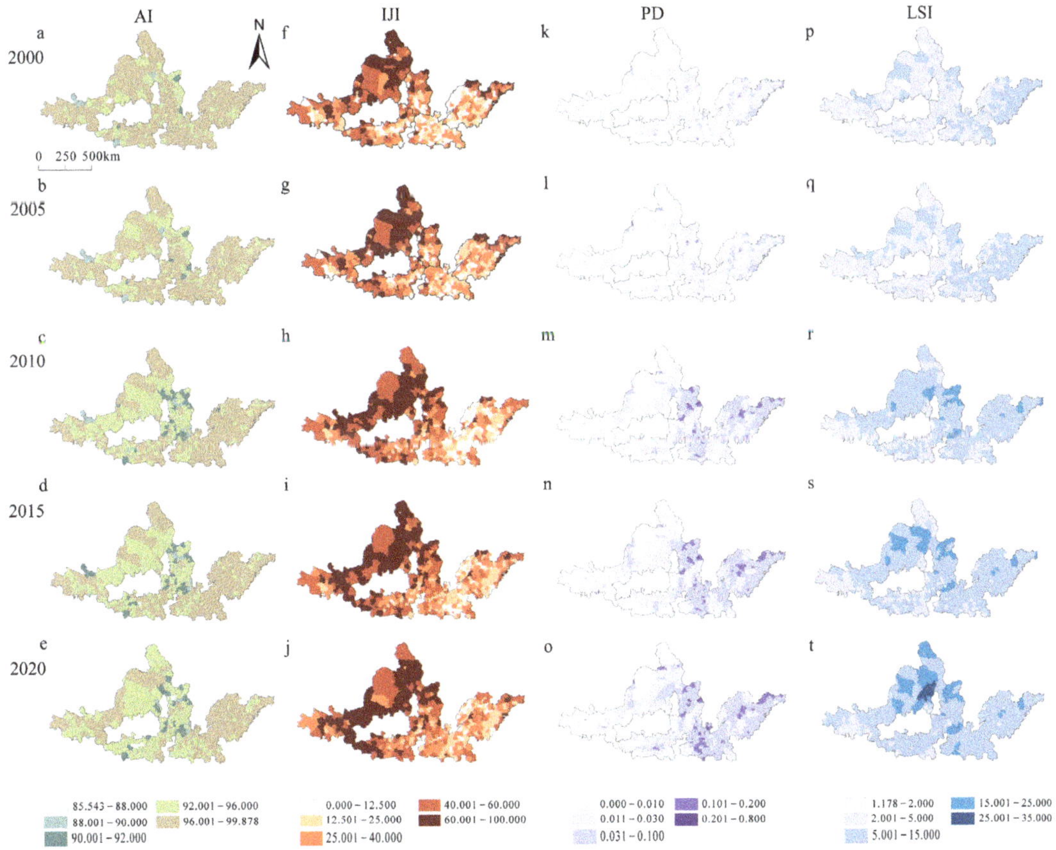

Figure 5. Spatial pattern of urban landscape metrics in the UAYRB during 2000–2020.

3.3. *Kernel Density of CEI and UE*

The kernel density contours of the normalized CEI and UE are shown in Figure 6. If the position of the contour lines is located near the 45° diagonal, it indicates that the study elements did not change drastically during the study period. Figure 6a and 6b show the kernel density results for CEI over a 5-year and 10-year period, respectively. It was found that the region of high values of CEI was continuously shifting to lower values, which also indicates the weakening of CEI in the UAYRB. Furthermore, this was further corroborated by the fact that the CEI shifted sharply to lower values over time. Figure 6c and 6d show the trend of UE transfer over a 5-year and 10-year time span, respectively. The results show that the kernel density contour peak was located near 45°, indicating the basic stability of the UE and the continuous slowing down of the UE rate in the UAYRB region. However, it is worth noting that, in some of the lower-value areas, a significant shift in UE to higher values may occur, which suggests that the potential for large-scale UE still exists in some counties.

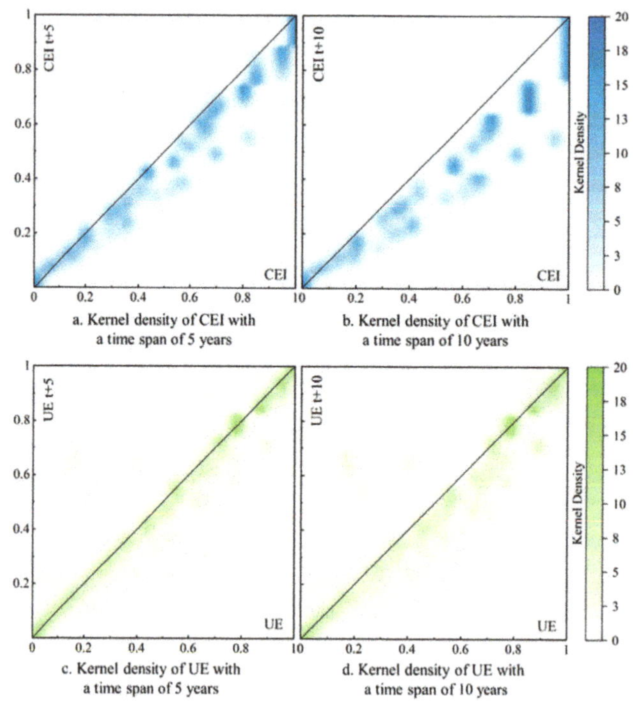

Figure 6. Kernel density contour map of CEI and UE. (**a**) is the kernel density of CEI with a time span of 5 years. (**b**) is the kernel density of CEI with a time span of 10 years. (**c**) is the kernel density of UE with a time span of 5 years. (**d**) is the kernel density of UE with a time span of 10 years.

3.4. Gini Coefficient of CEI and UE

The Gini coefficient of CEI and UE in the UAYRB during the study period is shown in Figure 7. The result of the Gini coefficient reveals the imbalance of CEI and UE. Generally speaking, the Gini coefficients of both CEI and UE were at high levels, indicating an obvious spatial imbalance of CEI and UE in the UAYRB region. As for the change in the Gini coefficient of CEI (Figure 7a), the CEI generally showed a downward trend, but the change was not dramatic. The Gini coefficient of CEI decreased by 3.357% during the study period, indicating that a slight decrease in the spatial imbalance of CEI occurred. The change in the Gini coefficient of UE showed an overall decreasing trend, and its decrease was greater than that of CEI, which is the same as in previous studies. Specifically, the decline in UE amounted to 13.506%. Urban land expansion has gradually taken on a regionally balanced pattern, partly attributed to China's strategy of balanced regional development.

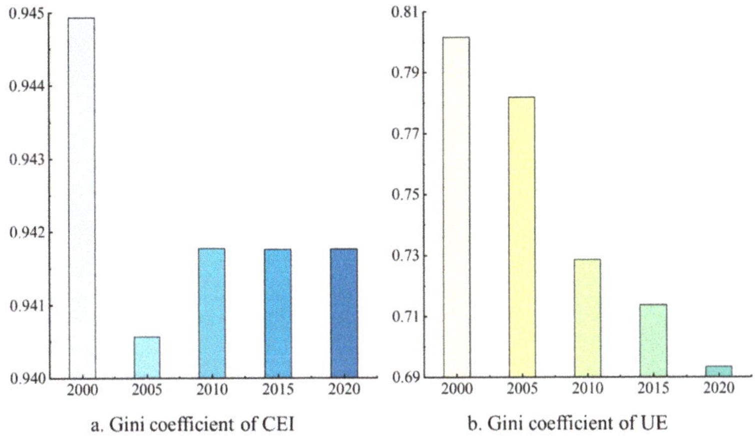

Figure 7. Gini coefficient of CEI and UE.

3.5. Markov Chain of CEI and UE

Markov chains are used to predict the future tendency of CEI and UE, and the constructed Markov transfer probability matrix is shown in Table 1. We broke CEI and UE into L (Low), ML (Middle-Low), MH (Middle-High), and H (High) using quartiles as interval points. The Markov transfer probability matrix for CEI with a period of five years showed that CEI growth will still occur over the next five years. However, when combined with the Markov transfer probability matrix for CEI with a period of 10 years, this tendency to converge to higher values kept weakening. This demonstrates the slowing growth rate of CEI as well. From the Markov transfer probability matrix of UE, the tendency to converge to higher values was more obvious than that of CEI for the five years, indicating the existence of the probability of larger-scale UE still occurring in the next five years. As the time span increases, the likelihood of UE shifting to higher values increases further, indicating that the UE remained in an increasing process.

Table 1. Markov transfer probability matrix of CEI and UE.

Variables	Time Span	Type	L	ML	MH	H	Number
CEI	5	L	0.881	0.119	0	0	530
		ML	0.013	0.812	0.175	0	527
		MH	0	0	0.889	0.111	496
		H	0	0	0	1	491
	10	L	0.811	0.184	0.005	0	419
		ML	0.012	0.650	0.337	0	412
		MH	0	0	0.777	0.223	354
		H	0	0	0	1	348
UE	5	L	0.808	0.183	0.007	0.002	558
		ML	0.006	0.691	0.291	0.013	540
		MH	0	0	0.859	0.141	467
		H	0	0	0.0167	0.983	479
	10	L	0.642	0.308	0.044	0.007	455
		ML	0.007	0.392	0.565	0.036	439
		MH	0	0	0.688	0.313	304
		H	0	0	0.027	0.973	335

Notes: L denotes low-level type; ML denotes medium-low-level type; MH denotes medium-high-level type; H denotes high-level type.

3.6. Spatial Heterogeneity of the Impact of UE on CEI

The regression coefficient of UE on CEI in the UAYRB in 2000, 2005, 2010, 2015, and 2020 was 1.111, 1.149, 1.210, 1.277, and 1.382, respectively, showing an increasing trend (Figure 8). Overall, counties with regression coefficients above 0 covered the majority, maintaining a relatively stable distribution pattern. Although areas with regression coefficients lower than 0 also exhibited similar stable distribution patterns, mainly concentrated west of the UAYRB, the proportion of the lowest regression coefficient shrunk to a few counties in the west of LXUA over time. The regression coefficient in the whole LXUA and NXUA, the majority of areas of HBOYUA, remained negative during the study period. By comparison, the regression coefficient in the JZUA and the SPUA was always positive. The regression coefficient stayed between 0.001 and 2.000. In 2020, the regression coefficient of two counties in the CPUA shifted from positive to negative. The spatial pattern of the negative regression coefficient in the west of the GZUA showed a decrease at first but then an increasing trend.

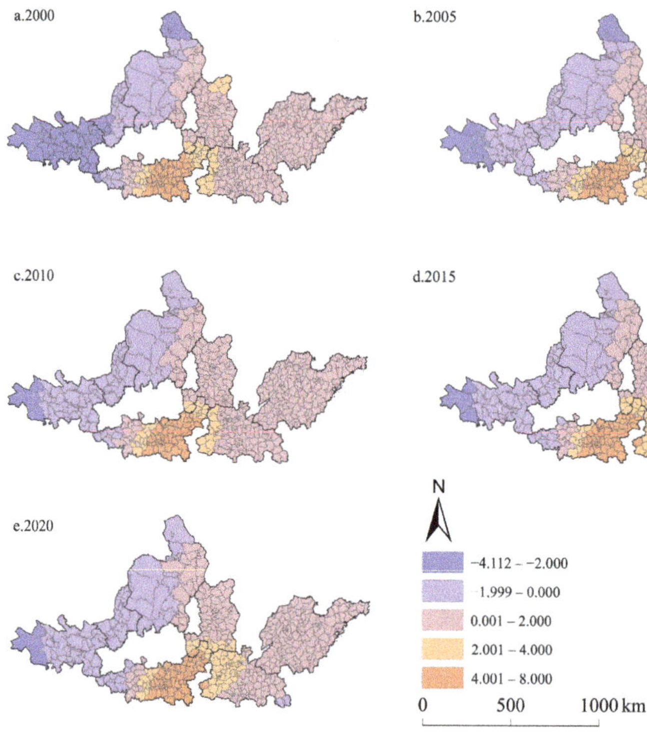

Figure 8. Regression coefficient of UE on CEI in the UAYRB during 2000–2020. (**a**–**e**) are the regression coefficient of UE on CEI in the UAYRB in 2000, 2005, 2010, 2015, and 2020, respectively.

The regression coefficients of different urban landscape metrics on CEI in the UAYRB varied greatly (Figure 9). Except for the LSI, the overall regression coefficients of the remaining three landscape metrics were positive. The regression coefficients of the AI on CEI in 2000, 2005, 2010, 2015, and 2020 were 14.500, 18.821, 22.325, 24.808, and 27.128, respectively, showing a growing trend. Spatially, similar to the distribution pattern of the regression coefficient of UE on CEI, the CPUA and SPUA concentrated the high regression coefficient. The proportion of the low regression coefficient in the west showed a tendency to shrink. The regression coefficients of the IJI on CEI in 2000, 2005, 2010, 2015, and 2020 were 0.358, 0.433, 0.475, 0.512, and 0.528, respectively, showing an increasing trend. However, the regression coefficients in some areas were negative, concentrated in the east of the SPUA and the whole GZUA. It is noteworthy that the regression coefficients in the whole GZUA was positive in 2000. Meanwhile, the spatial distribution of negative regression coefficients in the east of the SPUA and the north of the HBOYUA narrowed over time. The regression coefficient of PD on CEI showed a steady downward trend during the study period, but remained positive. However, its spatial distribution pattern changed significantly. The regression coefficient in the west of the UAYRB was generally higher than that in the east. The south of the CPUA remained the region with a low regression coefficient while the southwest of the NXUA remained the region with a high regression coefficient. The regression coefficient of the LSI on CEI remained negative. The most significant changes happened in the GZUA, while the remaining regions almost stayed the same with limited changes. Regions with the lowest regression coefficient tended to narrow while the high regression coefficient expanded to the east of the GZUA and the southwest of the CPUA.

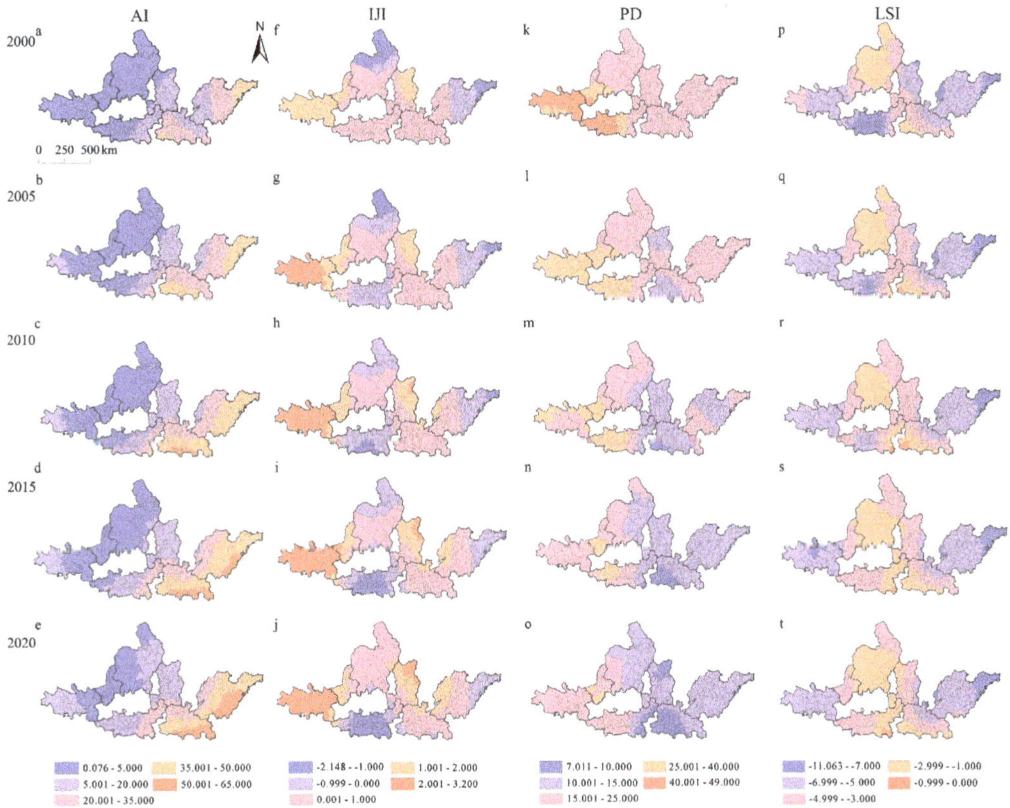

Figure 9. Regression coefficient of urban landscape metrics on CEI in the UAYRB during 2000–2020.

4. Discussion

4.1. Comparing with Previous Findings

Compared with previous studies, this study explored the impact of UE on CEs more comprehensively. It filled the research gap on the UAYRB, a region where current related studies have focused on specific scale units, such as global [53], national [54,55], provincial [56], and prefectural [57–60]. And, fewer scholars have studied the impact of UE on CEI using the county as a unit of study in the UAYRB. Some scholars also found that the CEs of the UAYRB increased sharply in general from 2000 to 2020 [61,62], and showed a pattern of high in the middle-east and low in the northwest [63], and that UE exacerbated CEs [53], which was in line with the results of this study. In addition, the urban spatial pattern influences the CEs status [64], and existing studies have focused on three aspects of urban landscape patterns: urban sprawl, urban form complexity, and urban compactness. It has been recognized that urban form (the spatial pattern and structural features of urban land use) is related to urban CEs; nevertheless, there are only a few studies that have empirically assessed the direct impacts and their relationship with CEs. Fang et al. (2015) found a positive association between the growth of urban areas and CO_2 in 30 provincial capitals in China, that an increase in urban continuity had a dampening effect on CEs, that an increase in the complexity of urban shapes positively affected CEs, that measures to make China's existing cities more compact may actually help to reduce CEs, and that increased fragmentation or irregularity of urban form may lead to increased CEs [60]. Ou et al. (2013) quantitatively analyzed the aspects of the impact of different urban shapes on CEs and found that increased fragmentation or irregularity of urban shapes may also lead to more CEs and that a compact development pattern of urban land use helps to reduce CEs [59]. The regression results of this study were the similar as those of the above studies. It was found that the regression coefficient between the LSI and CEI was negative, indicating that there was a negative association between the LSI and CEI, and the larger the value of the LSI, the more complex the landscape shape and the fewer CEs. This suggests that increased continuity in the urban landscape has a dampening effect on CEs. The positive regression coefficient between the AI and CEI indicates that CEs increase as urban form complexity increases. High LSI values may indicate a higher complexity of landscape patches, representing a richer ecosystem or green space environment within the city, helping to absorb and fix a large amount of CEs, thereby reducing CEs [65]. High AI values may indicate that urban interior landscape patches are more closely connected, reflecting the increasing degree of urbanization, resulting in more intensive and convenient transportation, concentration of industrial activities and intensification of land use, which may lead to more transportation emissions and energy consumption, as well as more industrial and commercial consumer electronics, thereby increasing CEs [66]. In addition, the positive regression coefficients between PD and the IJI and CEI also indicate that the increase in these urban landscape metrics may lead to increased CEs.

4.2. Spatiotemporal Differences in the Impact of UE on CEI

UE is a prevailing phenomenon in most parts of the world [67]. It is projected that urban areas are expected to triple between 2000 and 2030 [3]. Corresponding to UE, CEs in the past century have also witnessed a rocketing rise [68], leading to an increase in global average surface temperature [69]. It is estimated that such a global warming phenomenon will continue in the next century [70]. Urban areas account for 71% of global CEs, while that will increase to 76% by 2030 [71]. However, the situation in China is even more severe. It is reported that 84% of China's commercial energy usage comes from urban areas [72], while every 1% increase in urbanization in China is associated with a 1% increase in CEs [73,74]. Although research found that some counties in China have achieved the decoupling relationship of urbanization and CEs, such a status is not stable [75]. Much has been carried out to reveal the eco-environmental effects of UE. However, much less is known about the spatial heterogeneity of the impact of UE on CEs.

This study analyzed the impact of UE on CEs and their spatial heterogeneity in the UAYRB, China. The results show that both the CEI and UE of the UAYRB showed an obvious increasing trend during the study period. The spatial distribution of them also showed similar distribution patterns, both concentrated in the SPUA and CPUA. These two urban agglomerations have developed relatively rapidly, and their favorable location and vast plain area enable them to take the lead in the entire Yellow River Basin. Moreover, the kernel density contours of the normalized CEI and UE revealed the weakening of CEI and the potential for large-scale UE in some counties in the UAYRB. Xing et al. (2022) found the same evolution trend of CEI and a similar kind of UE in the urban agglomeration in south-central Liaoning province [76]. They found that, from 2010 to 2016, the average annual expansion rate of the urban agglomerations in south-central Liaoning province increased from 3.93% to 5.8%, with the expansion intensity increased from 0.211 to 0.525 [76]. However, the UE rate in the UAYRB region continuously slowed down. That is also the case in the Changsha–Zhuzhou–Xiangtan Urban Agglomerations [77–79]. Tian and Zhao (2024) found that UE in the Changsha–Zhuzhou–Xiangtan Urban Agglomerations decreased in 2015–2020 compared to the previous period [79]. Meanwhile, the finding in the three coastal agglomerations in China by Wen et al. (2019) also confirmed that UE is a major trend in the development of urban agglomerations [80]. Generally, affected by differences in urban resource endowment, development mode, and development direction, the UE and CEs of different cities also differ in the development process, showing spatial imbalance among regions. With the advance of time, the degree of CEs will change and the evolution trajectory of different cities will also change with different development processes. Meanwhile, in this study, the Gini coefficients of both CEI and UE were at high levels, indicating an obvious spatial imbalance of CEI and UE in the UAYRB region. It is noteworthy that UE has gradually taken on a regionally balanced pattern in the UAYRB during the study period, which can be partly attributed to the implementation of some development strategies in China. Such a phenomenon can also be found in the Changsha–Zhuzhou–Xiangtan Urban Agglomerations [79]. After the proposal of a "two-type society" in 2007, the CEs reduction of urban land in urban agglomerations has gradually achieved remarkable results [81]. Meanwhile, since the 18th National People's Congress, "ecological civilization construction" has been carried out, carbon-saving and emission reduction work have been continuously promoted, and the CEs of all districts and counties have been effectively controlled.

The synergistic relationship between UE and CEs has been confirmed in many studies. Tian and Zhao (2024) adopted Pearson correlation analysis and found that there was a high positive correlation between urban land area and CEs; that is, an increase in urban land area would lead to a simultaneous increase in CEs [79]. Xing et al. (2022) also confirmed the same synergistic expansion state of UE and CEs [80]. However, such synergies will change over time. As shown in this study, the Markov chain of CEI and UE revealed the slowing growth rate of CEI, while UE remained a continuing increasing process. This also proves to some extent that the effect of emission reduction policies implementation may be beginning to show. However, the spatial distribution of the impacts of UE on CEI may reveal that such policy effects may vary greatly among regions. In areas where UE and CEI values are high, the impacts of UE on CEI remain mostly positive with a lower regression coefficient, while, in places where UE and CEI values are low, the impacts of UE on CEI remain mostly negative with a higher regression coefficient. The difference in the response degree to the policy, the intensity of implementation, and the level of technology in different regions are the main reasons for the difference in the effectiveness of policy implementation. Areas with a relatively backward economic level, such as the LXUA and NXUA, have policies that restrict economic development but favor environmental protection that are often not well implemented. Meanwhile, policies are often ineffective when implemented due to the insufficient level of technology to support the low-carbon development of energy-consuming industries in the urbanization process in these regions. Therefore, the spatiotemporal differences in the impact of UE on CEs can be quite different, leading to significant spatial differences in the high-quality development level of the UAYRB, with the high-quality development index in the middle and lower reaches higher than that in the middle and upper reaches [82]. This conclusion is unanimously supported by Fang (2020) and Sun et al. (2022). Fang (2020) found that the higher upstream the Yellow River Basin, the lower the development of urban agglomerations [40], while Sun et al. (2022) revealed that the coordination degree of the upstream urban agglomerations is lower than that in the middle and lower reaches [39].

4.3. Policy Implications

Within the context of dual-carbon goals and new urbanization, ensuring stable economic growth while reducing CEs is one of the most important tasks for achieving high-quality urban development. UE is one of the most important manifestations of urban agglomerations development, and a reduction in the level of CEs in response to UE is conducive to the low-carbon development of urban agglomerations [79]. This study analyzed the impact of UE on CEs in the UAYRB region and proposed policy recommendations for low-carbon development in the following areas.

Although the UE rate in the UAYRB has slowed down, it is still in the expanding progress and some counties have the potential for large-scale UE. Once the process is out of control, it will inevitably lead to an unnecessary waste of resources and inefficient use of land. The expansion of urban land may also bring the possibility of occupying cultivated land, forest land, and other ecological land, which further aggravates the ecological and environmental problems of the city [83]. Therefore, it is highly recommended to formulate relevant legal policies to strictly control the urban sprawl. Meanwhile, strong penalties for illegal sprawl should be applied to enforce the policy to the end. In addition, urban planning is an important means of limiting urban sprawl [84], and construction land development needs to be carried out in strict accordance with the requirements of urban planning to ensure the sustainable development of cities. Local governments should abandon the idea of land finance and prioritize the healthy governance of cities and the realization of green and sustainable development to meet the needs of the people's happy lives [85].

Due to the needs of economic development, some regions are still at a stage of high CEs. In addition, although relevant energy-saving and emission reduction measures have been taken, they have not achieved the desired effect of emission reduction. For these areas, land resources should be used economically and intensively to avoid the disorderly expansion of urban land in the study area and realize the compact development of urban space. The proportion of tertiary industry should be increased, some high-polluting industries should be eliminated, advanced technologies should be introduced to improve CEs efficiency, and research and development of key technologies should be accelerated to achieve CEs reduction. At the same time, green areas should be increased to achieve carbon sink increase so as to achieve a strong decoupling as soon as possible and to enter into the stage of a low-carbon economy.

According to the spatial pattern of urban land use in the urban agglomerations of the UAYRB from 2000 to 2020 in the results of the study, including the AI, IJI, PD, and LSI, first of all, the overall decreasing trend of the AI indicates that the urban landscape of the UAYRB as a whole shows a certain degree of a discrete trend during the study period, which may be related to factors such as UE and land use changes. Second, the spatial pattern of the IJI shows a tendency to expand to the east. PD and the LSI continued to grow during the study period. According to the regression coefficients of urban landscape indicators on CEI in the research results, we can focus on areas with high regression coefficients, prioritize the formulation of targeted environmental protection policies and measures to cope with the trend of their growth in urban landscape indices, and design targeted environmental protection programs according to the distribution of the regression coefficients in different areas. For example, ecological protection and greening can be strengthened in CPUA and SPUA areas. In the process of policy formulation, it is necessary to consider the trends and spatial distribution patterns of the above indicators, strengthen the ecological protection and restoration of urban fringe areas, optimize the urban layout, and rationally guide urban development and low-carbon transformation.

4.4. Limitation and Future Directions

Although we tried to reveal the impact of UE on CEs in UAYRB with a series of analysis models, certain limitations still exist in this research. First of all, this study only analyzed the impact of UE on CEI, while the impacts of UE on CEs per GDP, CEs per person, or total CEs were still not clear. Different ways of quantifying CEs may produce more pronounced differences in results. In addition, this study explored the impact of the urban landscape pattern index on CEs with the GTWR model, while the internal mechanism of the impact of UE on CEs has not yet been clarified. Meanwhile, in this study, the impact of UE on CEs was analyzed only at the county scale, while a smaller scale can capture more regional differences. Therefore, the next step can be to conduct a smaller-scale analysis based on improved data accuracy to capture more features of regional differences. These limitations point to future research directions.

5. Conclusions

In this study, we analyzed the UE patterns and CEI in the UAYRB at the county level with kernel density, the Gini coefficient, and Markov chains, and revealed the spatial heterogeneity of the impact of UE on CEI with the GTWR model. The results show that CEI and UE in the UAYRB both showed a steady growing trend during the study period. The kernel density contours of the normalized CEI and UE revealed that the region with high values of CEI was continuously shifting to lower values, while the UE rate continuously slowed down in the UAYRB region. The Gini coefficients of both CEI and UE were at high levels, indicating obvious spatial imbalance in the UAYRB region. The Markov transfer probability matrix for CEI with a time span of five years showed that CEI growth will still occur over the next five years, while that of UE was more obvious, indicating the existence of the probability that larger-scale UE will still occur in the next five years. Counties with a regression coefficient of UE on CEI higher than 0 covered the majority, remaining as having a relatively stable distribution pattern. The regression coefficients of different urban landscape metrics on CEI in the UAYRB varied greatly; except for the LSI, the regression coefficients of the AI, IJI, and LSI remained positive. The study findings can enlighten policy implications for the high-quality development of the Yellow River Basin. The above analysis showed that, during the study period, CEs and UE in the UAYRB have continued to increase, showing obvious uneven spatial distribution. Looking forward to the next five years, the trend of CEs and UE will continue to grow, while the growth rate of CEs will gradually slow down and UE may be further accelerated in the following decade. The urban landscape showed a clear trend of discretization. Based on the urban landscape index and regression results, the characteristics of each region of the UAYRB must be fully considered in order to rationally guide urban development paths and realize low-carbon transformation.

Author Contributions: Conceptualization, Z.W. and W.C.; methodology, Z.W., Y.Z., T.G., X.W. and W.C.; software, Y.Z., T.G. and X.W.; validation, Z.W. and W.C.; formal analysis, Z.W. and W.C.; investigation, Z.W. and W.C.; resources, Z.W. and W.C.; data curation, Y.Z., T.G. and X.W.; writing—original draft preparation, Y.Z., T.G. and X.W.; writing—review and editing, Y.Z., T.G. and X.W.; visualization, Y.Z., T.G. and X.W.; supervision, Z.W. and W.C.; project administration, Z.W. and W.C.; funding acquisition, Z.W. and W.C. All authors have read and agreed to the published version of the manuscript.

Funding: This research was supported by the Ministry of education of Humanities and Social Science project (Granted No. 19YJC630179).

Data Availability Statement: The data that support the findings of this study are available from the corresponding author upon reasonable request. The data are not publicly available due to privacy restrictions.

Acknowledgments: The authors would like to thank the anonymous reviewers for their constructive comments on improving this paper.

Conflicts of Interest: The authors declare no conflicts of interest.

References

1. Berchin, I.I.; Valduga, I.B.; Garcia, J.; de Andrade Guerra, J.B.S.O. Climate change and forced migrations: An effort towards recognizing climate refugees. *Geoforum* **2017**, *84*, 147–150. [CrossRef]
2. Le Quéré, C.; Jackson, R.B.; Jones, M.W.; Smith, A.J.P.; Abernethy, S.; Andrew, R.M.; De-Gol, A.J.; Willis, D.R.; Shan, Y.; Canadell, J.G.; et al. Temporary reduction in daily global CO_2 emissions during the COVID-19 forced confinement. *Nat. Clim. Chang.* **2020**, *10*, 647–653. [CrossRef]
3. Seto, K.C.; Güneralp, B.; Hutyra, L.R. Global forecasts of urban expansion to 2030 and direct impacts on biodiversity and carbon pools. *Proc. Natl. Acad. Sci. USA* **2012**, *109*, 16083–16088. [CrossRef] [PubMed]
4. Liu, X.; Pei, F.; Wen, Y.; Li, X.; Wang, S.; Wu, C.; Cai, Y.; Wu, J.; Chen, J.; Feng, K.; et al. Global urban expansion offsets climate-driven increases in terrestrial net primary productivity. *Nat. Commun.* **2019**, *10*, 5558. [CrossRef] [PubMed]
5. DeFries, R.; Rosenzweig, C. Toward a whole-landscape approach for sustainable land use in the tropics. *Proc. Natl. Acad. Sci. USA* **2010**, *107*, 19627–19632. [CrossRef] [PubMed]
6. Huang, L.; Yan, L.; Wu, J. Assessing urban sustainability of Chinese megacities: 35 years after the economic reform and open-door policy. *Landsc. Urban Plan.* **2016**, *145*, 57–70. [CrossRef]
7. Tong, L.; Hu, S.; Frazier, A.E.; Liu, Y. Multi-order urban development model and sprawl patterns: An analysis in China, 2000–2010. *Landsc. Urban Plan.* **2017**, *167*, 386–398. [CrossRef]
8. Krayenhoff, E.S.; Moustaoui, M.; Broadbent, A.M.; Gupta, V.; Georgescu, M. Diurnal interaction between urban expansion, climate change and adaptation in US cities. *Nat. Clim. Chang.* **2018**, *8*, 1097–1103. [CrossRef]
9. van Vliet, J. Direct and indirect loss of natural area from urban expansion. *Nat. Sustain.* **2019**, *2*, 755–763. [CrossRef]
10. Liu, G.; Zhang, F. How do trade-offs between urban expansion and ecological construction influence CO_2 emissions? New evidence from China. *Ecol. Indic.* **2022**, *141*, 109070. [CrossRef]
11. Fang, C.; Zhou, C.; Gu, C.; Chen, L.; Li, S. Theoretical analysis of interactive coupled effects between urbanization and eco-environment in mega-urban agglomerations. *Acta Geogr. Sin.* **2016**, *71*, 531–550. [CrossRef]
12. Kii, M. Projecting future populations of urban agglomerations around the world and through the 21st century. *Npj Urban Sustain.* **2021**, *1*, 10. [CrossRef]
13. Bettencourt, L.M.; Lobo, J.; Helbing, D.; Kühnert, C.; West, G.B. Growth, innovation, scaling, and the pace of life in cities. *Proc. Natl. Acad. Sci. USA* **2007**, *104*, 7301–7306. [CrossRef] [PubMed]
14. Bettencourt, L.M.; Lobo, J.; Strumsky, D. Invention in the city: Increasing returns to patenting as a scaling function of metropolitan size. *Resour. Policy* **2007**, *36*, 107–120. [CrossRef]
15. Meerow, S. Double exposure, infrastructure planning, and urban climate resilience in coastal megacities: A case study of Manila. *Environ. Plan. A* **2017**, *49*, 2649–2672. [CrossRef]
16. Mi, Z.; Guan, D.; Liu, Z.; Liu, J.; Viguié, V.; Fromer, N.; Wang, Y. Cities: The core of climate change mitigation. *J. Clean. Prod.* **2018**, *207*, 582–589. [CrossRef]
17. Cengiz, S.; Görmüş, S.; Oğuz, D. Analysis of the urban growth pattern through spatial metrics; Ankara City. *Land Use Policy* **2022**, *112*, 105812. [CrossRef]
18. Seto, K.C.; Fragkias, M.; Güneralp, B.; Reilly, M.K. A Meta-Analysis of Global Urban Land Expansion. *PLoS ONE* **2011**, *6*, e23777. [CrossRef] [PubMed]
19. Beatriz, A.; Daunt, P.; Sanna, T.; Bürgi, M.; Hersperger, A. Urban expansion and forest reserves: Drivers of change and persistence on the coast of São Paulo State (Brazil). *Land Use Policy* **2021**, *101*, 105189.
20. Bennett, M.M.; Smith, L.C. Advances in using multitemporal night-time lights satellite imagery to detect, estimate, and monitor socioeconomic dynamics. *Remote Sens. Environ.* **2017**, *192*, 176–197. [CrossRef]
21. Jiao, L.; Liu, J.; Xu, G.; Dong, T.; Gu, Y.; Zhang, B.; Liu, Y.; Liu, X. Proximity Expansion Index: An improved approach to characterize evolution process of urban expansion. *Comput. Environ. Urban* **2018**, *70*, 102–112. [CrossRef]
22. Guo, R.; Wu, T.; Wu, X.; Luigi, S.; Wang, Y. Simulation of Urban Land Expansion Under Ecological Constraints in Harbin-Changchun Urban Agglomeration, China. *Chin. Geogr. Sci.* **2022**, *32*, 438–455. [CrossRef]
23. Lambin, E.F.; Meyfroidt, P. Land use transitions: Socio-ecological feedback versus socio-economic change. *Land Use Policy* **2010**, *27*, 108–118. [CrossRef]
24. Güneralp, B.; Reba, M.; Hales, B.U.; Wentz, E.A.; Seto, K.C. Trends in urban land expansion, density, and land transitions from 1970 to 2010: A global synthesis. *Environ. Res. Lett* **2020**, *15*, 044015. [CrossRef]
25. Mao, D.; Wang, Z.; Wu, J.; Wu, B.; Zeng, Y.; Song, K.; Yi, K.; Luo, L. China's wetlands loss to urban expansion. *Land Degrad. Dev.* **2018**, *29*, 2644–2657. [CrossRef]
26. Ren, Q.; He, C.; Huang, Q.; Shi, P.; Zhang, D.; Güneralp, B. Impacts of urban expansion on natural habitats in global drylands. *Nat. Sustain.* **2022**, *5*, 869–878. [CrossRef]
27. Hong, S.; Hui, E.C.M.; Lin, Y. Relationship between urban spatial structure and carbon emissions: A literature review. *Ecol. Indic.* **2022**, *144*, 109456. [CrossRef]
28. Liu, B.; Tian, C.; Li, Y.; Song, H.; Ma, Z. Research on the effects of urbanization on carbon emissions efficiency of urban agglomerations in China. *J. Clean. Prod.* **2018**, *197*, 1374–1381. [CrossRef]
29. Liu, X.; Xu, H.; Zhang, M. The effects of urban expansion on carbon emissions: Based on the spatial interaction and transmission mechanism. *J. Clean. Prod.* **2024**, *434*, 140019. [CrossRef]

30. Svirejeva-Hopkins, A.; Schellnhuber, H.J. Urban expansion and its contribution to the regional carbon emissions: Using the model based on the population density distribution. *Ecol. Model.* **2008**, *216*, 208–216. [CrossRef]
31. Cheng, L.; Mi, Z.; Sudmant, A.; Coffman, D. Bigger cities better climate? Results from an analysis of urban areas in China. *Energy Econ.* **2022**, *107*, 105872. [CrossRef]
32. Norman, J.; MacLean, H.L.; Kennedy, C.A. Comparing High and Low Residential Density: Life-Cycle Analysis of Energy Use and Greenhouse Gas Emissions. *J. Urban Plan. Dev.* **2006**, *132*, 10–21. [CrossRef]
33. Marshall, J.D. Energy-Efficient Urban Form. *Environ. Sci. Technol.* **2008**, *42*, 3133–3137. [CrossRef] [PubMed]
34. Angel, S.; Parent, J.; Civco, D.L.; Blei, A.; Potere, D. The dimensions of global urban expansion: Estimates and projections for all countries, 2000–2050. *Prog. Plan.* **2011**, *75*, 53–107. [CrossRef]
35. You, H.; Yang, X. Urban expansion in 30 megacities of China: Categorizing the driving force profiles to inform the urbanization policy. *Land Use Policy* **2017**, *68*, 531–551. [CrossRef]
36. Shen, X.; Wang, X.; Zhang, Z.; Lu, Z.; Lv, T. Evaluating the effectiveness of land use plans in containing urban expansion: An integrated view. *Land Use Policy* **2019**, *80*, 205–213. [CrossRef]
37. Zhao, P. Sustainable urban expansion and transportation in a growing megacity: Consequences of urban sprawl for mobility on the urban fringe of Beijing. *Habitat. Int.* **2010**, *34*, 236–243. [CrossRef]
38. Qiu, M.; Yang, Z.; Zuo, Q.; Wu, Q.; Jiang, L.; Zhang, Z.; Zhang, J. Evaluation on the relevance of regional urbanization and ecological security in the nine provinces along the Yellow River, China. *Ecol. Indic.* **2021**, *132*, 108346. [CrossRef]
39. Sun, J.; Cui, Y.; Zhang, H. Spatio-temporal pattern and mechanism analysis of coupling between ecological protection and economic development of urban agglomerations in the Yellow River Basin. *J. Nat. Resour.* **2022**, *37*, 1673–1690. [CrossRef]
40. Fang, C. Spatial Organization Pattern and High-Quality Development of Urban Agglomeration in the Yellow River Basin. *Econ. Geogr.* **2020**, *40*, 1–8.
41. Roten, D.; Wu, D.; Fasoli, B.; Oda, T.; Lin, J.C. An Interpolation Method to Reduce the Computational Time in the Stochastic Lagrangian Particle Dispersion Modeling of Spatially Dense XCO_2 Retrievals. *Earth Space Sci.* **2021**, *8*, e2020EA001343. [CrossRef]
42. Tan, R.; Liu, Y.; Liu, Y.; He, Q.; Ming, L.; Tang, S. Urban growth and its determinants across the Wuhan urban agglomeration, central China. *Habitat. Int.* **2014**, *44*, 268–281. [CrossRef]
43. Feng, F.; Wang, L.; Hou, W.; Yang, R.; Zhang, S.; Zhao, W. Analyzing the dynamic changes and causes of greenspace landscape patterns in Beijing plains. *Ecol. Indic.* **2024**, *158*, 111556. [CrossRef]
44. Zhao, F.; Li, H.; Li, C.; Cai, Y.; Wang, X.; Liu, Q. Analyzing the influence of landscape pattern change on ecological water requirements in an arid/semiarid region of China. *J. Hydrol.* **2019**, *578*, 124098. [CrossRef]
45. Xie, Y.; Yu, M.; Bai, Y.; Xing, X. Ecological analysis of an emerging urban landscape pattern—Desakota: A case study in Suzhou, China. *Landscape Ecol.* **2006**, *21*, 1297–1309. [CrossRef]
46. Liu, J.; Li, S.; Ji, Q. Regional differences and driving factors analysis of carbon emission intensity from transport sector in China. *Energy* **2021**, *224*, 120178. [CrossRef]
47. Quah, D.T. Empirics for Growth and Distribution: Stratification, Polarization, and Convergence Clubs. *J. Econ. Growth* **1997**, *2*, 27–59. [CrossRef]
48. Wu, J.; Wu, Y.; Cheong, T.S. New evidence on the convergence and regional clusters in China: A weighted continuous distribution dynamics approach. *Appl. Econ.* **2021**, *53*, 976–995. [CrossRef]
49. Yang, L.; Chen, W.; Zeng, J.; Pan, S.; Zhong, Y.; Gu, T. Regional differences and driving forces of ecosystem health in Yangtze River Basin, China. *Environ. Sci. Pollut. Res.* **2023**, *30*, 70985–71000. [CrossRef]
50. Meng, F.; Zhao, D.; Zhang, X. A fair consensus adjustment mechanism for large-scale group decision making in term of Gini coefficient. *Eng. Appl. Artif. Intel.* **2023**, *126*, 106962. [CrossRef]
51. Yang, L.; Fang, C.; Chen, W.; Zeng, J. Urban-rural land structural conflicts in China: A land use transition perspective. *Habitat. Int.* **2023**, *138*, 102877. [CrossRef]
52. du Toit, M.J.; Kotze, D.J.; Cilliers, S.S. Landscape history, time lags and drivers of change: Urban natural grassland remnants in Potchefstroom, South Africa. *Landsc. Ecol.* **2016**, *31*, 2133–2150. [CrossRef]
53. Sarkodie, S.A.; Owusu, P.A.; Leirvik, T. Global effect of urban sprawl, industrialization, trade and economic development on carbon dioxide emissions. *Environ. Res. Lett.* **2020**, *15*, 034049. [CrossRef]
54. Cheng, Z.; Hu, X. The effects of urbanization and urban sprawl on CO_2 emissions in China. *Environ. Dev. Sustain.* **2023**, *25*, 1792–1808. [CrossRef]
55. Hanif, I. Impact of fossil fuels energy consumption, energy policies, and urban sprawl on carbon emissions in East Asia and the Pacific: A panel investigation. *Energy Strateg. Rev.* **2018**, *21*, 16–24. [CrossRef]
56. Zhang, L.; Zhang, J.; Li, X.; Zhou, K.; Ye, J. The Impact of Urban Sprawl on Carbon Emissions from the Perspective of Nighttime Light Remote Sensing: A Case Study in Eastern China. *Sustainability* **2023**, *15*, 11940. [CrossRef]
57. Wu, Y.; Li, C.; Shi, K.; Liu, S.; Chang, Z. Exploring the effect of urban sprawl on carbon dioxide emissions: An urban sprawl model analysis from remotely sensed nighttime light data. *Environ. Impact Assess.* **2022**, *93*, 106731. [CrossRef]
58. Carpio, A.; Ponce-Lopez, R.; Lozano-García, D.F. Urban form, land use, and cover change and their impact on carbon emissions in the Monterrey Metropolitan area, Mexico. *Urban Clim.* **2021**, *39*, 100947. [CrossRef]
59. Ou, J.; Liu, X.; Li, X.; Chen, Y. Quantifying the relationship between urban forms and carbon emissions using panel data analysis. *Landsc. Ecol.* **2013**, *28*, 1889–1907. [CrossRef]

60. Fang, C.; Wang, S.; Li, G. Changing urban forms and carbon dioxide emissions in China: A case study of 30 provincial capital cities. *Appl. Energy* **2015**, *158*, 519–531. [CrossRef]
61. Xu, J.; Wang, H.; Zhao, K.; Li, Z. Evaluation of provincial carbon-neutral capacities in the Yellow River basin using DPSIR. *Sci. Rep.* **2022**, *12*, 18180. [CrossRef] [PubMed]
62. Ouyang, X.; Zhu, X. Spatiotemporal characteristics of urban land expansion in Chinese urban agglomerations. *Acta Geogr. Sin.* **2020**, *75*, 571–588.
63. Tian, M.; Chen, Z.; Wang, W.; Chen, T.; Cui, H. Land-use carbon emissions in the Yellow River Basin from 2000 to 2020: Spatio-temporal patterns and driving mechanisms. *J. Environ. Res. Public Health* **2022**, *19*, 16507. [CrossRef] [PubMed]
64. Zhang, Y.; Linlin, X.; Weining, X. Analyzing spatial patterns of urban carbon metabolism: A case study in Beijing, China. *Landsc. Urban Plan.* **2014**, *130*, 184–200. [CrossRef]
65. Zhang, X.; Zhang, X.; Li, D.; Lu, L.; Yu, H. Multi-Scenario Simulation of the Impact of Urban Land Use Change on Ecosystem Service Value in Shenzhen. *Acta Ecol. Sin.* **2022**, *42*, 2086–2097.
66. Peng, H.; Wang, Y.; Wang, Z. Temporal-spatial evolution of the integrated landscape patterns in Zhengzhou Metropolitan Area based on nighttime light data. *J. Appl. Ecol.* **2024**, 1–13.
67. Seto, K.C.; Sánchez-Rodríguez, R.; Fragkias, M. The New Geography of Contemporary Urbanization and the Environment. *Annu. Rev. Environ. Resour.* **2010**, *35*, 167–194. [CrossRef]
68. Wang, W.-Z.; Liu, L.-C.; Liao, H.; Wei, Y.-M. Impacts of urbanization on carbon emissions: An empirical analysis from OECD countries. *Energy Policy* **2021**, *151*, 112171. [CrossRef]
69. Qiao, W.; Lu, H.; Zhou, G.; Azimi, M.; Yang, Q.; Tian, W. A hybrid algorithm for carbon dioxide emissions forecasting based on improved lion swarm optimizer. *J. Clean. Prod.* **2020**, *244*, 118612. [CrossRef]
70. Sun, W.; Ren, C. The impact of energy consumption structure on China's carbon emissions: Taking the Shannon–Wiener index as a new indicator. *Energy Rep.* **2021**, *7*, 2605–2614. [CrossRef]
71. Shi, K.; Chen, Y.; Li, L.; Huang, C. Spatiotemporal variations of urban CO_2 emissions in China: A multiscale perspective. *Appl. Energy* **2018**, *211*, 218–229. [CrossRef]
72. Dhakal, S. Urban energy use and carbon emissions from cities in China and policy implications. *Energy Policy* **2009**, *37*, 4208–4219. [CrossRef]
73. Li, S.; Ying, Z.; Zhang, H.; Ge, G.; Liu, Q. Comprehensive Assessment of Urbanization Coordination: A Case Study of Jiangxi Province, China. *Chin. Geogr. Sci.* **2019**, *29*, 488–502. [CrossRef]
74. Li, S.; Zhou, C.; Wang, S. Does modernization affect carbon dioxide emissions? A panel data analysis. *Sci. Total Environ.* **2019**, *663*, 426–435. [CrossRef]
75. Pan, S.; Guo, J.; Ou, M. Exploring the coupling and decoupling relationship of urbanization and carbon emissions in China. *Environ. Sci. Pollut. Res.* **2023**, *30*, 96808–96826. [CrossRef] [PubMed]
76. Xing, Z.; Li, X.; Shi, Z.; Gulinaer, S.; Wu, H. Urban expansion and carbon emission effect of the urban agglomeration in south-central Liaoning Province. *Remote Sens. Nat. Resour.* **2022**, *34*, 272–279.
77. Li, Z.; Jiang, W.; Wang, W.; Lei, X.; Deng, Y. Exploring spatial-temporal change and gravity center movement of construction land in the Chang-Zhu-Tan urban agglomeration. *J. Geogr. Sci.* **2019**, *29*, 1363–1380. [CrossRef]
78. Ouyang, X.; Zhu, X.; He, Q.Y. Study of Spatio-temporal pattern and driving mechanism of urban land expansion in urban agglomeration: A case study of the Changsha-Zhuzhou-Xiangtan urban agglomeration. *Resour. Environ. Yangtze Basin* **2020**, *29*, 1298–1309.
79. Tian, Y.; Zhao, X. Simulation of construction land expansion and carbon emission response analysis of Changsha-Zhuzhou-Xiangtan Urban Agglomeration based on Markov-PLUS model. *Acta Ecol. Sin.* **2024**, *44*, 129–142.
80. Wen, J.; Chuai, X.; Li, S.; Song, S.; Li, Y.; Wang, M.; Wu, S. Spatial Heterogeneity of the Carbon Emission Effect Resulting from Urban Expansion among Three Coastal Agglomerations in China. *Sustainability* **2019**, *11*, 4590. [CrossRef]
81. Zhao, Y.N.; Xing, X.Q.; Wang, Q.; Gao, C.Y. Research on Environment Improvement with Development Strategies of Low Carbon Ecological Shijiazhuang City. *Adv. Mater. Res.* **2013**, *788*, 321–324. [CrossRef]
82. Ma, H.; Xu, X. High-Quality Development Assessment and Spatial Heterogeneity of Urban Agglomeration in the Yellow River Basin. *Econ. Geogr.* **2020**, *40*, 11–18.
83. Li, B.; Chen, D.; Wu, S.; Zhou, S.; Wang, T.; Chen, H. Spatio-temporal assessment of urbanization impacts on ecosystem services: Case study of Nanjing City, China. *Ecol. Indic.* **2016**, *71*, 416–427. [CrossRef]
84. Ho, C.S.; Matsuoka, Y.; Simson, J.; Gomi, K. Low carbon urban development strategy in Malaysia–The case of Iskandar Malaysia development corridor. *Habitat. Int.* **2013**, *37*, 43–51. [CrossRef]
85. Liu, Y.; Fan, P.; Yue, W.; Song, Y. Impacts of land finance on urban sprawl in China: The case of Chongqing. *Land Use Policy* **2018**, *72*, 420–432. [CrossRef]

land

MDPI

Article

Spatial and Temporal Changes in Ecological Resilience in the Shanxi–Shaanxi–Inner Mongolia Energy Zone with Multi-Scenario Simulation

Xinmeng Cai , Yongyong Song *, Dongqian Xue, Beibei Ma, Xianfeng Liu and Liwei Zhang

School of Geography and Tourism, Shaanxi Normal University, Xi'an 710119, China;
cxmdimon@snnu.edu.cn (X.C.); xuedq@snnu.edu.cn (D.X.); mabb@snnu.edu.cn (B.M.);
liuxianfeng7987@163.com (X.L.); zlw@snnu.edu.cn (L.Z.)
* Correspondence: syy2016@snnu.edu.cn

Abstract: The energy-driven expansion of artificial surfaces has resulted in severe ecological problems. Scientific evaluation of regional ecological resilience under different scenarios is crucial for promoting ecological restoration. This study chose the Shanxi–Shaanxi–Inner Mongolia Energy Zone (SEZ) and modeled an ecological resilience evaluation based on resistance, adaptability, and recovery. Land-use change and ecological resilience from 1980 to 2020 were then analyzed. Moreover, the SEZ land-use patterns and ecological resilience in 2030 were simulated under business as usual (BAU), energy and mineral development (EMD), and ecological conservation and restoration (ECR) scenarios. The results showed that (1) the SEZ was dominated by cultivated land, grassland, and unused land. (2) Ecological resilience showed a changing trend of decreasing and then increasing, with high ecological resilience areas mainly located in the Yellow River Basin, whereas low ecological resilience areas spread outward from the central urban areas. (3) The ecological resilience level was the lowest under the EMD scenario and the highest under the ECR scenario. This study not only expands the analysis framework of ecological resilience research but also provides scientific support for ecological conservation in ecologically fragile areas with intensive human activity worldwide.

Keywords: ecological resilience; PLUS model; multi-scenario simulation; land use

1. Introduction

Energy development zones, as a crucial energy supply to ensure sustainable global socio-economic development, contain an abundance of energy resources, including oil, natural gas, coal, and renewable resources. Energy development dominates production and construction activities in these zones, and energy-based industries play a major role in regional economic development [1]. Energy development zones are one of the areas with the most intense human activities [2]. For a long time, the irrational development of energy and mineral resources has been accompanied by the piecemeal expansion of construction land, occupying a large amount of ecological space [3], which poses a serious threat to the ecosystems. Therefore, scientific evaluation and prediction of the ability of ecosystems in energy development zones to withstand risks is an effective measure to address the growing unsustainability problems.

The concept of ecological resilience characterizes the ability of ecosystems to absorb, resist, adapt, and recover from disturbance [4], and is of great significance in guiding energy development zones to promote ecologically sustainable development. Resilience was first introduced into the field of ecology by the ecologist Holling [4], and research achievements such as resilience concept identification [5,6] and indicators for resilience evaluation [7,8] have provided theoretical support for ecological resilience research. However, in practical application, it is even more important to construct a reasonable quantitative methodological framework for regional ecological resilience. Current research is focused on three main aspects. Firstly, there is a focus on the impact of human activities on the ecological resilience process [9]. Research has shown that the expansion of construction areas and a development mode that purely pursues economic benefits can reduce the quality of ecosystem services [10]. Secondly, there have been studies combining the research methods of systems theory and ecology to evaluate the level of ecological resilience at different scales [11], explore its evolution process, and formulate development plans [12,13]. Lastly, attribution analysis of ecosystem change has been conducted to analyze the influencing factors of ecosystem change [14,15], and it was found that the level of urbanization, urban spatial patterns, and topographic conditions are key factors affecting ecological resilience. These studies provide a solid foundation and different perspectives for assessing ecological resilience.

As a hot research topic, the complex connotations of resilience have not been consistently finalized. For example, Macgillivray and Grime [16] believed that there may be a trade-off between resilience and resistance; Hodgson et al. [17] proposed that concepts of resistance and recovery can complement each other. Previous research [10,11] has illustrated that multiple processes influence resilience evaluation and that a single indicator cannot simply be used to measure resilience. This implies that it is necessary to establish appropriate evaluation frameworks based on different research objects. In 2019, Grafton et al. [18] proposed that social–ecological systems can be measured using the "3R model", which refines the dynamic process of resilience and has been initially explored in empirical case studies [19]. They explored resistance, recovery, and robustness as three important attributes of socio-ecological system resilience, and such definitions provide scientific and practical guidance on how different systems can achieve resilience. However, for energy development zones, which are vulnerable to human activities, the situation is somewhat different. It is crucial to focus on the adaptability of the ecosystems to disturbances [20] during ongoing activities like mining operations. Thus, this paper considers adding an "adaptability" index to modify the existing "3R" conceptual model to explore whether ecosystems can positively adapt to external disturbance by adjusting their internal processes when the disturbance continues to occur. In summary, this paper argues that the ecological resilience of energy development zones can be viewed as the ability of regional ecosystems to respond at different stages of disturbance. Ecological resilience is discussed comprehensively from three aspects: the resistance of the ecosystem in the transient or short-term period when the disturbance occurs, the adaptability during the continuation of the disturbance, and the recovery after the disturbance ends. These three are interconnected and indispensable and work together for ecological resilience, although their focus varies at different periods when disturbance occurs.

Some existing methods of constructing an ecological resilience evaluation index system through statistical data gradually show the drawbacks of different statistical calibers, missing data, and subjectivity [21,22]. In addition, most of the existing methods for measuring regional ecological resilience adopt the "scale-density-form" ecological resilience model based on administrative boundaries [21,23], which focuses on the impacts of human activities on the ecological environment, and less on natural factors such as land use, climate change, and topographic conditions. Land use, as a visual manifestation of the interaction between natural and human factors on the earth's surface [24,25], is an intrinsic driver of the evolution of ecological resilience in the SEZ [24], and land-use data are easier to access and collect than statistical data. Therefore, this paper uses land-use change as an endogenous drive to portray the changing law of ecological resilience. Furthermore, during the economic transformation and ecological civilization construction in energy development zones, conducting multi-scenario simulations can help clarify the future regional land-use expansion dynamics [26,27]. This, in turn, can offer valuable guidance in preparing for potential disturbances in regional ecosystems caused by uncertain risks [28]. Compared to traditional land-use simulation models [27,29,30], the PLUS model integrates the Land Expansion Analysis Strategy (LEAS) and CA model using a multi-class stochastic patch seed. This integration results in a more precise and efficient land-use simulation model [31]. The paper chose the PLUS model as a tool to simulate trends in land-use changes.

The issue of declining ecological resilience based on land-use changes must be brought to the forefront of global attention, particularly in typical energy zones [32,33]. Energy development zones are confronted with the critical problem of uncontrolled expansion of construction land and degradation of ecological land. This creates a conflict between economic development and ecological protection [34,35], which must be regulated and optimized in advance. This paper focuses on the SEZ, a critical energy security base in China with a delicate ecological environment, as a case study site. The objective is to deepen the basic meaning of ecological resilience and establish a practical and adaptable assessment framework for ecological resilience based on land-use change in energy development zones worldwide. The results could offer initial insights into protecting ecological resilience and optimizing land use in ecologically fragile energy development areas worldwide. The potential links among resistance, adaptability, and recovery are explained and used to develop the ecological resilience evaluation model. Local spatial autocorrelation and the PLUS model are then used with the aim of (1) elucidating spatial and temporal changes in land use and ecological resilience in the SEZ from 1980–2020; (2) exploring spatial differentiation patterns of ecological resilience and the spatial agglomeration pattern of energy development enterprises in the SEZ in 2020; (3) simulating evolution patterns of regional land use and ecological resilience in 2030 under the three scenarios of business as usual (BAU), energy and mineral development (EMD), and ecological conservation and restoration (ECR).

2. Research Framework

2.1. Conceptual Framework for Ecological Resilience Evaluation in the SEZ

Ecological resilience is a dynamic process that is constantly changing and challenging to characterize with just one variable [17]. A conceptual breakdown of ecological resilience is crucial. This paper considers a comprehensive portrayal of the ecological resilience of the SEZ in terms of resistance, adaptability, and recovery and needs to elucidate the interconnections among these three aspects.

As shown in Figure 1, the focus on the ability of ecosystems to respond to external disturbances may differ at different stages. When ecosystems located in energy development areas respond to external disturbances such as mining operations and land expansion for construction, their ability to resist the disturbances in the transient or short term by relying on their conditions [17], such as the natural environment and biodiversity, is called resistance. In reality, ecosystems often require a longer period to respond to external disturbances to assess their ability to adapt to such disturbances. Contrary to episodic natural hazards like floods and earthquakes that threaten the region's ecological resilience, human activities such as mining operations, while constituting external disturbances to the SEZ's ecological resilience, are crucial for sustaining and driving local and broader economic benefits. Therefore, more attention needs to be paid to the ability of ecosystems to continuously adapt themselves as disturbances continue. This paper introduces the index of "adaptability" as an important aspect of ecological resilience evaluation based on the 3R model [18]. Ecosystems are considered resilient if they can adapt positively to external disturbances over an extended period [20]. The ability of an ecosystem to transition from a disturbed state to a stable state is referred to as recovery [19], which is a specific manifestation of ecological resilience after the end of external disturbances. Ecosystems in energy development zones may have the three abilities of resistance, adaptability, and recovery described above, but with different emphases, in the overall process of facing external disturbances. Energy development zones' ecosystems may have a higher chance of recovering to a stable state through human interventions like land reclamation and reforestation. However, this does not necessarily mean that their ability to resist risks initially and adapt to disturbances in response to risks is also high. Thus, a complete assessment of regional ecological resilience requires considering all three aspects. Unlike the early prevention of natural disasters, this paper did not introduce the concept of early prevention into the SEZ ecological resilience evaluation model but expressed it by simulating future trends through multiple scenarios.

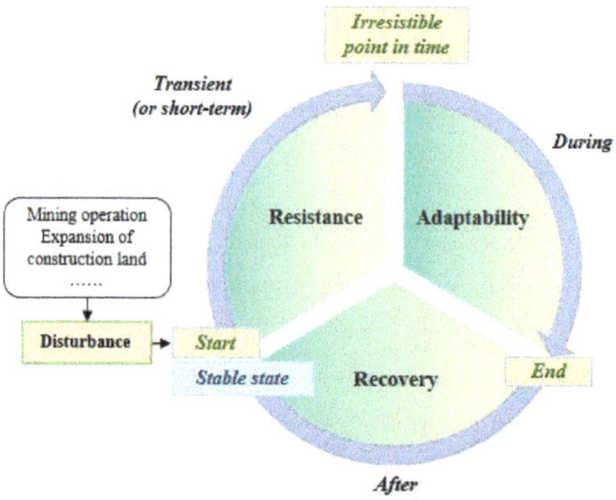

Figure 1. The conceptual framework for ecological resilience evaluation in the SEZ.

2.2. Framework for the Selection of Indices

Ecosystem resistance refers to the ability of an ecosystem to maintain its structural functions against short-term and transient external disturbances [17,36]. This aligns with the concept of ecosystem services. The benefits that ecosystems and ecological processes provide to sustain human survival and development are called ecosystem services [37]. From the perspective of human–nature coupling, healthy and resistant ecosystems should sustainably provide a range of valuable ecosystem services to fulfill human requirements while maintaining structural and functional integrity [38]. In the face of external perturbations like climate change, human activities, and natural disasters, these pressures can diminish both the quantitative and qualitative aspects of ecological services [39,40]. Consequently, ecosystems must be able to deliver sufficient and sustainable ecosystem services [41]. Ecosystems with limited services may become so fragile that they are unable to resist external disturbances and when disturbed struggle to provide ecosystem services and functions at the same level of value as before [40]. According to the Millennium Ecosystem Assessment, over 60% of ecosystem services are currently degraded. This degradation can lead to ecosystems losing the ability to support and protect themselves from external disturbances, which can ultimately pose a threat to both regional and global ecological security [42]. Hence, assessing the value of an ecosystem's services rendered at specific intervals over time can, to some extent, reflect the integrity of its structure and functioning, along with its resistance and ability to withstand risks.

This requires appropriate valuation of ecosystem services, which can be done using either value quantity or physical quantity assessments. Fixed-point observations facilitate the application of the physical quantity method in small-scale areas, yet obtaining data for large-scale areas poses a greater challenge [43]. Ecosystem service value (ESV) is a measure of the amount of value of ecosystem service functions using a monetization method [41], which is widely used in empirical studies due to its simplicity and practicality. The ecosystem service value approach was initially utilized in a study by Costanza et al. [44] for global ecosystems and biospheres, where the global ecosystem services' value was calculated. Then Xie et al. [45] developed a Chinese ESV equivalent factor table on this basis, which has been widely applied. Moreover, the ESV method, known for its ease of aggregating various services, proves suitable for comparative analyses of ecosystem service dynamics across different time frames and land-use scenarios. This paper employs ESV calculations to delineate ecosystem resistance.

Ecological adaptability is the ability of an ecosystem to continuously adjust itself in response to persistent external perturbations and internal demands [16,46]. The more stable the landscape structure of an ecosystem is, the more adaptable it is to external disturbances. The ecosystem landscape index was employed to characterize ecosystem adaptability, with landscape heterogeneity and connectivity reflecting the stability of the ecosystem landscape structure. Concerning these factors, the Shannon diversity index is sensitive to the non-equilibrium distribution status of each patch type in the landscape, and the area-weighted fractal dimension reflects the complexity of landscape patch shapes, both of which reflect landscape heterogeneity well; landscape fragmentation characterizes the degree of fragmentation in which the landscape has been segmented, and was used to measure landscape connectivity in this paper.

Ecosystem recovery reflects the ability and potential of an ecosystem to recover from a disturbed state to a stable state after an external disturbance has ended. Referring to the ecological resilience model and coefficients proposed by Peng et al. [38], this study believes that when an ecosystem encounters disturbances, unused land that has not been affected by human activities has a greater capacity for resilience, whereas the human-dominated construction land has a lower recovery and suffers from greater damage in the face of disturbance.

3. Materials and Methods

3.1. Study Area

The SEZ (37°21′–40°16′ N, 108°56′–111°29′ E) is located in the middle and upper reaches of the Yellow River in China, at the junction zone of Shanxi and Shaanxi Provinces and the Inner Mongolia Autonomous Region, and covers a total of three provinces (autonomous regions), five cities, and 13 counties (districts, banners) (Figure 2).

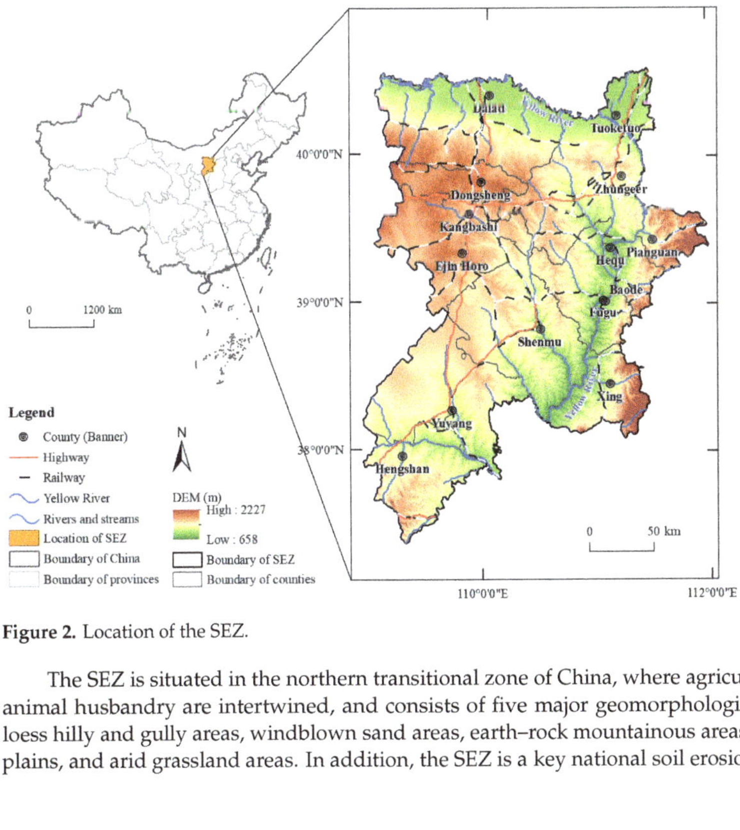

Figure 2. Location of the SEZ.

The SEZ is situated in the northern transitional zone of China, where agriculture and animal husbandry are intertwined, and consists of five major geomorphological types: loess hilly and gully areas, windblown sand areas, earth–rock mountainous areas, alluvial plains, and arid grassland areas. In addition, the SEZ is a key national soil erosion control area and a key supervision area for soil and water conservation, with abundant reserves of energy and mineral resources, such as coal, oil, and natural gas. In 2020, the region had a population of 3.97 million, a GDP of CNY 6157.46 billion, and a construction land area of 2247.22 km^2. With intensive human activity and a prominent ecological status, the region is a national energy security guarantee base and an area sensitive to global climate change, making it an ideal case study for regional ecological resilience.

3.2. Data Sources

The research data in this paper are mainly land-use data, the driving factor data needed for the PLUS model, and other socio-economic data. The specific research data are shown in Table 1.

Table 1. Detail of all data.

Category	Data	Years	Spatial Resolution	Data Sources
Land-use data	Land use	1980–2020	30 m	CAS (https://www.resdc.cn/, accessed on 19 November 2023)
Environmental data	Annual average temperature	2020	1000 m	CAS
	Annual average precipitation			
	Soil type	-	30 m	CAS
	DEM	2020	30 m	Geospatial Data Source (https://www.gscloud.cn/, accessed on 19 November 2023)
	Slope	2020	30 m	Calculated from DEM data in ArcGIS
Socio-economic data	GDP	2019	1000 m	CAS
	POP	2019	1000 m	CAS
	Distance to primary road	2020	30 m	OpenStreetMap (https://www.openstreetmap.org/, accessed on 20 November 2023)
	Distance to the secondary road			
	Distance to the tertiary road			
	Distance to railway			
	Distance to highway			
	Distance to open water			
	Distance to county (district/banner) governments			Baidu Maps (https://map.baidu.com/, accessed on 20 November 2023)

1. Land-use data. The land-use data were obtained from the Resource and Environment Science Data Center of the Chinese Academy of Sciences (CAS). According to the classification rules formulated by the CAS [47], land use was classified into six types: cultivated land, woodland, grassland, water body, construction land, and unused land;
2. Driving factor data. When using the PLUS model for multi-scenario simulations, driving factors must be produced, including environmental factors and socio-economic factors. In addition to those shown in Table 1, the distance to the county (district, banner) administrative center and individual road levels were calculated using the ArcGIS 10.3 Euclidean distance tool;
3. Other socio-economic data. Grain sowing area and production data in the SEZ were obtained from the 2001–2021 Statistical Yearbooks of each county and district. Grain prices were obtained from the "National Compendium of Agricultural Product Costs and Benefits". Energy development enterprises in the SEZ were selected from the

2020 list of key pollutant discharge units by the ecological environment bureaus of each county (district, banner), and the coordinate information was obtained from the Baidu coordinate picking platform and transformed into WGS84 coordinates.

3.3. *Methodology for the Calculation of Indices*

Based on clarifying the connotation of ecological resilience in the SEZ, this paper referred to the relevant studies [48,49], and chose suitable calculation methods for each index to precisely depict the resistance, adaptability, and recovery of ecosystems in the SEZ.

3.3.1. The Resistance Index

The ESV equivalent factor table was corrected by using one standard equivalent value equal to 1/7 of the average grain yield market value in the SEZ [50]. Taking the average grain yield of 3351.03 kg·hm^{-2} from 13 counties (districts and banners) in the SEZ from 2000 to 2020 as a substitute for the grain yield, and taking the national average grain selling price of CNY 2.26·kg^{-1} from 2011 to 2020 as the grain price, the ESV equivalent factor in the SEZ was calculated to be CNY 1081.90·hm^{-2}, and the ESV of land-use types in the SEZ was obtained (Table 2).

Table 2. Coefficients of ESV of different land-use types in the SEZ (CNY·hm^{-2}·a^{-1}).

Ecosystem Services	Ecosystem Sub-Services	Cultivated Land	Woodland	Grassland	Water Body	Construction Land	Unused Land
Supply services	Food production	919.62	313.75	108.19	865.52	0	0
	Raw material production	432.76	714.05	151.47	248.84	0	0
	Water supply	21.64	367.85	86.55	8968.95	0	0
Regulation services	Gas regulation	724.87	2347.72	551.77	833.06	0	21.64
	Climate regulation	389.48	7032.35	1449.75	2477.55	0	0
	Environmental purification	108.19	2088.07	476.04	6004.55	0	108.19
	Hydrological regulation	292.11	5128.21	1060.26	110,613.46	0	32.46
Support services	Soil conservation	1114.36	2867.04	670.78	1006.17	0	21.64
	Nutrient cycle maintenance	129.82	216.38	54.09	75.73	0	0
	Biodiversity protection	140.65	2607.38	605.86	2758.85	0	21.64
Cultural services	Aesthetic landscape	64.91	1146.81	270.48	2044.79	0	10.82

The calculation formula for the SEZ ecosystem resistance is as follows:

$$Resi = \sum_{i=1}^{6} A_i \times VC_{ij} \quad (1)$$

where *Resi* is the ecosystem resistance index characterized by the ESV function. A_i is the area of land-use type i, and VC_{ij} is the jth ESV of land-use type i.

3.3.2. The Adaptability Index

Landscape heterogeneity and connectivity are equally important in describing ecosystem landscape structure, so their weights can be set as equal [38]. The formula is as follows:

$$Adap = 0.25SHDI + 0.25AWM + 0.5C \quad (2)$$

where *Adap* is the ecosystem adaptability index, *SHDI* is the Shannon diversity index, *AWM* is the area-weighted average patch fractal dimension, and C is landscape fragmentation, all of which were calculated using Fragstats 4.2 software.

3.3.3. The Recovery Index

The calculation formula for the SEZ ecosystem recovery is as follows:

$$Reco = \sum_{i=1}^{6} A_i \times RC_i \quad (3)$$

where *Reco* is the ecosystem recovery index, A_i is the area of land-use type i, RC_i is the resilience coefficient of land-use type i, which was determined with reference to Peng et al. [38].

3.3.4. The Resilience Index

The ecosystem resistance, adaptability, and recovery were normalized to [0,1] using the extreme deviation standardization method, and then the ecological resilience index was calculated. The formula is as follows:

$$Resilience = (Resi \times Reco \times Adap)^{1/3} \quad (4)$$

where *Resilience* indicates the ecological resilience index.

3.4. *Spatial Autocorrelation Model*

The local autocorrelation LISA [51] was used to analyze the spatial distribution of ecological resilience and the energy development enterprise density agglomeration. The formula is as follows:

$$I = \frac{n(X_i - \overline{X}) \sum_{j=1}^{n} W_{ij}(X_j - \overline{X})}{\sum_{j=1}^{n} (X_j - \overline{X})^2} \quad (5)$$

where n is the number of grids into which the SEZ is divided or the number of energy development enterprises; X_i, X_j is the value of the ecological resilience or energy development enterprise kernel density for spatial locations i, j; W_{ij} is the weight matrix of the adjacency relationship between geographical units; and $\overline{X}$ is the average of the ecological resilience value or the kernel density of energy-developing enterprises.

3.5. *Land-Use Change Simulation Based on the PLUS Model*

3.5.1. PLUS Model

The PLUS model mainly consists of two parts: the LEAS and the CA model based on multiple random patch seeds (CARS) [31]; thus, the PLUS model can explore the driving factors of land expansion and better simulate the evolution of land-use patches. The LEAS module uses a random forest algorithm to sample the expansion portion of land use in different years and calculates the development probability of each land-use type and the driving factor contribution to the expansion of land use. The CARS module integrates random seeding and threshold-diminishing mechanisms to forecast future land-use distribution while considering development probability constraints.

In this paper, 14 driving factors (Table 1) were chosen for simulating land use in the SEZ in 2030. Moreover, water systems and nature reserves were set as restricted development areas. Before conducting the simulation, land-use distribution data for 2020 were simulated based on historical trends. In comparison with the actual land-use data in 2020, the results demonstrated high accuracy, with an overall accuracy of 89.2%.

3.5.2. Multi-Scenario Settings

Referring to the "Regulations on the Development and Construction of Soil and Water Conservation in the Border Region of Shanxi–Shaanxi–Inner Mongolia" and existing studies [10,26,52], in response to the actual situation of abundant mineral resources and the severe soil erosion phenomenon in the SEZ, this study set up three types of land-use change simulation scenarios: BAU, EMD, and ECR.

1. BAU scenario: Based on the actual development of the SEZ, according to the land-use change trend from 2010 to 2020, the area of each land-use type in 2030 was calculated using a Markov chain, which is the original 2030 prediction result generated by the PLUS model;
2. EMD scenario: Since the implementation of the Western Development Strategy in 2000, the large-scale development of energy and mineral resources and the construction of supporting facilities in the SEZ have led to an accumulation of waste soil and slag, which has blocked rivers. In addition, coal mining has damaged the natural ecosystem structure and changed landscape patterns and geomorphology. Accordingly, this study identified a 50% increase in the probability of conversion of cultivated land, woodland, grassland, and water bodies to unused land. Additionally, a 30% increase in the probability of conversion of cultivated land, woodland, grassland, and water bodies to construction land was determined. Moreover, there was a 30% decrease in the probability of conversion of construction and unused land to cultivated land, woodland, grassland, and water bodies, and a 20% increase in the probability of conversion of construction land to unused land.
3. ECR scenario: Under the promotion of a series of ecological restoration projects, such as returning farmland to woodland (grassland), the area of regional soil erosion has been significantly reduced, and the ecological construction results were remarkable. Therefore, this study designated the water system and nature reserve within the SEZ as a restricted development area. Simultaneously, it strictly limited the transfer of woodland, grassland, and water bodies, reducing the probability of conversion to construction and unused land by 50%. It also aimed to decrease the probability of cultivated land being converted to construction and unused land by 30% and increase the probability of unused and construction land being converted to woodland, grassland, and water bodies by 30%.

4. Results

4.1. Land-Use Change Characteristics

From 1980 to 2020, the land use in the SEZ was dominated by grassland (>48%), cultivated land (>23%), and unused land (>12%). During the 40-year study period, there was a significant increase in the area of construction land in the SEZ, with a rise of 156,332.34 hm^2. Notably, the construction land area remained relatively stable during the initial two decades but experienced a rapid expansion in the subsequent two decades. The figures highlight the swift expansion of construction land in tandem with China's rapid economic growth in the 21st century. Moreover, the area of ecological land has undergone significant changes in the past two decades. With the gradual implementation of economic development initiatives such as the "Western Development" policy, extensive deforestation and land clearing in the SEZ led to an 8.62% decline in woodland area from 2000 to 2010. Subsequently, from 2010 to 2020, as the Chinese economy transitioned from traditional rugged development practices to a greater focus on high-quality economic growth and ecological conservation, the woodland area exhibited a notable increase (Table 3 and Figure 3).

In this paper, a land-use transfer chord diagram was visualized by Origin for a more specific conversion relationship between each land-use type. There was no significant change in land-use transfer between 1980 and 1990. However, between 1990 and 2000, there was an expansion in the area of grassland. The largest transfer was observed from unused land to grassland, which accounted for 159,106.14 hm^2. As can be seen from the thickness of the chords, the interconversion between the three land types dominant in the SEZ was more pronounced and correlated during this decade. The period from 2000 to 2010 witnessed significant transformations, notably the noticeable transfer between construction land and other land-use types. The largest conversion occurred from grassland to construction land, amounting to 31,729.23 hm^2. Additionally, noteworthy conversions include cultivated land to woodland (25,783.2 hm^2) and to grassland (80,029.26 hm^2), reflecting the emphasis placed by local governments on initiatives such as the "Returning Cultivated Land to Woodland (Grassland)" policy. In the period from 2010 to 2020, there was a sharp increase in the area of construction land, primarily converted from grassland (77,831.55 hm^2) and cultivated land (21,482.28 hm^2). Notably, the woodland area also expanded during this decade, predominantly converted from unused land.

Table 3. Land-use area and proportion in the SEZ from 1980 to 2020.

Year	Index	Cultivated Land	Woodland	Grassland	Water Body	Construction Land	Unused Land
1980	Area (hm^2)	1,385,167.45	354,129.67	2,607,425.24	143,906.9	68,390.21	850,749.27
1990		1,384,814.49	355,127.44	2,594,057.03	138,313.8	69,435.17	868,012.62
2000		1,369,351.74	357,458.13	2,715,066.99	138,911.80	71,399.83	757,659.37
2010		1,332,817.65	326,925.14	2,828,018.35	127,356.69	130,247.36	664,509.28
2020		1,285,290.65	357,009.95	2,702,636.34	136,160.51	224,722.55	703,839.01
1980–1990	Variation (hm^2)	−352.96	997.77	−13,368.21	−5593.1	1044.96	17,263.35
1990–2000		−15,462.75	2330.69	121,009.96	598	1964.66	−110,353.25
2000–2010		−36,534.09	-30,532.99	112,951.36	-11,555.10	58,847.52	-93,150.09
2010–2020		−47,527.00	30,084.81	−125,382.01	8803.82	94,475.19	39,329.73
2000–2020		-84,061.09	-448.19	-12,430.65	-2751.29	153,322.72	-53,820.36

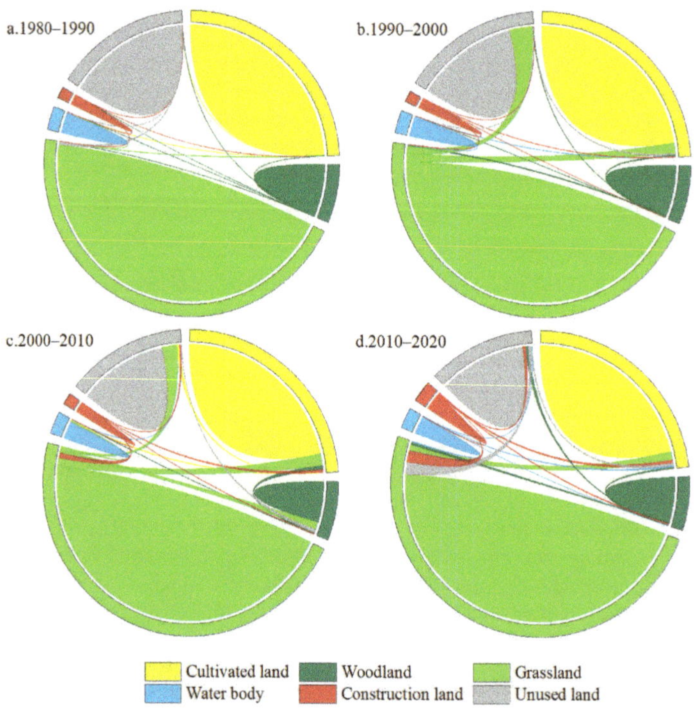

Figure 3. (**a–d**): Chord diagram of land-use transfer in the SEZ from 1980 to 2020. (The chord diagram is utilized to depict the correlation between multiple land-use types. The line segment that connects two points on a circle is referred to as a chord. Each chord represents the transformation between two land-use types, and the thickness of the chord represents the size of the transferred area.)

4.2. Ecological Resilience Spatial and Temporal Patterns

This study calculated the ecosystem resilience index in the SEZ from 1980 to 2020, using grid units, and divided the resilience index into three grades: low (0.00–0.23), medium (0.23–0.35), and high (0.35–0.80) by using the natural breaks (Jenks) method, to analyze the SEZ ecosystem resilience's spatial and temporal changes from 1980 to 2020 (Figure 4).

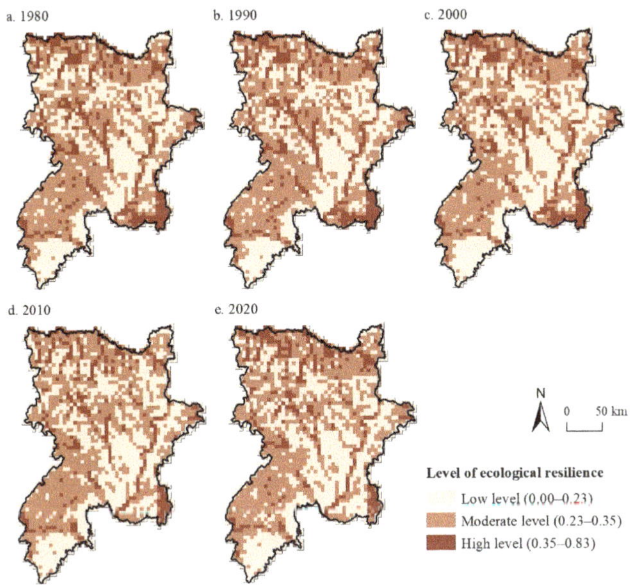

Figure 4. (**a**–**e**): Spatial and temporal changes in the ecological resilience level in the SEZ from 1980 to 2020.

Between 1980 and 2000, there was only a slight decrease in ecological resilience, with the average decreasing from 0.264 to 0.263. The spatial distribution did not change much during this period. However, between 2000 and 2010, the mean value of ecological resilience decreased significantly to 0.256. Specifically, the woodland area in the southern part of Xing County, Shanxi Province, diminished dramatically, leading to a significant decrease in ecological resilience. This is possibly because the region neglected the protection of ecological land while constructing production spaces and mineral resource transportation roads, resulting in a significant shrinkage in ecological land area. From 2010 to 2020, regional ecological restoration projects led to a significant increase in vegetation coverage, and the average ecological resilience value was 0.260 in 2020. The ecological recovery of the Yellow River Basin in the northern part of the SEZ was evident. However, a significant portion of land in the central towns of Shenmu City and Yuyang District in Shaanxi Province has been converted into production and habitable land, resulting in a decrease in ecological land area and an increase in landscape fragmentation. This has led to the formation of a low-value agglomeration area of ecological resilience that spreads outward from the central urban area.

4.3. Spatial Relationship between Ecological Resilience and Energy Development

In this study, a kernel density analysis of energy development enterprises in the SEZ was conducted using ArcGIS 10.3 software. In addition, univariate and bivariate local spatial autocorrelation analyses were performed on the regional ecological resilience level and the energy development enterprise kernel density for 2020 using Geoda 1.16 software (Figure 5).

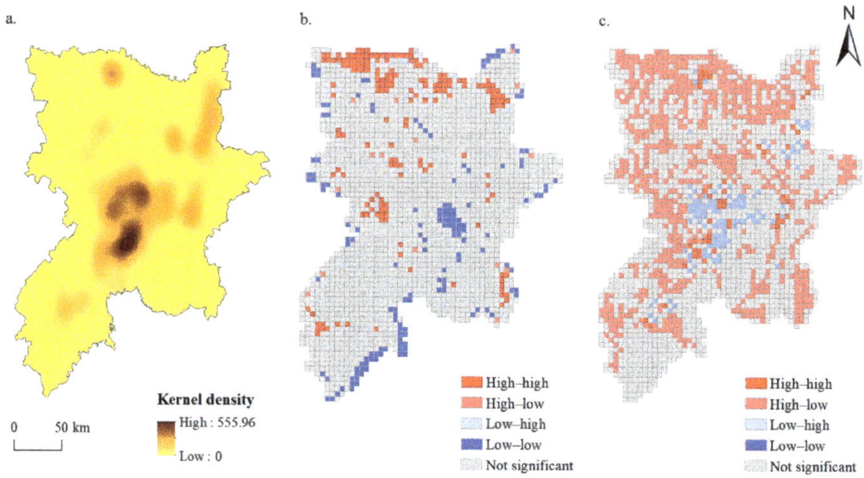

Figure 5. Spatial relationship between ecological resilience and energy development in the SEZ in 2020. ((**a**) Kernel density map of energy development enterprises; (**b**) local spatial autocorrelation map of ecological resilience level; (**c**) local spatial autocorrelation map of ecological resilience level—energy development enterprises).

Energy development enterprises were primarily concentrated in the central part of Shenmu City, forming a high-value circle of high kernel density levels spreading outward from the central urban area rich in energy and minerals (Figure 5a). Shenmu City is China's largest coal-producing county (city), with abundant resources and superior mining conditions; thus, the advantages of mineral resources have attracted a huge energy development industry chain and promoted the agglomeration of energy and mineral sewage enterprises.

The local spatial autocorrelation result of the ecological resilience level (Figure 5b) indicates that the areas with high ecological resilience values tended to cluster along the Yellow River in the northern part of the SEZ, illustrating that effective ecological management has significantly enhanced the ecological resilience of the Yellow River Basin in regards to sand and water problems. The bivariate spatial autocorrelation result of the analysis of the ecological resilience level and the kernel density of energy development enterprises (Figure 5c) reveals that energy development enterprises were concentrated in the central part of the SEZ, and there was a spatial clustering distribution of low resilience and high kernel density of energy development enterprises. The clustering of energy development enterprises is likely to have resulted in excessive discharge of wastewater and waste residues generated during production activity, contaminating the ecological environment. Furthermore, the concentration of enterprises has also led to the expansion of construction land, encroaching on ecological land, and decreasing the ecological resilience level in the surrounding areas. Concurrently, regions with a high intensity of human activity and a high risk of pollution where energy development enterprises were clustered were prone to ecological deterioration and a decrease in ecological resilience level, making them focal points that need to be regulated and controlled.

4.4. Multi-Scenario Simulation

4.4.1. Land Use under Different Scenarios

Because the SEZ is extensive, the land-use distribution map of the entire area cannot effectively illustrate the differences in land use in various scenarios. Therefore, selected areas within the three localities, along the Yellow River in the northern part of the SEZ, the downtown area of Shenmu City, and the woodlands of Xingxian County, are highlighted to provide detailed insights (Figure 6).

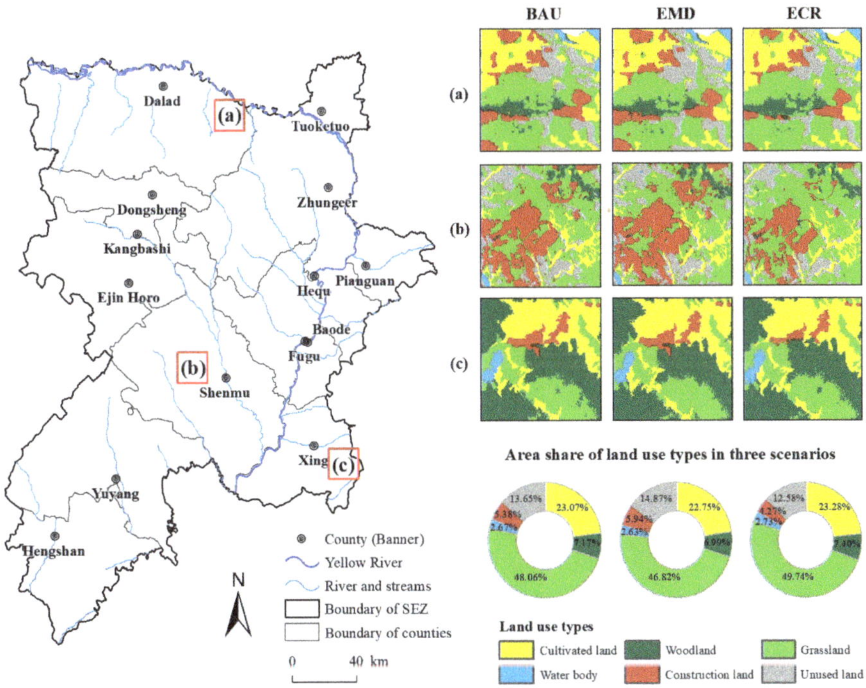

Figure 6. Prediction results of land use in the SEZ in 2030 under three scenarios.

In general (Table 4), under the BAU scenario, cultivated land area in the SEZ will decrease by 2.9% in 2030, and grassland area will decrease by 3.8% compared with those areas in 2020. The woodland and water body areas will increase by 8.6% and 6.1%, respectively. This indicates that under the BAU scenario, the SEZ will continue its basic land-use change trend from 2010 to 2020.

Table 4. Areas of land use in 2020 and various scenarios in 2030.

Scenario	Index	Cultivated Land	Woodland	Grassland	Water Body	Construction Land	Unused Land
2020	Area (hm^2)	1,285,290.65	357,009.95	2,702,636.34	136,160.51	224,722.55	703,839.01
BAU		1,248,048.81	387,605.88	2,599,639.65	144,434.97	291,278.43	738,682.29
EMD		1,230,900.15	377,966.50	2,533,097.43	142,015.19	321,362.89	804,347.85
ECR		1,259,151.26	400,574.91	2,690,466.04	147,884.74	230,908.65	680,704.40

Under the EMD scenario, cultivated land and grassland area will decrease, and construction land area in the SEZ in 2030 will rise significantly by 43% compared to the actual land-use area in 2020, indicating that under this scenario, energy exploitation and related construction activity will occupy the ecological space, exacerbating soil erosion and land desertification problems.

Under the ECR scenario, the woodland and water body areas in the SEZ in 2030 will rise significantly by 12.2% and 8.6%, respectively, compared with those of 2020. The increase in construction land area will be limited to 2.8% compared to 2020. This indicates that under this scenario, ecological land will be protected, construction land expansion will be suppressed, and ecological functions such as regional ecological protection and water conservation will be restored.

4.4.2. Ecological Resilience under Different Scenarios

The spatial distribution of the SEZ ecological resilience, resistance, adaptability, and recovery in 2030 under the BAU, EMD, and ECR scenarios is shown in Figure 7.

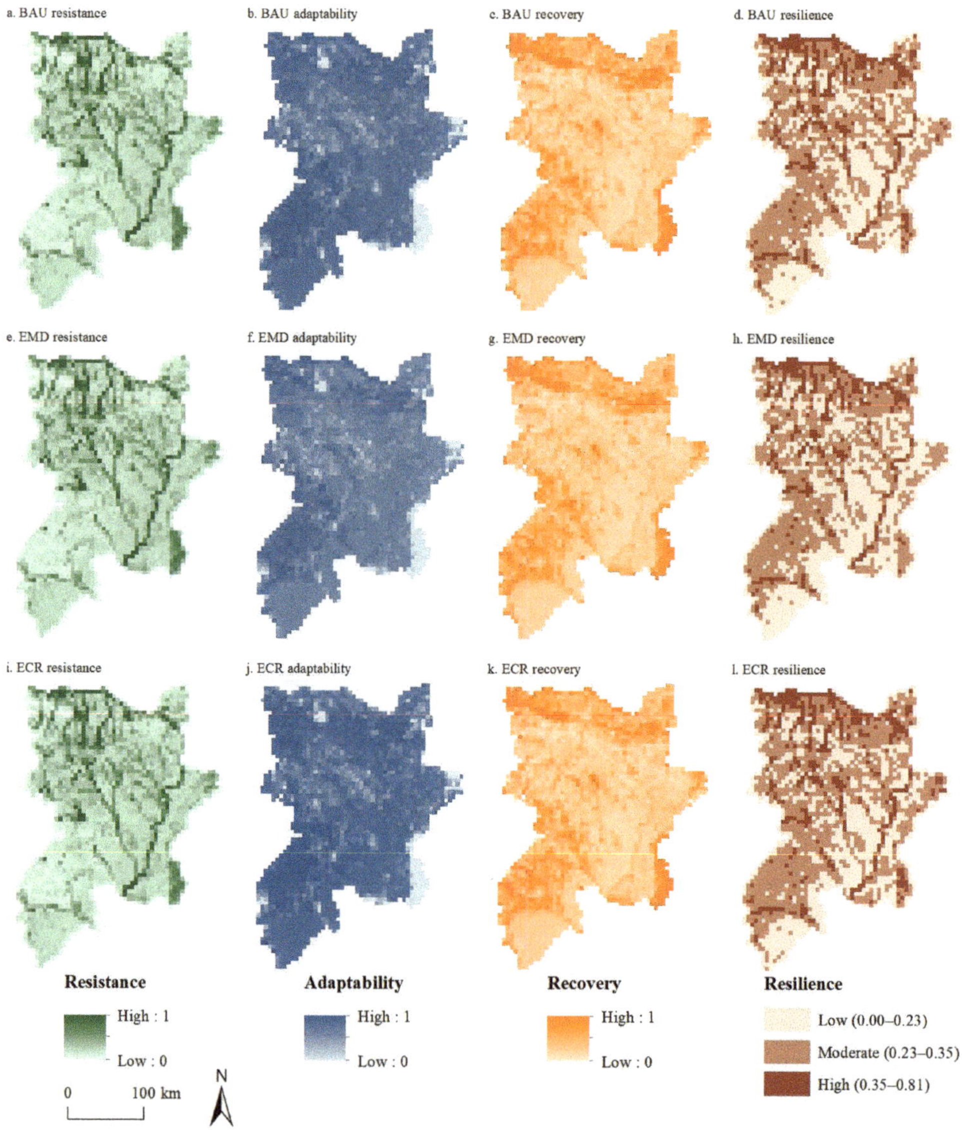

Figure 7. Spatial distribution of ecological resistance, adaptability, recovery, and resilience in the SEZ in 2030 under three scenarios. ((**a–d**) The BAU scenario; (**e–h**) the EMD scenario; (**i–l**) the ECR scenario).

1. BAU scenario. The proportions of low, medium, and high ecological resilience under the BAU scenario are 40.2%, 44.7%, and 15.1%, respectively. Among them, the proportion of low-level resilience increases by 1.8% compared to that of the actual situation in 2020, indicating that if not restricted, the SEZ's ecological resilience will decrease

further. The areas with high ecological resilience and resistance are located in the Yellow River Basin in the northern part of the SEZ, and the low-value adaptability area is located in the woodlands of the eastern part of Xing County in Shanxi Province. The high-value recovery areas are located in the western part of Yuyang District, Shaanxi Province, where unused land predominates, with less anthropogenic-dominated construction land, and a better recovery capacity;

2. EMD scenario. The proportions of low, medium, and high ecological resilience are 44.3%, 42.6%, and 13.1%, respectively. Ecological resilience is low in Shenmu City, located in the central part of the SEZ, and Hengshan County, in the southern part of the SEZ. The overall low level of ecological adaptability in the SEZ may be explained by the significant growth of construction land area under this scenario, resulting in landscape fragmentation and reduced landscape connectivity;
3. ECR scenario. The proportions of low, medium, and high ecological resilience are 37.5%, 46.4%, and 16.1%, respectively. Areas with high ecological resilience and resistance are mostly concentrated in the Yellow River Basin and the eastern part of Xing County, Shanxi Province. In addition, the overall ecological adaptability is high, indicating that under the ECR scenario, the SEZ emphasizes ecological land protection, reduces construction land encroachment on ecological land, and reduces landscape fragmentation. This will increase landscape heterogeneity and connectivity, and there will be a high level of ecological resilience in the future.

5. Discussion

5.1. Expansion Trend of Construction Land in the SEZ

To analyze the expansion trend of construction land in the SEZ more specifically, this paper divided the construction land into three types: urban land, rural land, and mining and transportation land, based on the classification rules of the land-use data used. Furthermore, the paper simulated the land use in 2030 under the BAU scenario. The expansion trends for the three land-use types are presented in Figure 8 from 2000 to 2030, as the changes from 1980 to 2000 were not apparent.

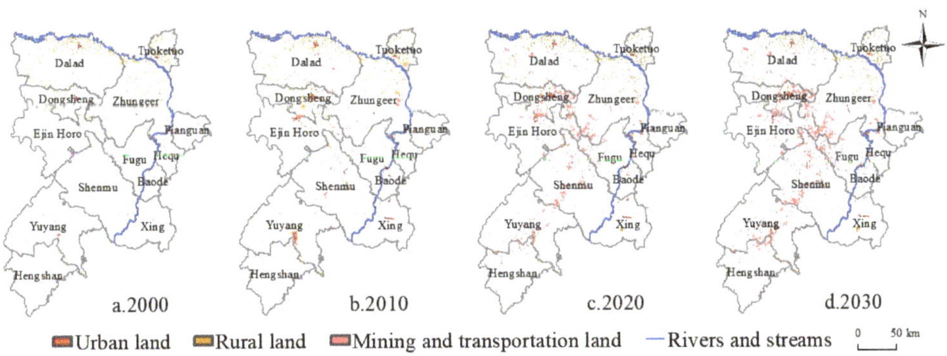

Figure 8. (**a**–**d**): Evolution of the three types of construction land between 2000 and 2030.

Since the 21st century, China's diversified development has brought about dramatic shifts, including changes in the spatial patterns of construction land expansion [53]. As is shown in Figure 8, from 2000 to 2010, the construction land in the SEZ was characterized by a dominance of rural land, primarily concentrated along the Yellow River in the northern areas. Meanwhile, during this period, there was a significant increase in urban land, amounting to 10,476.21 hm^2, which was clustered in the urban centers of Dalad Banner, Dongsheng District, and Yuyang District. This shift indicates that the acceleration of urbanization in the SEZ has led to the expansion of urban land. In 2020, the most notable feature was the rapid growth of mining and transportation land use, which became the dominant construction land in the region. Mining and transportation land was mainly concentrated in the central part of the SEZ, as well as in Dongsheng District and Shenmu City. Both of these areas are important coal resource bases in China, and energy development activities drive the rapid expansion of mining and transportation land use. According to the simulated 2030 BAU scenario, the expansion of mining and transportation land will continue and dominate the regional construction land in the Coal Resource Zone. In addition, as can be seen in Figure 9, the overall area of construction land in the SEZ grows rapidly between 2000 and 2030; however, this growth is accompanied by a rapid decline in cultivated land and ecological land (woodland, grassland, water body, and unused land), which will pose a significant threat to the ecological resilience of the region.

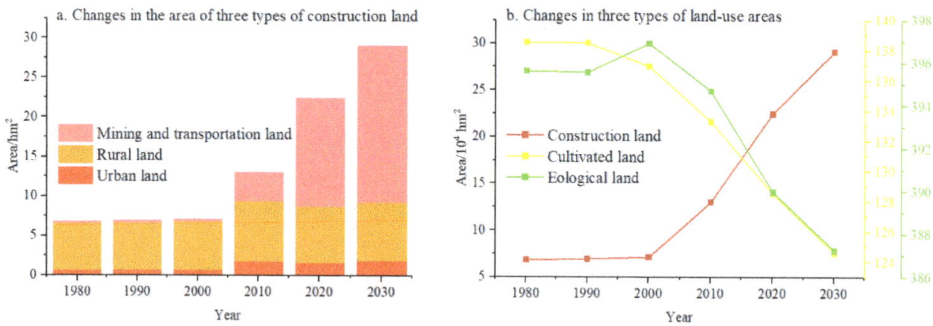

Figure 9. Evolution of different types of land use between 1980 and 2030. ((**a**) Changes in the area of mining and transportation land, rural land, and urban land; (**b**) Trends in the area of construction land, cultivated land, and ecological land).

Energy development projects and construction activities within energy development zones play a crucial role in driving regional economic growth. However, as the analysis presented above shows, it is evident that such resource development activities have adversely impacted the regional ecosystem. The expansion of construction land, mainly for mining and transportation, has caused a decline in the region's ecological resilience. China's economic progress has been challenged by the need for ecological conservation, especially in energy development zones. Balancing energy extraction, economic growth, and ecological protection has led to a focus on the gradual allocation of construction land as a potential solution to this dilemma. Based on the simulated results for 2030, it is expected that the main land-use change within the SEZ will be the continuous growth of construction land, driven by energy development. In light of this, the SEZ may need to prioritize the effective use of construction land, rather than solely focusing on controlling its growth. For instance, implementing comprehensive management strategies for coal mining subsidence areas and strengthening land reclamation efforts can help to effectively restore the use of construction land resources.

5.2. Comparison of Multi-Scenario Ecological Resilience Simulation Results

Compared to the BAU scenario, the ecological resilience in the SEZ in 2030 under the EMD scenario declines significantly, with a 4.1% increase in the low-level proportion and a 2% decrease in the high-level proportion. This suggests that energy extraction must be accompanied by a focus on protecting and restoring ecosystems. In the eastern part of the SEZ, the ecological resilience of Pianguan, Baode, and Xing counties in Shanxi Province declines with the shrinkage of woodland and grassland areas, illustrating that the construction land area encroaches upon ecological land during production and construction activity. Furthermore, the ecological adaptability in the EMD scenario is significantly lower. This could be attributed to the substantial increase in construction land area under the EMD scenario, leading to sporadic accumulation and infiltration into other land-use types. This, in turn, causes landscape fragmentation and reduces landscape connectivity. Therefore, the energy zones must take decisive measures to prohibit private mining of iron ore resources and other activities that have a detrimental impact on the environment. It is imperative to adopt and implement advanced technology to modernize traditional mining practices, promote sustainable mining, and enhance mineral resource recovery rates and comprehensive utilization rates. Furthermore, it is necessary to reduce the dependence on highly polluting resources and ensure a sustainable supply of energy and raw materials for various industries.

Under the ECR scenario, the ecological resilience in 2030 is higher than that in the BAU and EMD scenarios, with a high level of ecological resilience accounting for 16.1% of the area. This shows that the woodland and water bodies in the SEZ are effectively protected under the ECR scenario, resulting in an increase in ecological resilience in woodlands and along water bodies. Ecological conservation projects are essential for protecting regional ecology, improving ecological quality, enhancing ecological resilience, and promoting sustainable regional development. The SEZ is a key area for the ecological protection and restoration of the Yellow River Basin. In March 2000, the State Council implemented a policy of returning cultivated land to woodland (grassland) to increase vegetation cover and control soil erosion. In this context, the SEZ has successfully implemented ecological projects such as the construction of silt embankments and terraces, the "Three Norths" protective forests, and Yellow River Basin water loss and soil erosion comprehensive management, effectively reducing the degree of erosion of the regional slopes and channels, further promoting regional ecological restoration, and improving the ecological resilience. Additionally, the ecological resistance and adaptability in the ECR scenario are higher, indicating that construction land expansion in the ECR scenario is effectively suppressed and the ecological land is restored, which contributes to the enhancement of ecosystem service functions and a decrease in landscape fragmentation in the SEZ.

The above analyses demonstrate that in energy zones like the SEZ, ecological conservation is essential for improving regional ecological resilience. In 2017, the concept of "high-quality development" introduced by the state shifted from traditional rough development to prioritizing green development for the harmonious coexistence of humans and nature. Therefore, it is crucial to prioritize ecological principles and pursue moderate development in actual development efforts. This may involve focusing on the development of ecological industries that align with the primary goal of ecological conservation, particularly in the counties and central towns within the region. Additionally, ongoing efforts to enhance ecological construction should be prioritized to ensure the integrity of the ecosystem is maintained to the fullest extent possible.

5.3. Research Insights

Regarding the SEZ, ecological resilience is generally low in the intensive areas of energy development enterprises and their surroundings, indicating that industrial production activities such as energy extraction have seriously affected the surrounding land-use types. Extraction of energy and mineral resources is temporary [54]; therefore, it is necessary to strengthen land restoration in these areas by comprehensively integrating and rehabilitating land that has been damaged by energy and mineral extraction. In 2011, the State Council issued the Regulations on Land Reclamation, which required the remediation of land destroyed by production and construction activity to make it available for use. Under the EMD scenario established in this study, construction and unused land areas increased by 43% and 14.3%, respectively, compared with those of 2020. Thus, ecological restoration of industrial and abandoned energy and mineral development production land is required to reduce landscape fragmentation caused by the occupation of ecological land for production and construction to improve regional ecological resilience. Moreover, governments should set up and implement the "Who Destroys, Who Compensate" concept of ecological compensation in energy-rich regions, to make full compensation for the economic and societal losses. Moreover, it is essential to enhance the efficiency of productive and habitable land to mitigate the adverse effects of construction land expansion [55,56].

It is a fact that mining activities cause environmental disturbances. Nevertheless, it is necessary to take into account the evolution of the Chinese economy and to observe the rules of economic and social development. Given that coal remains a primary component of China's energy structure, and electricity plays a central role, it is imperative to pursue a balanced approach that integrates protection and development. This entails enhancing the level and efficiency of resource utilization. The results of this paper underscore the importance of ecological conservation in preserving the ecological resilience of energy zones. However, relying solely on technical governance measures may not be sufficient to establish a resilient ecological civilization in these zones. Higher government authorities must implement management measures that guarantee the adoption of ecological conservation and governance techniques. This can be accomplished by incentivizing social funds to invest in ecological conservation within energy zones and promoting ecological restoration in mineral and resource-based areas through preferential policies, including tax incentives and financial subsidies.

5.4. Shortcomings and Prospects

First, in Section 3, this paper chose the simpler method of ESV to measure ecological resistance, which may have some limitations. Due to the complexity of the economic valuation of ecosystem services, the methodology of ESV has been questioned in practical applications [57,58]. Criticisms include concerns regarding the mechanistic nature of value assignment and the unilateral nature of evaluation criteria. Research on the ecological significance of ecosystem services may be neglected due to excessive focus on calculating their economic value. For example, the economic value of unused land is difficult to account for, which may result in a portion of the ecological value of its regulation and support not being captured. In addition, ecosystem change is a dynamic process, and the area-weighted approach to calculating the total value of ecosystem services may cause functional de-differentiation and impede the identification of faster-changing ecosystem services [42], leading to less accurate construction of ecosystem resistance.

Second, in the previous study, to ensure that the results of the simulation are comparable to the actual land use, we classified the land use into six types, which may limit the reliability of specific land-use types. For example, previous research [59] classified open-pit coal mines as a land-use type based on specific research needs, which can improve the analysis accuracy of coal mine land-use types. Although we discussed the refinement and modeling of the construction land, we did not proceed to calculate the ecological resilience of the refined land types due to the indicator calculation method chosen. However, we believe that the different types of construction land have varying levels of resistance, recovery, and adaptability, and this should serve as a basis for future research to investigate the impact on ecological resilience after refining the land-use types. Furthermore, the development of energy zones was not limited to the three scenarios set in this study. In the future, we can consider scenarios that coordinate economic development and ecological protection, seeking a balance between the two, rather than just considering one aspect. Simulation results can better support planning policies when the scenario setting is coupled with multi-objective optimization functions.

6. Conclusions

This study analyzes land-use change characteristics of the SEZ from 1980 to 2020, explores the spatial and temporal variation laws of regional ecological resilience, reveals the spatial agglomeration pattern of regional ecological resilience and energy development enterprises and their correlations, and finally simulates the spatial pattern of land use and ecological resilience of the SEZ in 2030. The results show that the land-use type of the SEZ is dominated by cultivated land, grassland, and unused land, and the construction land area increased dramatically in the 21st century. The ecological resilience level in the SEZ showed a trend of first decreasing and then increasing. From a spatial perspective, the high ecological resilience value area was distributed in a belt shape in the Yellow River Basin, whereas the central urban area of the city spread outward to form a low ecological resilience agglomeration area. The ecosystem resilience level was high in southeastern Xing County, Shanxi Province, whereas the ecological resilience level was low in Fugu County, Shenmu City, and Hengshan County, Shaanxi Province. The expansion of energy-based enterprises and resource-based cities drove construction land expansion, resulting in generally low ecological resilience in these areas.

Although the BAU scenario showed an increasing trend in woodland and water bodies, the decrease in regional ecological resilience was still mainly due to the expansion of the construction land area. The increase in construction land area under the EMD scenario was particularly prominent, leading to a low level of ecological resilience, which does not support sustainable regional development. However, the ECR scenario was more conducive to improving ecological resilience and realizing sustainable development in the SEZ. Therefore, during energy zone development, it is necessary to focus on ecological environmental conservation and sustainable ecosystem management to continuously improve regional ecological resilience.

Author Contributions: Conceptualization, X.C.; Formal analysis, X.C.; Funding acquisition, Y.S.; Methodology, X.C.; Supervision, Y.S., D.X., B.M., X.L. and L.Z.; Writing—original draft, X.C.; Writing—review and editing, Y.S. and D.X. All authors have read and agreed to the published version of the manuscript.

Funding: This research was funded by the National Natural Science Foundation of China, grant number 42001251; the Fundamental Research Funds for the Central Universities, grant number GK202201008.

Data Availability Statement: The data presented in this study are available on request from the corresponding author. The data are not publicly available due to the data is designed to be used in other ongoing research and should be protected before official publication.

Conflicts of Interest: The authors declare no conflicts of interest.

References

1. Zhang, H.L.; Shen, L.; Zhong, S.; Elshkaki, A. Coal resource and industrial structure nexus in energy-rich area: The case of the contiguous area of Shanxi and Shaanxi Provinces, and Inner Mongolia Autonomous Region of China. *Resour. Policy* **2020**, *66*, 101646. [CrossRef]
2. Zhang, M.; Wang, J.; Li, S. Tempo-spatial changes and main anthropogenic influence factors of vegetation fractional coverage in a large-scale opencast coal mine area from 1992 to 2015. *J. Clean. Prod.* **2019**, *232*, 940–952. [CrossRef]
3. Aizizi, Y.; Kasimu, A.; Liang, H.; Zhang, X.; Zhao, Y.; Wei, B. Evaluation of ecological space and ecological quality changes in urban agglomeration on the northern slope of the Tianshan Mountains. *Ecol. Indic.* **2023**, *146*, 109896. [CrossRef]
4. Holling, C.S. Resilience and Stability of Ecological Systems. *Annu. Rev. Ecol. Syst.* **1973**, *4*, 1–23. [CrossRef]
5. Delettre, O. Identity of ecological systems and the meaning of resilience. *J. Ecol.* **2021**, *109*, 3147–3156. [CrossRef]
6. Folke, C. Resilience: The emergence of a perspective for social-ecological systems analyses. *Glob. Environ. Chang. Hum. Policy Dimens.* **2006**, *16*, 253–267. [CrossRef]
7. Quinlan, A.E.; Berbes-Blazquez, M.; Haider, L.J.; Peterson, G.D. Measuring and assessing resilience: Broadening understanding through multiple disciplinary perspectives. *J. Appl. Ecol.* **2016**, *53*, 677–687. [CrossRef]
8. Ingrisch, J.; Bahn, M. Towards a Comparable Quantification of Resilience. *Trends Ecol. Evol.* **2018**, *33*, 251–259. [CrossRef]
9. Cao, T.G.; Yi, Y.; Liu, H.; Xu, Q.; Yang, Z. The relationship between ecosystem service supply and demand in plain areas undergoing urbanization: A case study of China's Baiyangdian Basin. *J. Environ. Manag.* **2021**, *289*, 112492. [CrossRef]
10. Gao, L.; Tao, F.; Liu, R.; Wang, Z.; Leng, H.; Zhou, T. Multi-scenario simulation and ecological risk analysis of land use based on the PLUS model: A case study of Nanjing. *Sustain. Cities Soc.* **2022**, *85*, 104055. [CrossRef]
11. Zuniga-Teran, A.A.; Gerlak, A.K.; Mayer, B.; Evans, T.P.; Lansey, K.E. Urban resilience and green infrastructure systems: Towards a multidimensional evaluation. *Curr. Opin. Environ. Sustain.* **2020**, *44*, 42–47. [CrossRef]
12. Berkes, F.; Jolly, D. Adapting to climate change: Social-ecological resilience in a Canadian Western Arctic community. *Conserv. Ecol.* **2002**, *5*, 18. [CrossRef]
13. Meerow, S.; Newell, J.P. Spatial planning for multifunctional green infrastructure: Growing resilience in Detroit. *Landsc. Urban Plan.* **2017**, *159*, 62–75. [CrossRef]
14. Xiu, C.; Cheng, L.; Song, W.; Wu, W. Vulnerability of large city and its implication in urban planning: A perspective of intra-urban structure. *Chin. Geogr. Sci.* **2011**, *21*, 204–210. [CrossRef]

15. Zhang, Y.; Yang, Y.; Chen, Z.; Zhang, S. Multi-criteria assessment of the resilience of ecological function areas in China with a focus on ecological restoration. *Ecol. Indic.* **2020**, *119*, 106862. [CrossRef]
16. Macgillivray, C.W.; Grime, J.P. Testing predictions of the resistance and resilience of vegetation subjected to extreme events. *Funct. Ecol.* **1995**, *9*, 640–649. [CrossRef]
17. Hodgson, D.; McDonald, J.L.; Hosken, D.J. What do you mean, 'resilient'? *Trends Ecol. Evol.* **2015**, *30*, 503–506. [CrossRef] [PubMed]
18. Grafton, R.Q.; Doyen, L.; Béné, C.; Borgomeo, E.; Brooks, K.; Chu, L.; Cumming, G.S.; Dixon, J.; Dovers, S.; Garrick, D.E.; et al. Realizing resilience for decision-making. *Nat. Sustain.* **2019**, *2*, 907–913. [CrossRef]
19. Yu, S.; Kong, X.; Wang, Q.; Yang, Z.; Peng, J. A new approach of Robustness-Resistance-Recovery (3Rs) to assessing flood resilience: A case study in Dongting Lake Basin. *Landsc. Urban Plan.* **2023**, *230*, 104605. [CrossRef]
20. Folke, C.; Carpenter, S.R.; Walker, B.; Scheffer, M.; Chapin, T.; Rockstrom, J. Resilience Thinking: Integrating Resilience, Adaptability and Transformability. *Ecol. Soc.* **2010**, *15*, 20. [CrossRef]
21. Wang, S.; Cui, Z.; Lin, J.; Xie, J.; Su, K. Coupling relationship between urbanization and ecological resilience in the Pearl River Delta. *Acta Geogr. Sin.* **2021**, *76*, 973–991. [CrossRef]
22. Nan, G.; Han, L.; Wang, Y. Assessing vulnerability of coupled human–environment systems and its influence factors in Yan'an City, China. *Hum. Ecol. Risk Assess. Int. J.* **2022**, *28*, 783–801. [CrossRef]
23. Zhang, T.; Sun, Y.; Zhang, X.; Yin, L.; Zhang, B. Potential heterogeneity of urban ecological resilience and urbanization in multiple urban agglomerations from a landscape perspective. *J. Environ. Manag.* **2023**, *342*, 118129. [CrossRef] [PubMed]
24. Zhou, W.; Yu, W.; Qian, Y.; Han, L.; Pickett, S.T.A.; Wang, J.; Li, W.; Ouyang, Z. Beyond city expansion: Multi-scale environmental impacts of urban megaregion formation in China. *Natl. Sci. Rev.* **2022**, *9*, nwab107. [CrossRef] [PubMed]
25. Jin, X.; Jin, Y.; Mao, X. Ecological risk assessment of cities on the Tibetan Plateau based on land use/land cover changes—Case study of Delingha City. *Ecol. Indic.* **2019**, *101*, 185–191. [CrossRef]
26. Lin, Z.; Peng, S. Comparison of multimodel simulations of land use and land cover change considering integrated constraints—A case study of the Fuxian Lake basin. *Ecol. Indic.* **2022**, *142*, 109254. [CrossRef]
27. Shirmohammadi, B.; Malekian, A.; Salajegheh, A.; Taheri, B.; Azarnivand, H.; Malek, Z.; Verburg, P.H. Scenario analysis for integrated water resources management under future land use change in the Urmia Lake region, Iran. *Land Use Policy* **2020**, *90*, 104299. [CrossRef]
28. Xiong, M.Q.; Li, F.J.; Liu, X.H.; Liu, J.F.; Luo, X.P.; Xing, L.Y.; Wang, R.; Li, H.Y.; Guo, F.Y. Characterization of Ecosystem Services and Their Trade-Off and Synergistic Relationships under Different Land-Use Scenarios on the Loess Plateau. *Land* **2023**, *12*, 2087. [CrossRef]
29. Kang, J.; Fang, L.; Li, S.; Wang, X. Parallel Cellular Automata Markov Model for Land Use Change Prediction over MapReduce Framework. *ISPRS Int. J. Geo-Inf.* **2019**, *8*, 454. [CrossRef]
30. Liu, X.; Liang, X.; Li, X.; Xu, X.; Ou, J.; Chen, Y.; Li, S.; Wang, S.; Pei, F. A future land use simulation model (FLUS) for simulating multiple land use scenarios by coupling human and natural effects. *Landsc. Urban Plan.* **2017**, *168*, 94–116. [CrossRef]
31. Liang, X.; Guan, Q.; Clarke, K.C.; Liu, S.; Wang, B.; Yao, Y. Understanding the drivers of sustainable land expansion using a patch-generating land use simulation (PLUS) model: A case study in Wuhan, China. *Comput. Environ. Urban Syst.* **2021**, *85*, 101569. [CrossRef]
32. Xu, W.X.; Wang, J.M.; Zhang, M.; Li, S.J. Construction of landscape ecological network based on landscape ecological risk assessment in a large-scale opencast coal mine area. *J. Clean. Prod.* **2021**, *286*, 125523. [CrossRef]
33. Campbell, M.; Nel, V.; Mphambukeli, T.N. A thriving coal mining city in crisis? The governance and spatial planning challenges at Witbank, South Africa. *Land Use Policy* **2017**, *62*, 223–231. [CrossRef]
34. Li, S.C.; Xiao, W.; Zhao, Y.L.; Lv, X.J. Incorporating ecological risk index in the multi-process MCRE model to optimize the ecological security pattern in a semi-arid area with intensive coal mining: A case study in northern China. *J. Clean. Prod.* **2020**, *247*, 119143. [CrossRef]
35. Song, Y.-Y.; Xia, S.; Xue, D.Q.; Ma, B.-B.; Liu, X. Spatio-temporal differences and influencing factors of carbon emission equity in the Loess Plateau based on major function-oriented zones. *J. Geogr. Sci.* **2023**, *33*, 1245–1270. [CrossRef]
36. Nimmo, D.G.; Mac Nally, R.; Cunningham, S.C.; Haslem, A.; Bennett, A.F. Vive la resistance: Reviving resistance for 21st century conservation. *Trends Ecol. Evol.* **2015**, *30*, 516–523. [CrossRef] [PubMed]
37. Daily, G.C. *Nature's Services: Societal Dependence on Natural Ecosystems*; Island Press: Washington, DC, USA, 1997.
38. Peng, J.; Liu, Y.X.; Wu, J.S.; Lv, H.L.; Hu, X.X. Linking ecosystem services and landscape patterns to assess urban ecosystem health: A case study in Shenzhen City, China. *Landsc. Urban Plan.* **2015**, *143*, 56–68. [CrossRef]
39. Cairns, J.J.; Pratt, J.R. The Relationship Between Ecosystem Health and Delivery of Ecosystem Services. In Proceedings of the NATO Advance Research Workshop, Montebello, QC, Canada, 10–15 October 1993.
40. Xie, Z.L.; Li, X.Z.; Chi, Y.; Jiang, D.G.; Zhang, Y.Q.; Ma, Y.X.; Chen, S.L. Ecosystem service value decreases more rapidly under the dual pressures of land use change and ecological vulnerability: A case study in Zhujiajian Island. *Ocean Coast. Manag.* **2021**, *201*, 105493. [CrossRef]
41. Ma, K.; Kong, H.; Guan, W.; Fu, B. Ecosystem health assessment: Methods and directions. *Acta Ecol. Sin.* **2001**, *12*, 2106–2116.
42. Cai, G.; Xiong, J.F.; Wen, L.S.; Weng, A.F.; Lin, Y.Y.; Li, B.Y. Predicting the ecosystem service values and constructing ecological security patterns in future changing land use patterns. *Ecol. Indic.* **2023**, *154*, 110787. [CrossRef]

43. Yin, Z.; Feng, Q.; Zhu, R.; Wang, L.; Chen, Z.; Fang, C.; Lu, R. Analysis and prediction of the impact of land use/cover change on ecosystem services value in Gansu province, China. *Ecol. Indic.* **2023**, *154*, 110868. [CrossRef]
44. Costanza, R.; D'arge, R.C.; Groot, R.d.; Farber, S.B.; Grasso, M.; Hannon, B.M.; Limburg, K.E.; Naeem, S.; O'Neill, R.V.; Paruelo, J.M.; et al. The value of the world's ecosystem services and natural capital. *Nature* **1997**, *387*, 253–260. [CrossRef]
45. Xie, G.; Zhang, C.; Zhang, C.; Xiao, Y.; Lu, C. The value of ecosystem services in China. *Resour. Sci.* **2015**, *37*, 1740–1746.
46. Carpenter, S.R.; Brock, W.A. Adaptive Capacity and Traps. *Ecol. Soc.* **2008**, *13*, 1740–1746. [CrossRef]
47. Lai, S.; Sha, J.; Eladawy, A.; Li, X.; Wang, J.; Kurbanov, E.; Lin, Z.; Wu, L.; Han, R.; Su, Y.-C. Evaluation of ecological security and ecological maintenance based on pressure-state-response (PSR) model, case study: Fuzhou city, China. *Hum. Ecol. Risk Assess. Int. J.* **2022**, *28*, 734–761. [CrossRef]
48. Xia, C.; Dong, Z.; Chen, B. Spatio-temporal analysis and simulation of urban ecological resilience: A case study of Hangzhou. *Acta Ecol. Sin.* **2022**, *42*, 116–126.
49. Xue, F.; Zhang, N.; Xia, C.; Zhang, J.; Wang, C.; Li, S.; Zhou, J. Spatial evaluation of urban ecological resilience and analysis of driving forces: A case study of Tongzhou District, Beijing. *Acta Ecol. Sin.* **2023**, *43*, 6810–6823.
50. Xie, G.; Lu, C.; Leng, Y.; Zheng, D.; Li, S. Ecological assets valuation of the Tibetan Plateau. *J. Nat. Resour.* **2003**, *18*, 189–196.
51. Anselin, L. Local Indicators of Spatial Association—LISA. *Geogr. Anal.* **1995**, *27*, 93–115. [CrossRef]
52. Liao, Z.; Zhang, L. Spatio-temporal analysis and simulation of urban ecological resilience in Guangzhou City based on the FLUS model. *Sci. Rep.* **2023**, *13*, 7400. [CrossRef]
53. Liu, T.; Liu, H.; Qi, Y.J. Construction land expansion and cultivated land protection in urbanizing China: Insights from national land surveys, 1996–2006. *Habitat Int.* **2015**, *46*, 13–22. [CrossRef]
54. Cao, X. Regulating mine land reclamation in developing countries: The case of China. *Land Use Policy* **2007**, *24*, 472–483. [CrossRef]
55. Pan, N.; Guan, Q.; Wang, Q.; Sun, Y.; Li, H.; Ma, Y. Spatial Differentiation and Driving Mechanisms in Ecosystem Service Value of Arid Region:A case study in the middle and lower reaches of Shule River Basin, NW China. *J. Clean. Prod.* **2021**, *319*, 128718. [CrossRef]
56. Ye, S.; Ren, S.; Song, C.; Cheng, C.; Shen, S.; Yang, J.; Zhu, D. Spatial patterns of county-level arable land productive-capacity and its coordination with land-use intensity in mainland China. *Agric. Ecosyst. Environ.* **2022**, *326*, 107757. [CrossRef]
57. Serafy, S.E. Pricing the invaluable: The value of the world's ecosystem services and natural capital. *Ecol. Econ.* **1998**, *25*, 25–27. [CrossRef]
58. Opschoor, J.B. The value of ecosystem services: Whose values? *Ecol. Econ.* **1998**, *25*, 41–43. [CrossRef]
59. Gao, Y.; Wang, J.; Zhang, M.; Li, S. Measurement and prediction of land use conflict in an opencast mining area. *Resour. Policy* **2021**, *71*, 101999. [CrossRef]

land

MDPI

Article

Exploring Integrative Development of Urban Agglomeration from the Perspective of Urban Symbiosis and Production–Living–Ecological Function

Sijia Lin [1], Chun Li [2,*], Yanbo Li [1] and Liding Chen [2,3,4]

1 Institute of International Rivers and Eco-Security, Yunnan University, No. 2 North Cuihu Road, Kunming 650091, China; linsijia@stu.ynu.edu.cn (S.L.); yblee@ynu.edu.cn (Y.L.)
2 School of Ecology and Environmental Science, Yunnan University, No. 2 North Cuihu Road, Kunming 650091, China; liding@ynu.edu.cn
3 State Key Laboratory of Urban and Regional Ecology, Research Center for Eco-Environmental Sciences, Chinese Academy of Sciences, Beijing 100085, China
4 University of Chinese Academy of Sciences, Beijing 100049, China
* Correspondence: lichun@ynu.edu.cn

Abstract: Integrative development is an effective way to enhance urban potential and implement resource-optimal relocation, especially in urban agglomeration regions. Conventionally, the evaluation of urban integration is usually studied from one aspect of urban interaction intensity or urban functional similarity, but considering both together can better reflect the integrative condition of urban agglomeration. This paper introduces the symbiosis theory into the exploration of urban integration. The production–living–ecological function is taken to analyze urban function, and the improved radiation model is adopted to measure urban interaction. Under the framework of symbiosis theory, we integrate urban function and urban interaction to indicate the integrative condition of urban agglomeration from a production–living–ecological aspect. Urban agglomeration in the Central Yunnan Urban Agglomeration is taken as the study area. The results show that (1) spatial variations occur in high-value areas with distinct functions. The east emphasizes production and living, while the west leans towards ecology. (2) Urban agglomeration is in its early developmental stages without stable symbiosis. Interactions among counties mostly show sporadic point symbiosis, lacking stability. It mainly radiates outward from the central area, with more stable interactions in high-value areas, often causing inter-city competition. (3) Urban agglomeration integration is generally low, with distinct high-value production and ecological areas. The central, eastern, and southern regions exhibit strong production and living interactions, while the west benefits from ecological interactions. These findings can offer some insights for informing relevant policies and fostering the integrated development of urban agglomerations.

Keywords: integrative development; production–living–ecological function; symbiosis theory; interactive relationship; urban agglomeration

1. Introduction

The acceleration of economic globalization and urbanization has led to the rapid growth of urban populations, resulting in urban expansion and restructuring of urban areas [1]. The regional development model has transitioned from independent development of a single city to integrated development facilitated by resource interaction across multiple cities [2]. The regional development model has shifted from the independent development of individual cities to a city agglomeration or group with the core city at its center, driving the collective development of surrounding cities. Urban agglomeration develops as a unified entity, with each city mutually interacting and progressing together. It has been viewed as the predominant model for regional development in China. The integrated development of urban agglomerations is crucial for stabilizing internal dynamism and enhancing county competitiveness. Simultaneously, the integrated development of urban agglomerations is a crucial component of sustainable development. When examining the sustainable development level of urban agglomerations, it is essential to consider interactions and spillover effects between towns within the agglomerations, along with variations in the developmental levels of individual towns [3]. The goal of integrated development is to achieve rapid resource circulation and optimal allocation, including industrial specialization, market expansion, distribution of living resources, and synergistic CO_2 reduction effects [4–7]. Such integration allows the entire urban agglomeration system to not only generate income from individual cities but also realize economies of scale when organized as a whole. Integrative development within urban agglomeration is crucial for enhancing regional competitiveness, optimizing regional resource allocation, and fostering regional sustainable development [8].

In summarizing scholars' research on the integration of urban agglomeration, two research frameworks are generally involved. The first framework examines the interaction between counties, with a focus on the element flow among inner cities, such as economic interactions [9,10], production transportation [11], population flow [12], information flow [13,14], etc. Quantification of inter-county flow is often achieved through the use of methods such as the gravity model and its extensions [15,16], the field strength model [17] and the radiation model [18]. Additionally, the structure of the urban network is analyzed using social network analysis methods and complex network models to examine the interweaving of flow elements. The second framework focuses on urban functions. These researches typically explore urban functional similarities by analyzing economic complementarities [19], industrial structure similarity [8], commodity price convergence [20], etc. New economic geography models and trade theories are commonly employed for analysis within this research framework. Furthermore, regarding urban elements involved in related research, urban economy and industry have been widely concerned in the exploration of urban integration. Transportation, as a foundation for urban connectivity, has also been extensively studied. As research progresses, scholars have begun to address other integration issues, such as green development [21], ecosystem service value [22], coordinated emission reduction [7], culture tourism industries and culture development [23], etc.

Previous research has made significant contributions to exploring the integrated development of urban agglomerations. However, there are still unresolved issues that require in-depth investigation. First, the research's focus on urban connection cannot well capture the features of individual cities, and the research conducted based on the individual function features analysis may ignore the interactive condition among member cities. Combining urban interaction and urban function to comprehensively delineate urban development processes requires further research. Second, the integrated degree of the urban area is mostly assessed by synthesizing multiple indicators into a comprehensive measure. However, cities within an urban agglomeration have diverse functional positions, and the overall indicator might lack specificity in describing urban connections. While the overall degree of closeness is generally assessed, the specific modes of city connectivity remain underexplored.

To fix the gap, this paper endeavors to (1) introduce the symbiosis theory into the study of urban agglomeration integration. The symbiosis theory was originally defined as the coexistence of different species according to a certain relationship [24]. Counties are viewed as symbiotic units to observe how they co-exist in a symbiotic environment. The application of symbiosis theory in exploring the integrative development of urban agglomerations can be an effective approach to consider both urban interaction and urban function. (2) Utilizing the functions of production, living, and ecology to elaborate the integrated development of urban areas. Economy, society, and ecology are commonly regarded as the three subsystems of cities, which encompass the functions of production, living, and ecology [25]. These three functions comprehensively summarize the different functions provided by counties within urban agglomeration and serve as a suitable framework for distinguishing diverse functions among different urban agglomerations.

In this paper, urban function is summarized into the aspects of production, living, and ecology. An index system is built to establish the three functions of each town in urban agglomeration. Based on the developed condition of the urban production–living–ecology function, urban interaction and urban functional distance are separately measured by the improved radiation model and functional distance model. Adopting the framework of symbiosis theory, urban interaction and urban function are integrated to explore the integrative degree of urban agglomeration from the aspects of productive integration, living integration, and ecological integration. The Central Yunnan Urban Agglomeration (CYUA), which lies in southwestern China, is taken as a study area. The CYUA serves as the central hub of China's radiance toward South Asia and Southeast Asia. It strategically intersects the "Belt and Road" and the "Yangtze River Economic Belt" initiatives. Positioned as a significant economic growth pole in the western region of China, research on this area can promote prosperity and stability in border areas. We analyze the urban production–living–ecology function, urban interaction condition, urban symbiotic modes, and integrated development condition in the CYUA from 2000 to 2020. Some policy suggestions are proposed to guide the integrative development of the CYUA. This study provides a research framework for exploring the integrated development of urban agglomerations, not only for the CYUA but also for other counterpart areas.

2. Materials and Methods

2.1. Study Area

We chose the Central Yunnan Urban Agglomeration (CYUA) as the study area. The CYUA is located in the southwest mountainous plateau region of China, covering Kunming City, Qujing City, Yuxi City, Chuxiong Prefecture, and 7 counties in the northern Honghe Prefecture, a total of 49 counties, as shown in Figure 1. It has a land area of 111,400 square kilometers, a permanent population of 21.27 million, and a GDP of CNY 1.02 trillion, accounting for 28.3%, 44.1%, and 61.6% of the province, respectively. It has the most intensive transportation facilities, the highest degree of development, and the highest level of development in Yunnan Province. It is also the most important industrial agglomeration area in Yunnan Province [26]. However, compared with other urban agglomerations in China, the flow of regional elements in the inner CYUA is not frequent. According to the standards of the seventh national census results, the urban agglomeration has only 1 mega-city (5–10 million people), 2 medium-sized cities (1–3 million people), 5 type I small cities (200,000–500,000 people), and 33 type II small cities (Less than 200,000 people). Furthermore, researchers have pointed out that the urban function of member cities of the CYUA represents an extent of similarity. Investigating the integrative condition of the CYUA with regard to urban interaction and function is imperative not only to steer its own progress but also to be pivotal for the advancement of Yunnan province.

2.2. Data Source and Processing

The data used in this study mainly include socioeconomic statistical data, basic geographic information, road network data, and land use data. We use socioeconomic cross-sectional data from the CYUA every five years from 2000 to 2020. These data are collected from the Yunnan Statistical Yearbook (2000–2020) and the China Statistical Yearbook (county-level) (2000–2020). Basic geographic information is acquired from the National Geomatic Center of China. Road networks are obtained from the Open Street Map. Based on the traffic situation in different years, Google Earth is compared to adapt to the actual situation of the region at a specific time. Land use classification data with 30 m precision are obtained from the Resource and Environmental Science Data Center of the Chinese Academy of Sciences. The spatialization processing of all the GIS data is performed in ArcGIS 10.2.

Figure 1. Location of study area.

2.3. *Methodology*

In order to establish a reasonable system to classify urban functions and combine the characteristics of a single county and the interaction between multiple counties to jointly evaluate the integrated development of urban agglomerations, four issues should be considered: firstly, an index system is constructed to measure the production, living, and ecological functions of each town within the CYUA from the year 2000 to 2020. Secondly, based on the production, living, and ecological functions of each town, the improved radiation model is applied to obtain interaction intensity and interaction direction between towns; the functional distance model is employed to identify the specific interaction functions between towns. Thirdly, a symbiotic model is proposed to identify the organizational mode and behavioral pattern of towns based on interaction intensity, interaction direction, and interaction function. Finally, the urban integration index is constructed to explore the degree of coordinated participation of the CYUA and each town within it.

2.3.1. Evaluation Index System of Production–Living–Ecological Function

Based on the connotation of production, living, and ecological functions, we establish an index system of production–living–ecological functions. First, data from different standards need to be normalized. Second, the critic weight method is used to determine the weight of each index in the index system. Finally, by utilizing the aforementioned approach, we calculate the production, living, and ecological function indices of each county within the urban agglomeration, which allows us to better characterize the attributes of each county. The specific steps are as follows:

2.3.2. Establishment of Index System

We refer to the index systems used in previous studies [27,28] to select and consider the rationality and availability of data and establish an index system based on the status of the study area. Twenty-two indicators are selected to establish the evaluation index system presented in Table 1.

Table 1. Index system of the production–living–ecological function.

Level 1 Indicators	Secondary Indicators	Serial Number	Weight
Production function	Average fiscal revenue per land	X1	0.091
	Economic density	X2	0.11
	Grain production	X3	0.181
	The total output value of agriculture, forestry, animal husbandry, and fishery	X4	0.177
	Total tourism revenue	X5	0.101
	Number of units of industrial enterprises above designated size	X6	0.135
	The proportion of secondary and tertiary industries	X7	0.206
Living function	The total retail sales of social consumer goods	X8	0.11
	Population density	X9	0.128
	Average salary of employees in each city and county	X10	0.227
	Number of beds in health institutions per 10,000 people	X11	0.164
	Number of students in primary and secondary schools	X12	0.187
	Number of beds per 10,000 people in social welfare adoption units	X13	0.184
Ecological function	Forest cover rate	X14	0.176
	Bio abundance index	X15	0.164
	NDVI mean	X16	0.12
	Carbon storage based on invest	X17	0.16
	Ecosystem service value	X18	0.069
	Water coverage	X19	0.312

2.3.3. The CRITIC Method

The CRITIC (Criteria Importance Through Inter-criteria Correlation) method is used to calculate weights. It is an objective weighting method [29]. This method uses two fundamental weights to evaluate the relative importance of different indicators based on the strength of contrast and conflict between them. The strength of contrast is evaluated using the standard deviation to measure the size of the gap between the same index in the indexing scheme, while the conflict is reflected by the correlation coefficient between the indicators. Based on these concepts, the method constructs an indicator that contains both types of information, which indicates the amount of information contained in the j_{th} indicator. The formula for calculating this indicator is as follows:

$$C_j = \sigma_j \sum_{i=1}^{n} (1 - r_{ij}). \tag{1}$$

The larger the C_j, the greater the amount of information contained in the j_{th} evaluation indicator, so the objective weight of the j_{th} indicator is as follows:

$$\theta_j = \frac{C_j}{\sum_{j=1}^{m} C_j} \quad j = 1, 2 \ldots m. \tag{2}$$

2.3.4. The Improved Radiation Model for Urban Interaction

The radiation model, applied in urban studies, draws on the analysis of material dispersion and absorption processes in physics [30]. Considering the existence of intervening opportunities in urban interaction processes [18], the model characterizes the interaction intensity, which is influenced by the background conditions of both interacting parties. These background conditions include urban interaction potential, urban external capacity, urban interaction channels, and urban interference intermediaries. The specific model framework is shown below:

$$\mathrm{T_{ij}} = \mathrm{T_i}\frac{\mathrm{m_i m_j}}{(\mathrm{m_i} + \mathrm{s_{ij}})(\mathrm{m_i} + \mathrm{m_j} + \mathrm{s_{ij}})}, \tag{3}$$

where $\mathrm{T_{ij}}$ is the size of the interaction from location *i* to location *j*; $\mathrm{m_i}$ and $\mathrm{m_j}$ represent the interaction potential of location i and location j, which refers to the self-foundation of interaction between towns, characterized by the production, living, and ecological functions of towns in this study; $\mathrm{T_i}$ characterizes the external function of towns, measured using the location entropy model of the three functions of towns, where the size of the location entropy reflects whether the function has an external function. The specific formula is as follows:

$$L_{qij} = \frac{\mathrm{P}_{ij} \div \mathrm{P}_i}{\mathrm{P}_j \div \mathrm{P}}, \tag{4}$$

$$L_{ij} = \begin{pmatrix} 1, L_{qij} > 1 \\ 0, L_{qij} < 1 \end{pmatrix}, \tag{5}$$

$$T_i = \sum L_{ij}\left[\mathrm{P}_{ij} - \mathrm{P}_i(\mathrm{P}_j/P)\right], \tag{6}$$

where P_{ij}, P_i, P_j, and P represent the magnitude of the *j* functional quantity of town *i*, the sum of all functional quantities of town *I*, the weighted sum of all urban energy in the urban agglomeration, and the sum of all functional quantities in the urban agglomeration. When the location quotient > 1, it is considered that the function has an outward function; that is, the category function exceeds the average level in the region and can provide the category of services to other regions.

s_{ij} represents the potential size of intervening opportunities that may exist in the interaction between town *i* and town *j*. Based on the shortest path analysis of the road network, the interaction potential of the towns passed through is summed up and represented as the interfering intermediary between town *i* and town *j*. The formula is as follows:

$$\mathrm{s_{ij}} = \sum_{\mathrm{i=1}}^{\mathrm{n}} \mathrm{Z_k}, \tag{7}$$

where s_{ij} represents the potential interference opportunities that may exist in the interaction process between town *i* and town *j*, and *n* represents the number of towns passed through in the shortest path between town *i* and town *j* during the interaction process. Z_k represents the interaction potential of town *k*.

2.3.5. The Evaluation of Urban Functional Distance

This study uses the urban functional distance model to analyze the functional differences between counties. Based on urban production, living, and ecological function, the functional similarity between counties is evaluated. Generally, the larger the functional distance between towns, the lower the functional similarity between counties and the stronger the complementarity; otherwise, the higher the similarity, the stronger the competition. The specific formula is as follows:

$$D_{ij} = \sqrt[2]{(x_i - x_j)^2 + (y_i - y_j)^2 + \ldots + (z_i - z_j)^2}, \tag{8}$$

$$FD_{ij} = (D_{ij} - D_i)/\sqrt{\sum (D_{ij} - D_i)^2/n}, \tag{9}$$

where D_{ij} represents the functional distance between town i and town j. $x, y \ldots z$ represents the functions of the county and FD_{ij} is the functional distance index between town i and town j, which this paper characterizes by the production, living, and ecological functions of the county. When $FD_{ij} \geq 0.5$, it is considered that the functional heterogeneity between cities and towns is large and complementary. When $FD_{ij} \leq 0.5$, it is considered that the functional homogeneity between cities and towns is large and competition is formed. When $-0.5 < FD_{ij} < 0.5$, it is considered that the functional attributes between towns are not clear.

2.3.6. The Identification of Urban Symbiotic Pattern

Symbiosis, as a concept in ecology, emphasizes cooperation among populations. Symbiotic units, symbiotic patterns, and symbiotic environments are the three fundamental elements of the symbiosis theory framework. In urban symbiosis research, cities are usually taken as the basic unit, and symbiotic patterns refer to the ways in which cities interact with each other. The intermediary required for interaction is called the symbiotic interface, and other elements that affect symbiosis besides the symbiotic interface are referred to as the symbiotic environment. The symbiotic patterns between cities are based on their own attributes and are influenced by the symbiotic interface and environment. Symbiotic evolution is inherent in symbiotic systems, and local measures affecting city conditions can destabilize the region, prompting changes in the urban symbiosis system. The evolution of the symbiosis system bears similarities to the integration process of urban agglomerations, making the application of symbiosis theory a valuable approach to evaluate regional integration.

There are two symbiotic modes in the urban symbiosis system: organizational models and behavioral models. We use symbiosis patterns classification [31] to classify the organizational patterns and behavioral patterns of urban agglomerations. The organizational model gauges the size and intensity of interactions between cities, emphasizing the "absolute value" of these interactions. Behavioral patterns, on the other hand, capture the manner and nature of urban interactions, collectively forming a comprehensive symbiotic relationship. Symbiosis entails both positive and negative effects on the entities involved. Symbiotic units engage in both struggle and cooperation, driven by a mechanism of competition and cooperation that propels mutual advancement and individual excellence.

The organizational pattern is identified based on the interaction intensity and direction between cities, including four categories: point symbiosis, intermittent symbiosis, continuous symbiosis, and integrated symbiosis, which characterize the connections between cities from accidental and occasional interactions to gradually evolving into selective interactions with continuity, and finally reaching a stable state of integrated symbiosis driven by endogenous forces, as shown in Table 2. In this study, the global natural breakpoint method is used to classify the interaction intensity.

Table 2. Symbiotic organizational pattern type.

Organizational Pattern	The Degree of Urban Interaction	Urban Development Model
Point symbiosis	An interaction and connection occur by chance	Independent development under self-organization
Intermittent symbiosis	Have some selective interaction	External forces gradually influence urban development
Continuous symbiosis	There is a continuous interaction	Town-linked development
Integrated symbiosis	Interactions with continuity, stability, continuity, and endogenous dynamics	Integrated and coordinated development of cities and towns

The behavioral pattern is identified from the functional differences of towns, and through the comparison of the positive and negative functions between towns, it is divided into six categories: mutual benefit symbiosis, parasitism, competition symbiosis, partial harm symbiosis, partial benefit symbiosis, and irrelevant symbiosis, as shown in Table 3.

Table 3. Types of symbiotic behavioral patterns.

F_{dji} \ F_{dij}	[0.5, +∞)	(−0.5,0.5)	(−∞,0.5]
[0.5, +∞)	mutualism	partial benefit symbiosis	parasitism
(−0.5,0.5)	partial benefit symbiosis	irrelevant symbiosis	partial harm symbiosis
(-∞,0.5]	parasitism	partial harm symbiosis	competition symbiosis

2.3.7. The Evaluation of Integrative Degree of Urban Agglomeration

The degree of urban agglomeration integration is an important indicator of the degree of urban agglomeration development. We devise an Urban Agglomeration Integration Index by leveraging the organizational and behavioral models of urban agglomeration. This index assesses the extent of active participation of each county within the urban agglomeration integration framework, based on a benign symbiosis model, throughout the urban agglomeration's development. The fundamental framework is outlined as follows:

$$\mathrm{Int-Index_i} = \omega \mathrm{UI_{ij+}} / \mathrm{SUM_{UI}}. \tag{10}$$

Among them, $\mathrm{Int-Index_i}$ represents the integration index of town i, $\omega \mathrm{UI_{ij+}}$ represents the weighted interaction intensity from town i to town j under the benign symbiosis mode, and $\mathrm{SUM_{UI}}$ represents the weighted total value of all interaction intensities of town i.

Throughout the development of an urban agglomeration, it undergoes a non-linear, spiraling process. Cities within the agglomeration, at different stages, work towards a common goal, engaging in varying degrees of coordinated action. We adopt the coordinated development standard of urban agglomeration to identify the integrative condition of urban agglomeration exhibited by the $\mathrm{Int\text{-}Index_i}$ [32], as delineated in Table 4. In instances where the integration level falls below 10%, it is referred to as the assist stage, with cities operating independently and engaging in limited collaboration. Upon reaching an integration degree of 10–20%, it is called the cooperative stage. Cities are still relatively independent of each other but are beginning to strengthen their ties, cooperation, and contact. Upon reaching an integration degree of 20–30%, it is called the coordination stage, accompanied by an escalation in conflicts between cities, passing multiple conflicts; only the joint cooperation of cities can solve the problem. Subsequently, when the integration degree reaches 30–40%, the coordination level of urban agglomerations has entered the concordant stage, marked by a substantial increase in inter-city conflicts; cities begin to form alliances to form a community with a shared future. Advancing to an integration degree of 40–50%, it is referred to as the collaborative stage; at this stage, the total number of problems and conflicts faced by each city and the conflicts resolved through collaborative methods are basically the same, and cities exhibit a heightened inclination for mutual development. Upon attaining an integration degree of 50–70%, it is referred to as the co-vibration stage. At this stage, there are far more regional problems and conflicts resolved through collaborative methods among cities than through non-collaborative methods. At the 70–85% integration range, cities establish a closely-knit cooperation mechanism, signifying the attainment of the integration stage. This phase is characterized by a high level of integration, reflecting cities' strong commitment to collaborative development. Finally, when the integration degree surpasses 85–100%, urban agglomerations achieve a state akin to a single city; this is referred to as the urbanization stage.

Table 4. Urban agglomeration integration stage table.

Serial Number	Stage Name	Degree of Collaboration %
1	Assist stage	0–10
2	Cooperative stage	10–20
3	Coordination stage	20–30
4	Concordant stage	30–40
5	Collaborative stage	40–50
6	Co-vibration stage	50–70
7	Integration stage	70–85
8	Urbanization stage	85–100

3. Result

3.1. The Production–Living–Ecological Functions in the CYUA

Based on the evaluation index system of production–living–ecological function, the quantitative and spatiotemporal features of production function (PF), living function (LF), and ecological function (EF) in the CYUA were calculated. Figure 2 shows the average number of these three functions in each county of the urban agglomeration in 2000, 2005, 2010, 2015, and 2020. From the perspective of the entire urban agglomeration, three functions continuously increased from 2000 to 2020. Among them, the living function had the fastest growth, from 0.08 to 0.38. The production function increased from 0.13 to 0.30, and the ecological function index was in a growing state; however, the increase was not significant, rising from 0.32 to 0.34.

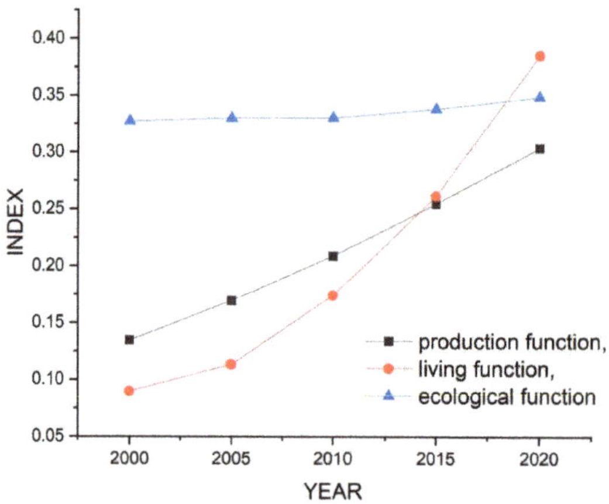

Figure 2. Changes of the production–living–ecological functions in the CYUA.

From the perspective of each county in the urban agglomeration, we used the global natural breakpoint method to divide the functional index into five levels: low, low, medium, high, and high, and observe the quantitative changes and spatial distribution. Figure 3 visually displays how these changes occurred spatially.

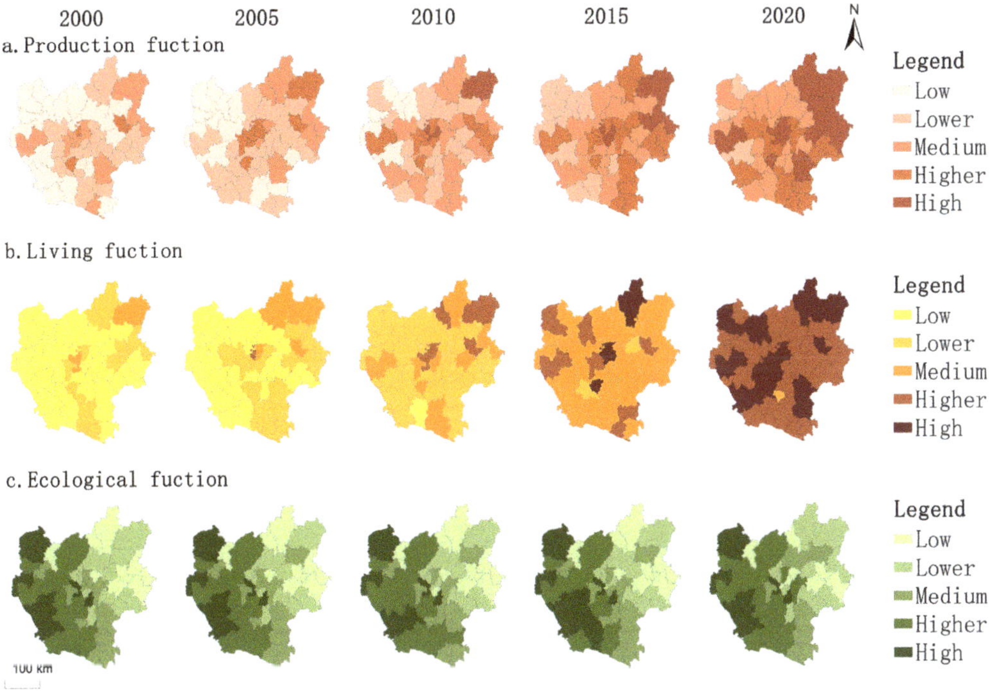

Figure 3. Spatiotemporal features of production–living–ecological functions.

Figure 3a illustrates the spatiotemporal fluctuation of the production function. High production functions are primarily distributed in the central and eastern regions of the CYUA, with low-value concentrations prevalent in its western reaches. As temporal and spatial changes, the production functions of all counties demonstrated notable improvements between 2000 and 2020. The distribution of high-value areas of the living function index is relatively sporadic, mainly in the central part of the urban agglomeration. As shown in Figure 3b, the change process presents a radial structure in which the central part drives the surrounding areas. As shown in Figure 3c, the distribution of ecological functions is high in the west and low in the east. Between 2000 and 2020, the functional spatial pattern of ecological function remained essentially constant, with only marginal changes in quantity.

3.2. The Symbiotic Mode of Cities in the CYUA

Based on the identification of counties' production, living, and ecological functions, this study carried out urban interaction measurement based on an improved radiation model and analyzed 2352 interactive relationships among 49 counties in the CYUA from 2000 to 2020. It was found that the total interaction strength, average interaction strength, and maximum interaction strength of production functions in CYUA were 1.69, 0.000144, and 0.026, respectively; life functions were 1.22, 0.000104, and 0.035; ecological functions were 2.49, 0.000212, and 0.028. At the same time, this study calculated the functional distance of towns based on the functional distance model. During the study period, the maximum functional distance reached 3.38. Based on the evaluation results of interaction strength and urban function distance, the symbiotic modes of towns can be identified under the framework of symbiosis theory.

3.2.1. The Organized Mode of Urban Symbiosis in the CYUA

The organized mode of urban symbiosis focuses on the intensity and direction of interaction among counties. Based on the calculation of the improved radiation model, we employed the global natural breakpoint method to classify urban interaction into four classes, which separately represent point symbiosis, intermittent symbiosis, continuous symbiosis, and integrated symbiosis. Figure 4 illustrates the distribution and evolution of the organizational mode of urban symbiosis in the CYUA from 2000 to 2020.

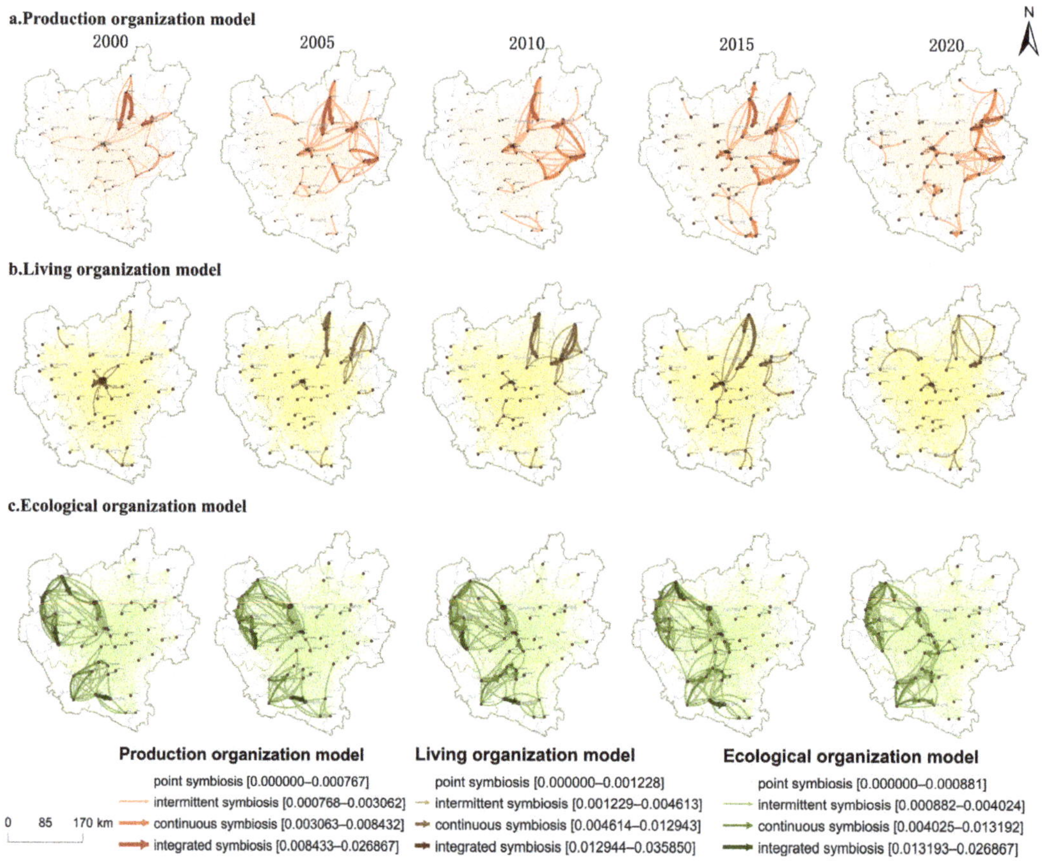

Figure 4. The organized mode of urban symbiosis.

The production function organization model, depicted in Figure 4a, is primarily characterized by point symbiosis, which constitutes more than 96% of all organizational models. Point symbiosis is commonly observed among secondary counties or those located far apart. Intermittent symbiosis is prevalent between primary and secondary counties within urban agglomerations, whereas continuous and integrated symbioses tend to occur more consistently among primary counties.

The living function organization model is depicted in Figure 4b. The evolution and distribution of the living function organizational model are somewhat dispersed, primarily demonstrated by the central urban area of Kunming acting as a growth pole that extends outward and stimulates reciprocal growth among adjacent sub-poles. This model predominantly follows the pattern of pole-driven development in surrounding counties. Point symbiosis is the dominant organizational model for living functions across the entire urban agglomeration, accounting for over 98% of all living symbiotic models. Integrated symbiosis is mainly observed in the middle and northwest regions of the CYUA, while continuous and intermittent symbiotic patterns tend to occur in high-value areas of living functions scattered throughout the southern, central, and eastern parts of the urban agglomeration.

The ecological function organization model is depicted in Figure 4b. High values of ecological function symbiotic organization span across the western and southern portions of the urban agglomeration. The organizational model is dominated by point symbiosis in up to 95% of all models. The direction of high ecological function symbiotic organization is opposite to that of production and living, which is mainly from west to east.

In general, urban agglomeration is still at a low level of symbiotic organization, and the three functions are all in a low-level interaction range. A stable organizational model with endogenous driving forces has not been formed.

3.2.2. The Behavior Mode of Urban Symbiosis in the CYUA

Combining the functional parity among counties, which is assessed by the functional distance model, the behavior mode of urban symbiosis can be identified based on both urban interaction and urban function. The quantitative characteristics of symbiotic behavior patterns in the CYUA are presented in Table 5. The dominant symbiotic patterns in urban agglomeration include competitive symbiosis, favorable symbiosis, and irrelevant symbiosis. Competitive symbiosis represents the largest proportion, while parasitic symbiosis accounts for the smallest proportion. Competitive symbiosis occurs between symbiotic units.

Table 5. Quantitative characteristics of symbiotic behavior patterns in the CYUA.

Year	Irrelevant Symbiosis	Partial Benefit Symbiosis	Mutualism	Parasitism	Partial Harm Symbiosis	Competition Symbiosis
2000	211 (17.94%)	258 (21.94%)	210 (17.86%)	24 (02.04%)	167 (14.20%)	306 (26.02%)
2005	177 (15.05%)	285 (24.23%)	194 (16.50%)	37 (03.15%)	156 (13.27%)	327 (27.81%)
2010	182 (15.48%)	268 (22.79%)	212 (18.03%)	32 (02.72%)	171 (14.54%)	311 (26.45%)
2015	181 (15.39%)	270 (22.96%)	213 (18.11%)	35 (02.98%)	164 (13.95%)	313 (26.62%)
2020	240 (20.41%)	240 (20.41%)	203 (17.26%)	26 (02.21%)	166 (14.12%)	301 (25.60%)

Figure 5 shows the spatial distribution of the three main behavioral modes of the CYUA, namely competition symbiosis, mutualism, and partial benefit symbiosis. In this paper, the cost of travel time was used to represent the interaction distance between counties. Among them, the average time cost between competition symbiosis was 247.42 min, the average time cost between mutualism was 333.44 min, and the time cost between partial benefit symbiosis was 302.87 min, which represents the distance of interaction between counties. Competition symbiosis < partial benefit symbiosis < mutualism. This shows that, in the CYUA, competition symbiosis tends to occur between counties that are close together, while mutualistic symbiosis occurs between counties that are far apart.

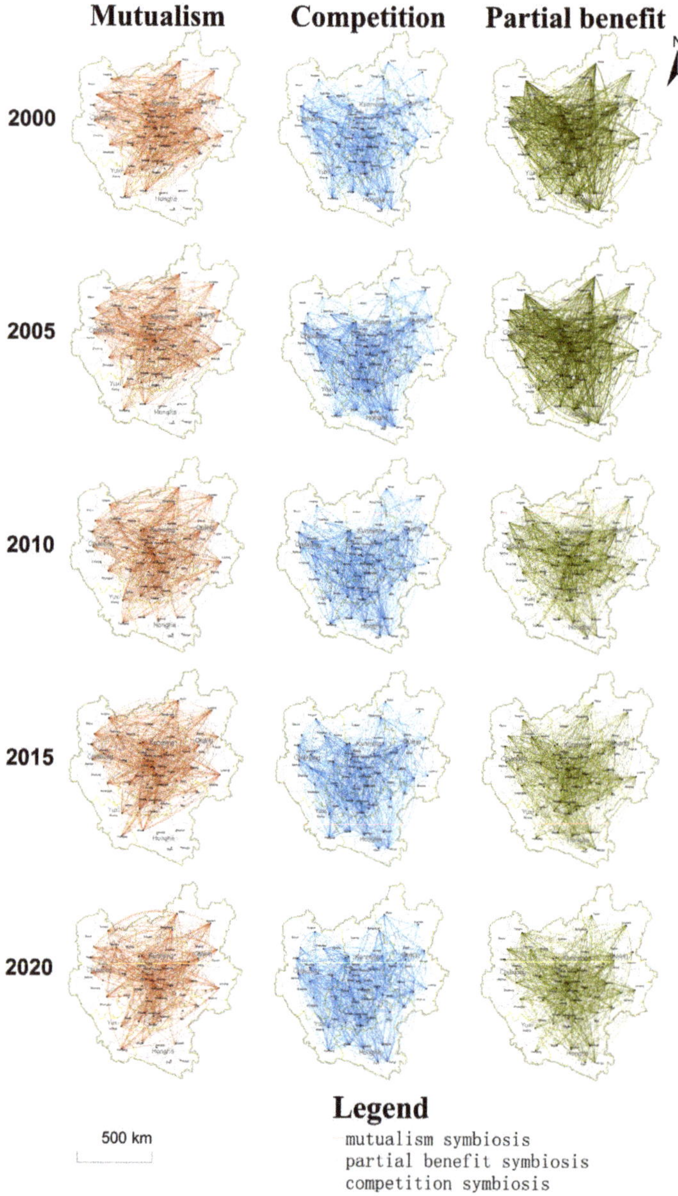

Figure 5. Distribution diagram of the three main behavioral patterns.

3.2.3. The Organized Mode and the Behavior Mode of Urban Symbiosis in the CYUA

We combined the organizational model and the behavioral model to demonstrate the current state of symbiosis between counties in the CYUA. Figure 6 shows the symbiosis model of the CYUA in 2020. Figure 6a shows the production function symbiosis model, in which counties are mainly concentrated in the central and eastern regions. The counties with strong interaction in the central part are mainly based on mutualism, while the eastern part is mainly based on competition symbiosis and partial-harm symbiosis. It is necessary to adjust the industrial structure to achieve a mutually beneficial development model. Figure 6b clearly shows the symbiosis model of living functions. It can be seen that the interaction intensity of living functions is generally low, the radiation range is small, and the distribution of behavioral patterns is relatively even. Figure 6c shows the ecological functions. In the symbiosis model, counties are mainly concentrated in the central and western regions. The western and central regions mainly adopt mutually beneficial symbiosis or partial-benefit symbiosis, while the counties in the western region mainly adopt competitive symbiosis and partial-harm symbiosis.

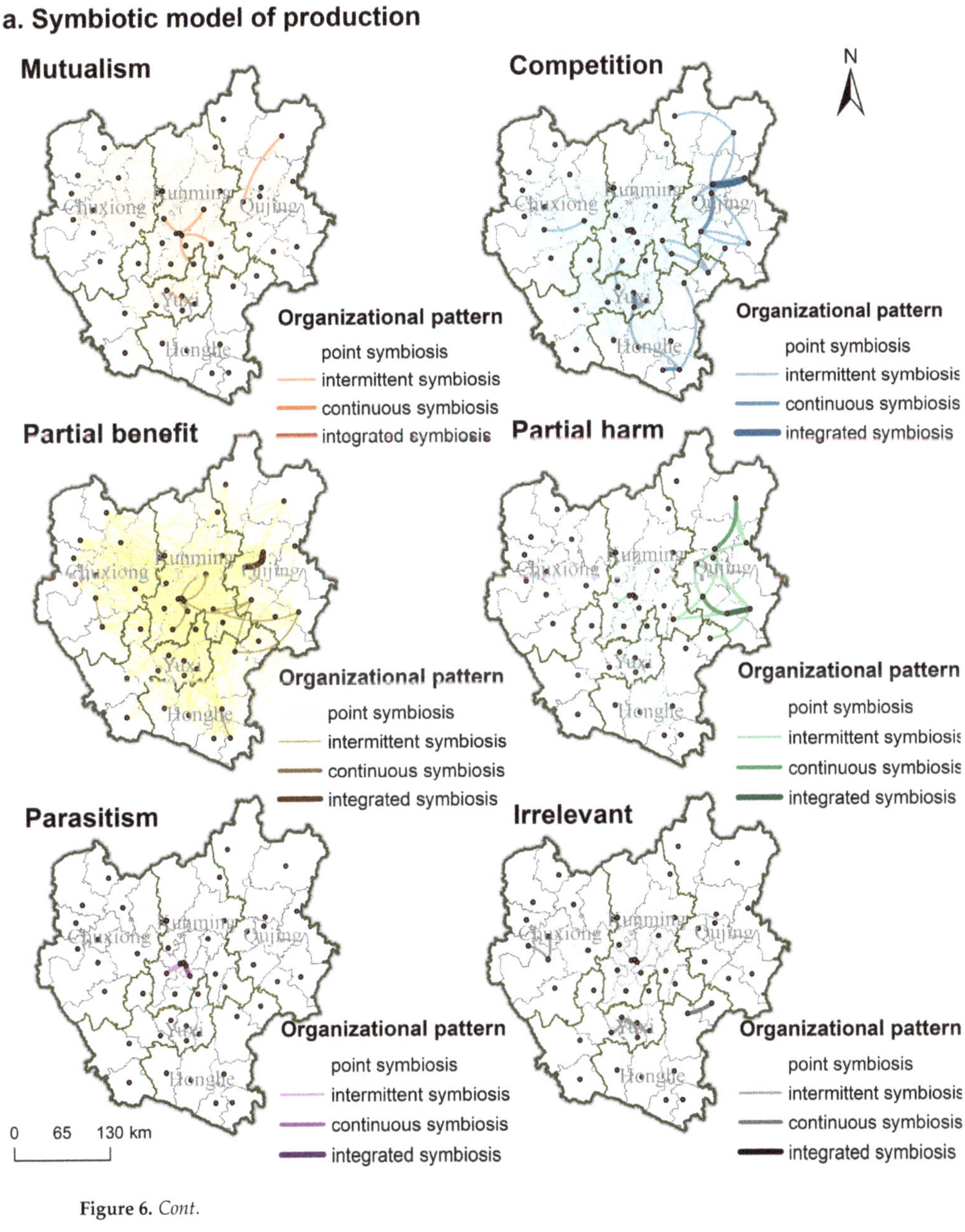

Figure 6. *Cont.*

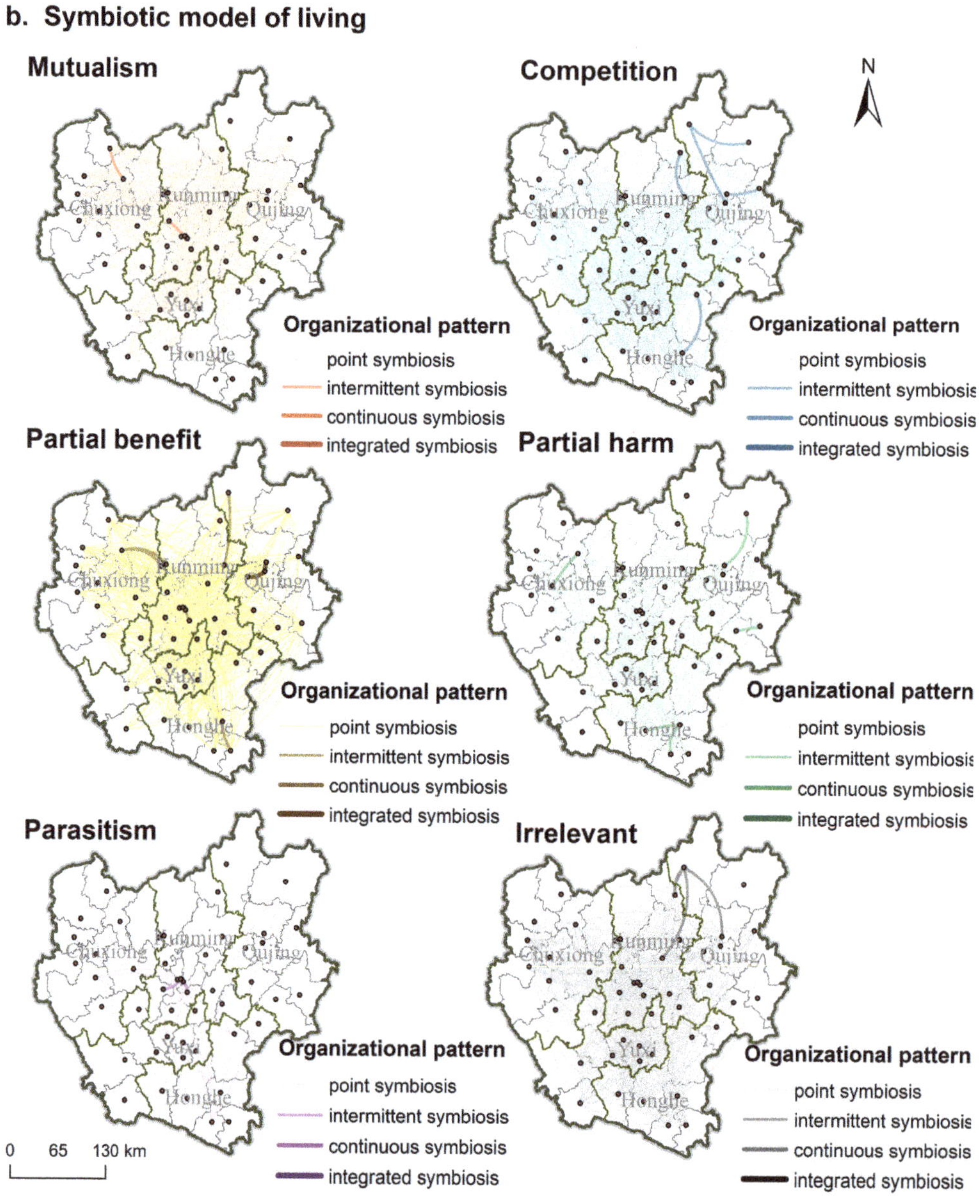

Figure 6. *Cont.*

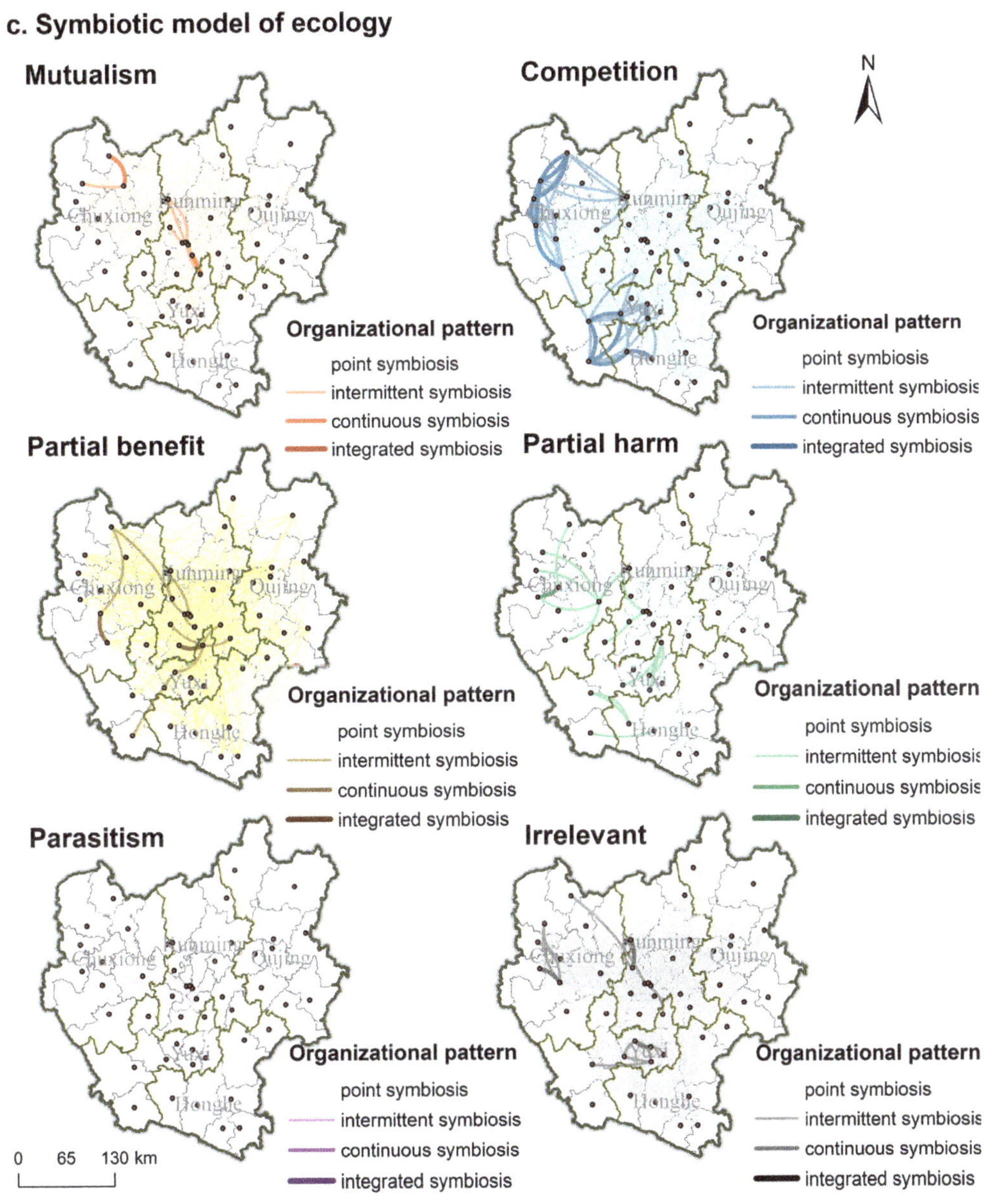

Figure 6. (**a**) Spatial distribution of production symbiosis models in 2020. (**b**) Spatial distribution of living symbiosis models in 2020. (**c**) Spatial distribution of ecological symbiosis models in 2020.

The CYUA is an urban agglomeration developed with central cities at its core. However, the dynamics within Kunming's urban area are predominantly competitive. On the contrary, positive interactions can be observed with counties in other areas, such as Yuxi. Yuxi, located in the six-river ecological zone, prominently develops living and ecological functions, fostering a positive symbiotic relationship with Kunming, which primarily focuses on the industrial sector. This phenomenon underscores that, in terms of functional planning, the primary focus remains on city areas. There is a deficiency in functional division planning within the counties, resulting in positive interactions between counties in neighboring cities but competitive dynamics and other symbiotic models within the city itself.

3.3. The Integrated Degree of the CYUA from the Perspective of Production–Living–Ecological Symbiosis

The level of integration was evaluated by analyzing the positive interaction generated by each county. As shown in Figure 7, we calculated changes in the quantitative characteristics of the integrated degree of the CYUA. Figure 8 shows changes in the spatial distribution of the integrated degree of the CYUA.

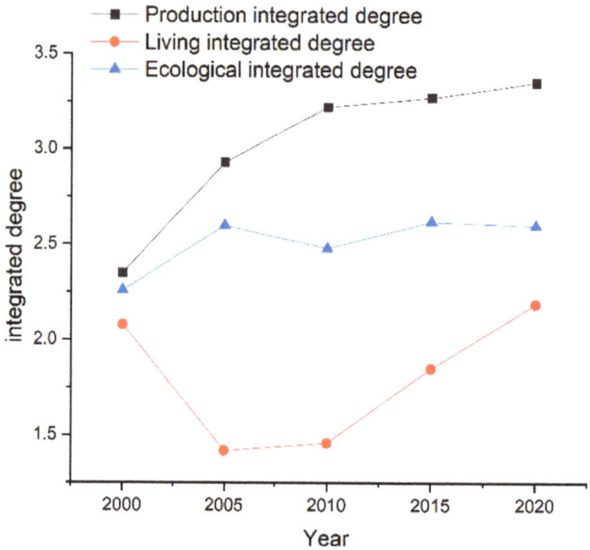

Figure 7. Changes in the quantitative characteristics of the integrated degree of the CYUA from 2000 to 2020.

As shown in Figure 8, the integrated level of the production function shows a continuous upward trend from 2000 to 2020. The spatial distribution is shown in Figure 8. Figure 8a shows the spatial distribution of the production integrated degree. It can be seen that the degree of participation in the production integration of urban agglomerations gradually increased. The cities entering the early stage of development increased from 12% to 26% regarding the distribution of urban production integration participation. The situation also developed from discrete distribution to gradually contiguous development. However, it was mainly concentrated below the coordination stage, and the development of integrated production in urban agglomerations was in its early stages. Figure 8b shows the spatial distribution of the living integrated degree. The degree of participation in the living integration of urban agglomerations is low. From 2000 to 2020, the degree of increase was not obvious, from the original 12% to 14%, and development was in the early stage, mainly concentrated in the central region, with scattered urban distribution and no stable development. Figure 8c shows the spatial distribution of the ecological integrated degree. The overall ecological integration of urban agglomerations was also in its early stages. From a distribution perspective, it has not formed a stable development model. The ecological integration of the eastern and southern regions was involved to a lesser extent.

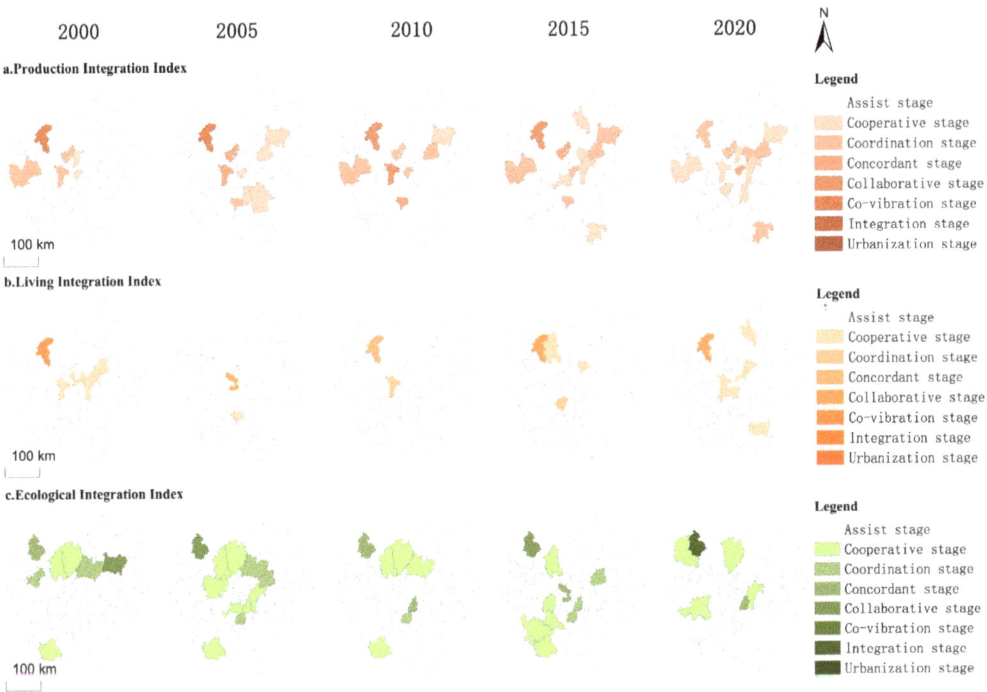

Figure 8. Changes in the spatial distribution of the integrated degree of the CYUA from 2000 to 2020.

In general, from 2000 to 2020, urban agglomeration was still in its initial stage of development, with few counties capable of participating in integrated development to a significant extent.

4. Discussion

4.1. Symbiotic Modes of Towns Derived from Production–Living–Ecology Conditions in the CYUA

The symbiotic mode of counties from the perspective of production, living, and ecology is continuously improved but still at a low level. Reasonable organizational and behavioral modes among counties are conducive to coordinated development. In order to promote urban agglomeration development, the following aspects must be considered: (1) similar functions tend to cluster regionally, leading to competitive dynamics among counties in areas with high functional organization. In the CYUA, central cities can form a good interactive relationship with the surrounding areas, but the surrounding areas are mainly in a competitive state. It is necessary to clarify the functional positioning of small cities and strengthen the driving ability of sub-poles. (2) The city's own resource endowment and interaction intensity complement each other, and the stable interaction between counties is rooted in a certain degree of endogenous forces. (3) The development of transportation enhances the level of production organization while hindering the level of ecological organization. The development of transportation within the CYUA is primarily centered around Kunming and radiates outward. This can be observed through the organization models of production and living functions. During the period from 2010 to 2015, Honghe Prefecture, located in the southern part of the urban agglomeration, was officially incorporated into the urban agglomeration planning. The establishment of the Yuxi–Mengzi Railway connected Honghe Prefecture with Yuxi. It is evident that the degree of production organization in the southern and central regions experienced a significant increase during 2010–2015, while the degree of ecological organization was subdued. In the development of urban agglomeration, the improvement of transportation deepens communication among counties. Urban expansion may lead to the decline of the carbon sequestration capacity of vegetation [33] and the loss of biodiversity [34], thus affecting the decline of the overall ecological function interaction of urban agglomeration.

4.2. Integration Development of the CYUA from the Perspective of Production–Living–Ecology

Figure 9 shows the development plan of the CYUA. The CYUA plans to build the Kunming metropolitan area as its center, the integrated development area of Qilin, Zhanyi, and Malone districts as its sub-core, and the Yuxi, Chuxiong, and Honghe Prefectures as three sub-polar areas, including Hongta, the Jiangchuan Group development area, Chuxiong, Nanhua, the Muding Group development area, and the Mengzi Old opening source group development area. This is mainly based on the ecological pattern and relies on the transportation axis to form a two-way "T" shape pattern; that is, a point-axis linkage development pattern with Chuxiong, Kunming, and Qujing as the horizontal axis and Kunming, Yuxi, and Honghe as the vertical axis. Figure 10 shows the current distribution of major functional areas in urban agglomerations.

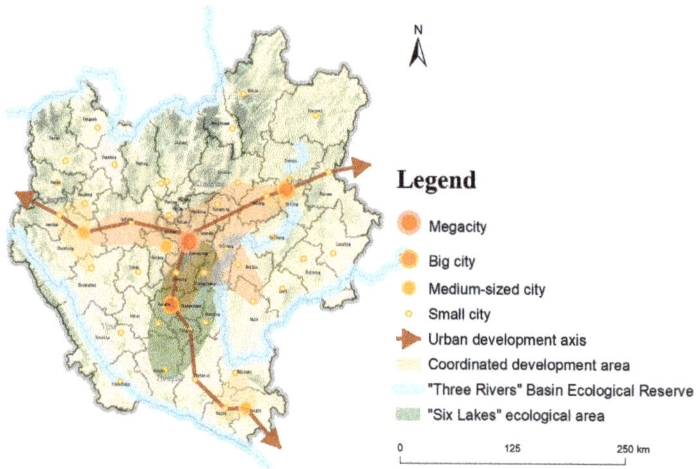

Figure 9. Development plan of the CYUA.

According to the integration degree of urban agglomerations calculated in this study, from the perspective of production function integration, the overall integration degree of the CYUA is on the rise, and the integration degree of counties on the Kunming–Qujing development axis is gradually deepening, while the integration degree of cities and towns on the Kunming–Chuxiong development axis has no obvious change. From the perspective of living function integration, the overall participation of the living integration degree is low, and the urban development pattern is mainly independent development. From the perspective of ecological function integration, the counties participating in ecological integration are mainly concentrated in ecological function areas, as shown in Figure 10. The integration degree of the central region has been improved under the governance of the six lakes area, while the ecological integration degree of the ecological protection area in the eastern Pearl River Basin needs to be strengthened. On the whole, existing planning mainly focuses on the central provincial capital cities and spreads to the surrounding areas. In the process of urban development in China, the gap in social and economic development in prefecture-level cities is more obvious than that in provincial capital cities [35]. How to coordinate and organize among these cities is also the focus of the coordinated development of the entire urban agglomeration.

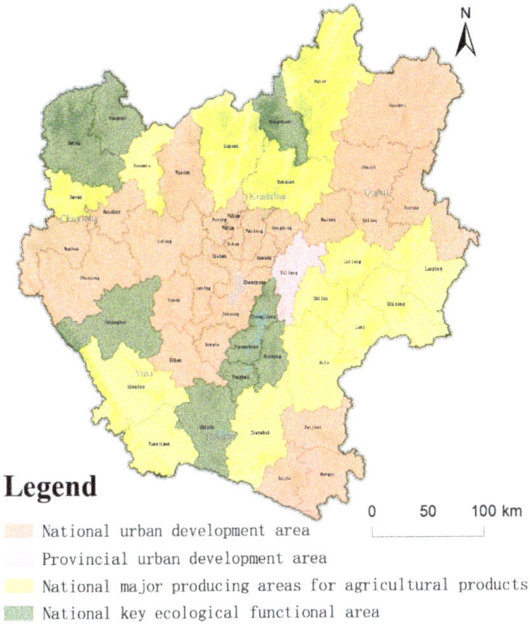

Figure 10. Distribution of main functional areas in the CYUA.

Based on the current development situation, this article proposes the following recommendations for strengthening urban agglomeration in the future: (1) regarding production, there is an imbalance in development between the east and west. It is essential to enhance connectivity between the central and western regions, bolstering the industrial functions of the western part and reinforcing its radiation ability to the surrounding areas. (2) In terms of living, currently, participation in the integration of life functions in urban agglomerations is not substantial, with most focusing on isolated development. It is crucial to increase investment in the construction of life functions and expand the investment capacity in living functions. (3) Regarding the ecological aspect, the involvement in ecological integration primarily centers around western cities, with insufficient participation from eastern cities. There is a need to enhance ecological zone governance in eastern cities and strengthen ecological integration efforts in the eastern region.

4.3. The Advantage of Refined Analysis of Integration Development from Production–Living–Ecology Angle

In the course of urban agglomeration development, the interaction among cities is inevitable. The focal point in urban research is how to achieve mutually beneficial outcomes in urban development through these interactions. Conventional research on urban agglomeration development typically employs a comprehensive indicator to investigate interconnection among cities. For instance, by utilizing socioeconomic statistics, traffic data, and network big data, the Urban Connectivity Index (UCI) was developed to gauge the strength of urban connections and assess the correlation between the closeness of urban connections and urban development [36]. Although this method employs comprehensive indicators to depict the intensity among cities, further exploration of the urban symbiosis mechanism requires refinement due to variations in functional positioning among towns. For example, employing spatial big data to examine city connections from a network perspective [37] or investigating the impact of urban agglomeration integration on green development from a market integration viewpoint provides more nuanced insights into urban closeness [21]. These studies delve into specific aspects based on the foundation of urban integrated development. This article sought to go a step further by subdividing the diverse functional positioning of cities beyond a general depiction of city connections. When discussing the integration process of urban agglomerations, we assessed it from three perspectives: production, living, and ecology. China has adopted the practices of production, living, and ecological functions as key measures to attain sustainable urban development. On the one hand, this provides a suitable overview of different functions and illustrates the interaction among towns with different functions. On the other hand, it avoids excessive detail, ensuring an overarching understanding.

Second, this article utilized the radiation model to gauge the interaction between towns. According to spatial interaction theory, key conditions for town interaction encompass complementarity, accessibility, and intermediary interference effects. In comparison with other spatial interaction models, the radiation model is better equipped to consider the impact of intermediary interference effects, leading to a more accurate measurement of spatial interactions between towns.

Third, we also considered the impact of urban functions and interactions. We identified the two-way interaction status between counties under the benign symbiosis model to prevent the overestimation of urban, comprehensive development caused by ineffective interactions. This approach ensured a more accurate measure of the degree of integration within urban agglomeration.

5. Conclusions

5.1. Conclusions

This study focused on the integrative development of urban agglomeration, assessing urban functions and urban interaction across three dimensions: production, living, and ecology. Each county's functional scale was quantified using an index system, while county relationships were examined utilizing both the improved radiation model and functional distance model. Building upon the calculations mentioned above, we drew from symbiosis theory, delving into the study of organizational and behavioral models from 2000 to 2020. Through these analyses, we evaluated the integration status of urban agglomeration. The results show that (1) the overall integration of production functions of the CYUA is in the primary stage. Examining the developmental trajectory over the past two decades, the production integration value of cities within the agglomeration demonstrated an ascending trend. The number of counties participating in the construction of production integration increased, and spatial contiguous development gradually formed. In spatial distribution, the counties actively involved in the integrated development of production within the urban agglomeration shifted from fragmented states to cohesive and coordinated development patterns emanating outward from the center. (2) The overall integration of living functions of the CYUA is in the primary stage. The degree of integration of life functions was the lowest among the three functions. Meanwhile, the overall integration trend showed an upward trend. Fewer counties are participating in the integration of living, mainly in the central and northern regions, and their distribution is sporadic, without forming a stable development model. (3) The overall integration of ecological functions of the CYUA is in the primary stage. Counties participating in ecological integration are mainly concentrated in the west and north of the CYUA. During the development process of the urban agglomeration, the distribution of ecological integration gradually changed from the original contiguous form to fragmentation. This shows that urban development has a certain inhibitory effect on the integration of ecological functions.

This study exclusively examined the developmental model of urban agglomeration integration, focusing on three dimensions: production, living, and ecology. The objective was to investigate the urban agglomeration development model within the framework of these three functions. However, there is still a need to delve into how these three functions interact during the integration process. Additionally, further research is required to explore external influences on urban agglomerations and other pertinent aspects.

5.2. Limitations and Prospects

This article employed symbiosis theory to examine the integration of urban agglomerations, focusing on three dimensions: production, living, and ecology. This has contributed to the advancement of urban agglomeration development. Nevertheless, certain limitations are acknowledged. Firstly, the exploration of symbiosis within the urban agglomeration employs the county as a unit but neglects the urban–rural relationship, necessitating further exploration in subsequent research. Secondly, the article concentrated on the symbiosis model within the urban agglomeration, recognizing that the region is not a closed system but an open one. As the urban agglomeration interacts internally, it also engages with surrounding cities, requiring additional investigation into the nature of this interaction. Thirdly, the paper utilized the radiation model to simulate the internal interaction of urban agglomeration. In subsequent research, the model can be refined, and its effectiveness enhanced with the support of geographical big data and other relevant information. Fourthly, the interaction between the production, living, and ecological functions of the urban agglomeration was discussed separately; however, understanding how these functions influence each other warrants further exploration. Finally, the CYUA discussed in this article represents an urban agglomeration in its early developmental stages. Future research will continue to explore urban agglomerations at various stages of development.

Author Contributions: Conceptualization, S.L. and C.L.; methodology, S.L. and C.L.; validation, S.L. and C.L.; formal analysis, S.L.; investigation, S.L.; resources, S.L., C.L., Y.L. and L.C.; data curation, S.L.; writing—original draft preparation, S.L.; writing—review and editing, S.L. and C.L.; visualization, S.L. and C.L.; supervision, C.L., Y.L. and L.C.; project administration, C.L.; funding acquisition, C.L. All authors have read and agreed to the published version of the manuscript.

Funding: This research was financially supported by the National Natural Science Foundation of China (No. 42201290) and the Applied Basic Research Programs of Science and Technology Department of Yunnan Province (No. 202001BB050027, No. 202301AT070335).

Data Availability Statement: The data presented in this study are available upon request.

Conflicts of Interest: The authors declare no conflict of interest.

References

1. Liu, Y.; Gao, H.; Cai, J.; Lu, Y.; Fan, Z. Urbanization Path, Housing Price and Land Finance: International Experience and China's Facts. *Land Use Policy* **2022**, *113*, 105866. [CrossRef]
2. Fang, C.; Yu, D. Urban Agglomeration: An Evolving Concept of an Emerging Phenomenon. *Landsc. Urban Plan.* **2017**, *162*, 126–136. [CrossRef]
3. Du, X.; Zhou, J.; Xiao, C. Spatial Effects and Influencing Factors of Urban Sustainable Development: An Analysis of Urban Agglomerations in China. *Econ. Anal. Policy* **2024**, *81*, 556–575. [CrossRef]
4. Florida, R.; Gulden, T.; Mellander, C. The Rise of the Mega-Region. *Camb. J. Reg. Econ. Soc.* **2008**, *1*, 459–476. [CrossRef]
5. Liu, Y.; Zhang, X.; Pan, X.; Ma, X.; Tang, M. The Spatial Integration and Coordinated Industrial Development of Urban Agglomerations in the Yangtze River Economic Belt, China. *Cities* **2020**, *104*, 102801. [CrossRef]
6. Fu, C.; Xu, Y.; Zhou, F. Environmental Collaborative Governance of Urban Agglomeration in China: Influencing Factors and Drivers. *Environ. Sci. Pollut. Res.* **2022**, *30*, 38363–38379. [CrossRef]
7. Shen, W.; Liang, H.; Dong, L.; Ren, J.; Wang, G. Synergistic CO2 Reduction Effects in Chinese Urban Agglomerations: Perspectives from Social Network Analysis. *Sci. Total Environ.* **2021**, *798*, 149352. [CrossRef]
8. Ye, C.; Zhu, J.; Li, S.; Yang, S.; Chen, M. Assessment and Analysis of Regional Economic Collaborative Development within an Urban Agglomeration: Yangtze River Delta as a Case Study. *Habitat Int.* **2019**, *83*, 20–29. [CrossRef]
9. Gao, X.; Zhang, A.; Sun, Z. How Regional Economic Integration Influence on Urban Land Use Efficiency? A Case Study of Wuhan Metropolitan Area, China. *Land Use Policy* **2020**, *90*, 104329. [CrossRef]
10. Huang, T.; Xi, J.-C.; Ge, Q.-S. Spatial Differentiation and Integration Optimization of an Urban Agglomeration Tourism System under the Influence of High-Speed Railway Network Evolution. *Appl. Spat. Anal. Policy* **2019**, *12*, 349–376. [CrossRef]
11. Yang, Y.; Lu, X.; Chen, J.; Li, N. Factor Mobility, Transportation Network and Green Economic Growth of the Urban Agglomeration. *Sci. Rep.* **2022**, *12*, 20094. [CrossRef] [PubMed]
12. Li, Y.; Xiong, C.; Song, Y. How Do Population Flows Promote Urban–Rural Integration? Addressing Migrants' Farmland Arrangement and Social Integration in China's Urban Agglomeration Regions. *Land* **2022**, *11*, 86. [CrossRef]

13. Li, H.; Liu, Y.; He, Q.; Peng, X.; Yin, C. Simulating Urban Cooperative Expansion in a Single-Core Metropolitan Region Based on Improved CA Model Integrated Information Flow: Case Study of Wuhan Urban Agglomeration in China. *J. Urban Plan. Dev.* **2018**, *144*, 05018002. [CrossRef]
14. Lin, Q.; Xiang, M.; Zhang, L.; Yao, J.; Wei, C.; Ye, S.; Shao, H. Research on Urban Spatial Connection and Network Structure of Urban Agglomeration in Yangtze River Delta—Based on the Perspective of Information Flow. *Int. J. Environ. Res. Public. Health* **2021**, *18*, 10288. [CrossRef]
15. Ma, W.; Yu, Y.; Meng, W.; Fan, J. Comparative Analysis of City Connection Network Based on Gravity Model and Baidu Index in Wuhan Urban Agglomeration. *IOP Conf. Ser. Earth Environ. Sci.* **2020**, *446*, 022050. [CrossRef]
16. Zhao, Y.; Zhang, G.; Zhao, H. Spatial Network Structures of Urban Agglomeration Based on the Improved Gravity Model: A Case Study in China's Two Urban Agglomerations. *Complexity* **2021**, *2021*, 6651444. [CrossRef]
17. He, J.; Li, C.; Yu, Y.; Liu, Y.; Huang, J. Measuring Urban Spatial Interaction in Wuhan Urban Agglomeration, Central China: A Spatially Explicit Approach. *Sustain. Cities Soc.* **2017**, *32*, 569–583. [CrossRef]
18. Li, C.; He, J.; Duan, X. Modeling the Collaborative Evolution of Urban Land Considering Urban Interactions under Intermediate Intervention, in the Urban Agglomeration in the Middle Reaches of the Yangtze River in China. *Land* **2020**, *9*, 184. [CrossRef]
19. van Oort, F.; Burger, M.; Raspe, O. On the Economic Foundation of the Urban Network Paradigm: Spatial Integration, Functional Integration and Economic Complementarities within the Dutch Randstad. *Urban Stud.* **2010**, *47*, 725–748. [CrossRef]
20. You, S.; Chen, X. Regional Integration Degree and Its Effect on a City's Green Growth in the Yangtze River Delta: Research Based on a Single-City Regional Integration Index. *Clean Technol. Environ. Policy* **2021**, *23*, 1837–1849. [CrossRef]
21. Liao, B.; Li, L. How Can Urban Agglomeration Market Integration Promote Urban Green Development: Evidence from China's Yangtze River Economic Belt. *Environ. Sci. Pollut. Res.* **2022**, *29*, 10649–10664. [CrossRef]
22. Guo, X.; Chuanglin, F.; Xufang, M.; Dan, C. Coupling and Coordination Analysis of Urbanization and Ecosystem Service Value in Beijing-Tianjin-Hebei Urban Agglomeration. *Ecol. Indic.* **2022**, *137*, 108782. [CrossRef]
23. Shen; Chen Zhong; Hou Lulu Temporal-Spatial Evolution and Dynamic Mechanism of the Culture-Oriented Urban Agglomerations:A Focus on the Eastern Guangdong. *Trop. Geogr.* **2022**, *42*, 1628–1639.
24. Douglas, A.E. *Symbiotic Interactions*; Oxford University Press: Oxford, UK, 1994; ISBN 0-19-854286-0.
25. Chen, K.; Long, H.; Liao, L.; Tu, S.; Li, T. Land Use Transitions and Urban-Rural Integrated Development: Theoretical Framework and China's Evidence. *Land Use Policy* **2020**, *92*, 104465. [CrossRef]
26. Zhang, H.Z.; Wu, Y.D.; Wang, Y.C. Research of the Match State between the Economic Contact and Geo-Economics Relationship in Yunnan Central Urban Agglomeration. *Areal Res Dev* **2014**, *33*, 16–19.
27. Wang, D.; Jiang, D.; Fu, J.; Lin, G.; Zhang, J. Comprehensive Assessment of Production–Living–Ecological Space Based on the Coupling Coordination Degree Model. *Sustainability* **2020**, *12*, 2009. [CrossRef]
28. Yang, Y.; Bao, W.; Liu, Y. Coupling Coordination Analysis of Rural Production-Living-Ecological Space in the Beijing-Tianjin-Hebei Region. *Ecol. Indic.* **2020**, *117*, 106512. [CrossRef]
29. Diakoulaki, D.; Mavrotas, G.; Papayannakis, L. Determining Objective Weights in Multiple Criteria Problems: The Critic Method. *Comput. Oper. Res.* **1995**, *22*, 763–770. [CrossRef]
30. Yan, X.-Y.; Wang, W.-X.; Gao, Z.-Y.; Lai, Y.-C. Universal Model of Individual and Population Mobility on Diverse Spatial Scales. *Nat. Commun.* **2017**, *8*, 1639. [CrossRef] [PubMed]
31. Zhang, H.; Wu, Y. Research on Urban Symbiosis and Co evolutionA Case Study of Central Yunnan Urban Agglomeration. *Resour. Dev. Mark.* **2016**, *32*, 1078–1082.
32. Fang, C. Theoretical foundation and patterns of coordinated development ofthe Beijing Tianjin Hebei Urban Agglomeration. *Prog. Geogr.* **2017**, *36*, 15–24.
33. Zhuang, Q.; Shao, Z.; Li, D.; Huang, X.; Li, Y.; Altan, O.; Wu, S. Impact of Global Urban Expansion on the Terrestrial Vegetation Carbon Sequestration Capacity. *Sci. Total Environ.* **2023**, *879*, 163074. [CrossRef] [PubMed]
34. Li, F.; Wu, S.; Liu, H.; Yan, D. Biodiversity Loss through Cropland Displacement for Urban Expansion in China. *Sci. Total Environ.* **2024**, *907*, 167988. [CrossRef] [PubMed]
35. Ning, Y.; Liu, S.; Smith, A.R.; Qiu, Y.; Gao, H.; Lu, Y.; Yuan, W.; Feng, S. Dynamic Multi-Dimensional Scaling of 30+ Year Evolution of Chinese Urban Systems: Patterns and Performance. *Sci. Total Environ.* **2023**, *863*, 160705. [CrossRef]
36. Yin, J.; Liu, H.; Shi, P.; Zhang, W. Exploring Coupling Relationship between Urban Connection and High-Quality Development Using the Case of Lanzhou-Xining Urban Agglomeration. *Complexity* **2021**, *2021*, e9933582. [CrossRef]
37. Fang, C.; Yu, X.; Zhang, X.; Fang, J.; Liu, H. Big Data Analysis on the Spatial Networks of Urban Agglomeration. *Cities* **2020**, *102*, 102735. [CrossRef]

land

MDPI

Article

Spatiotemporal Analysis of the Impacts of Land Use Change on Ecosystem Service Value: A Case from Guiyang, China

Qinglan Li [1], Liu Yang [1,2,*], Hongzan Jiao [3,4] and Qing He [5]

1 School of Public Administration, Guizhou University, Guiyang 550025, China; qinglanli0@outlook.com
2 ASEAN Research Institute, Guizhou University, Guiyang 550025, China
3 Department of Urban Planning, School of Urban Design, Wuhan University, Wuhan 430072, China; jiaohongzan@whu.edu.cn
4 Digital City Research Center, Wuhan University, Wuhan 430072, China
5 College of Big Data and Information Engineering, Guizhou University, Guiyang 550025, China; qhe@gzu.edu.cn
* Correspondence: lyang3@gzu.edu.cn

Abstract: The significance of ecosystem services and land use for human well-being and sustainable development cannot be understated. Scientifically assessing the ecosystem service value (ESV) and studying the relationship between land use change and the ESV can provide a theoretical groundwork for land use planning and ecological administration in Guiyang. In this study, gradient analysis was utilized to explore the changes of ESV at district level of Guiyang. Then, the synergistic relationship and the strength of the interaction between land use intensity (LUI) and ESV were explored by using a coupled coordination model and spatial autocorrelation analysis. Furthermore, polynomial fitting was carried out for the LUI index and its linked coordination index in relation to the ESV. The results showed that (1) the areas of farmland, forest, grassland, and unused land in Guiyang decreased from 2000 to 2020, while the areas of construction land and water body increased conversely. (2) The expansion of the construction land and water body was the main cause of the ESV change pattern in Guiyang, which first moved downward and then upward. (3) The ESV and LUI had a low overall coupling coordination degree (CCD). Spatial autocorrelation studies showed that low–to–low aggregation and high–to–high aggregation dominated the spatial patterns of essential regions. (4) The LUI and CCD indexes exhibited an inverted U-shaped curve correlation.

Keywords: ecosystem service value; land use change; land use intensity; coupling coordination; polynomial fit

1. Introduction

An ecosystem comprises living organisms and their surrounding conditions [1]. The products and services ecosystems offer people are known as ecosystem services. These services are categorized into four categories: providing, regulating, supporting, and cultural. Humans may derive direct or indirect benefits from the functions of an ecosystem [2,3]. Ecosystems impact economic growth and human welfare, and ecosystem changes may offer notable benefits to humans. Still, the accompanying costs involve the degradation of ecosystem services and an increase in their potential risks [4,5]. The rapid industrialization and urbanization, economic expansion, and extensive deterioration of ecosystems all strain ecosystems increasingly [6]. The concept of ecosystem service value has gained significance in addressing the challenges of ecosystem services more crucial [7]. As the LUI can alter notable elements of ecosystem functioning, including the conflicts between ecosystem functions and services, as well as their benefits, it is imperative to consider how the LUI impacts ecosystem services and functions [8]. The relationship between land use and ESV needs to be explored to improve human well-being and provide theoretical support for the creation of regional development policies on land use and ecosystem services.

ESV measurements are essential for social development [9,10] because they necessitate balancing several factors that impact the welfare of people in various circumstances in which decisions are made [11]. Since the 1990s, significant research has been conducted on the quantification of ESVs [12–14]. The financial ESV can be determined by comparing the relative values of various indicators and by measuring them in labor or time units [15]. Further research on expressing ESVs in monetary terms is warranted, as doing so would support decision making. Monetary valuation methods for ESV usually include ecological modeling, economic valuation methods based on unprocessed data, and value transfer methods based on land use in terms of unit values [16]. There are two steps involved in the primary estimation process: the first step is to quantify ecosystem services and processes with the use of ecological models (such as water conservation models) or indicators (such as land utilization) and then evaluate them using economic valuation methods [17–19]. Because these methods are computationally demanding and require many parameters, harmonizing and standardizing the assessment parameters for every ecosystem function is more challenging when methods that rely on unprocessed data are used. Consequently, these methods are suited for assessing ESV for specific services within a particular ecosystem or at small spatial scales [20]. The unit value technique calculates ESVs using the economic worth of a unit area inside a given environment. It is appropriate for large spatial scales and integrated ESV evaluations [21]. Costanza updated the worldwide ESV using data from an ESV database that De Groot created [7,22], which included the ESV for ten key biomes. In order to calculate the ESV in China based on prior research, Xie constructed an equivalency factor approach based on the unit value method and consulted 500 Chinese ecological specialists [23,24]. In China, the equivalency factor approach is frequently applied, particularly for research that assesses the ESV of land use change (LUC) [19].

Since terrestrial ecosystem services are susceptible to LUC, land use planning must consider these effects [25,26]. Scholars have demonstrated the relationship between LUC and ESV by analyzing how land use patterns can alter how ecosystem services are given [27,28]. LUC significantly affects the form and functions of significant ecological systems and ecosystem services and their values [29]. Ecological services and land use are influenced and constrained by each other [30]. The speed at which human society has developed has increased human influence on the natural environment, and this has caused environmental harm and underscored the vulnerability of ecosystems [31]. Research has shown that land degradation brought on by LUC may obstruct the delivery of ecosystem services in a particular location and impede the ability of ecosystems to develop sustainably [32,33]. Different land use types can provide various ecosystem services. For instance, farmland can provide more food than forest, but forest has higher carbon reserves and provides more services for the production of lumber [34]. Land use changes include modifications to the kinds of land utilized and adjustments to the intensity and spatial patterns of land usage [35]. In analyzing the connection between ecosystem services and LUC, more researchers have recently looked at variations in land use types [36–38]. For instance, environmental degradation and excessive human encroachment on ecological lands, such as wetland, grassland, and forest, are the reasons behind the decline in associated ecosystem services [27,39]. Proactive measures such as reforestation and tree planting improve some ecosystem functions [40,41]. Although the connection between LUI changes and ecosystem services has not received much attention [42], changes in LUI also affect the services and functions of ecosystems [43]. Xu [44] conducted a variance and correlation analysis to determine how LUI affected ecological services and human well-being. He discovered that the increasing LUI improved food production, soil conservation, and climate control. According to Chillo's [45] research on the effects of LUI on ecosystem functions, it can be observed that the indirect effects of LUI on ecosystems were of significant importance. Lucia's research indicated that while ecosystem services and ecosystem functions can be enhanced at moderate LUI levels. However, the positive correlation between ecosystem services and ecosystem functions decreased, at higher LUI levels [8].

Guiyang is a typical karst city in Southwest China, characterized by high fragmentation and heterogeneity in its landscape, as well as fragile ecosystems and ecological environments [46]. Due to the weak and unstable ecological restoration capacity of karst ecosystems, limited ecological carrying capacity, and the high ecological sensitivity, the ecological system of Guiyang City is easily influenced by external pressures [47]. Although there has been an overall improvement in ecological quality in Guiyang since the 1990s, primarily as a result of the implementation of national ecosystem restoration and other policy projects, the degradation of land still persists in the city center area, and the high rate of urbanization is associated with low environmental quality [48]. Guiyang, with its growing economic prosperity, is experiencing continuous urban expansion and damage to its fragile ecological environment. As LUC can alter notable elements of ecosystem functioning, including the conflicts and benefits between ecosystem functions and services, it is necessary to unveil how LUC influences ecosystem functions and services. However, few studies have concentrated on ESV changes in karst areas, and few have addressed the relationship between LUI and ESV. Karst areas have more fragile ecosystems and need to pay more attention to their land use–ecosystem relationship. Theoretical frameworks can be proposed for land use planning and ecological regulation in Guiyang by evaluating ESVs scientifically and examining the connection between LUC and ESV.

The coupled coordination model and polynomial fit were employed in this work to explore the patterns and changes in land use and ecosystem services in Guiyang, both spatially and temporally, from the standpoint of linking land use with ecosystem services. This model attempted to explain how urbanization has evolved with the relationship between land use and ecological services. The following are the goals of this study: (1) quantitative assessments of these variables in Guiyang in 2000, 2005, 2010, 2015, and 2020, as well as an assessment of the spatiotemporal distribution aspects of LUC and ESV; (2) assessment of the features of the temporal and spatial distributions of various forms of coupled coordination using the coupled coordination model to explore the connection between ecosystem services and land use; and (3) examination of the degree of coupled coordination in the LUI–ESV pattern of change, as well as the trend in the change between the two indicators and the possibility of a turning point in the change process.

2. Materials and Methods

2.1. Study Area

Guiyang (106°07′ E–107°17′ E; 26°11′ N–26°55′ N) is the provincial center of Guizhou Province and has a karst landscape. The city of Guiyang is located on a watershed that separates the tributaries of the Chishui River of the Pearl River system and the Wujiang River of the Yangtze River system. The jurisdiction of Guiyang includes six main urban areas and "one city and three counties", and its main urban areas include Nanming, Yunyan, Huaxi, Wudang, Guanshanhu, and Baiyun, and the "one city and three counties" refers to Qingzhen, Kaiyang, Xifeng, and Xiuwen (Figure 1). Different degrees of impact of increasing urbanization in Guiyang on the land use structure and ecological services have been observed. The years 2000, 2005, 2010, 2015, and 2020 were selected as study periods to expose changes in land use and the ESV of Guiyang. This study explained the changes in land use dynamics, LUI, and ESV in terms of space and time, explored the coupling and coordination of LUI and ESV, and provided references for ecological security and sustainable development in Guiyang.

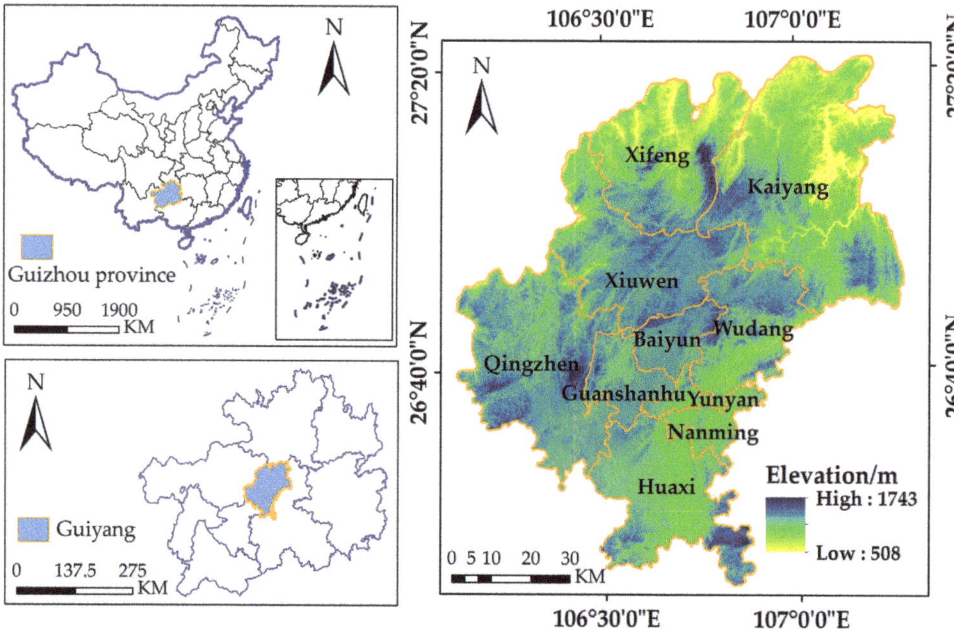

Figure 1. Location of Guiyang in China.

2.2. Data Sources

Land use data with a resolution of 30 m raster data was sourced from the Resource and Environment Science Data Center of the Chinese Academy of Sciences. Six primary categories exist for land use statistics: farmland, forest, grassland, water body, construction land, and unused land [49]. Digital elevation model data was acquired from the Geographic Data Spatial Cloud. Administrative divisions originated from the National Catalogue Service for Geographic Information. The area, production, and selling price of major grains in Guiyang were obtained from the Guizhou Statistical Yearbook, the Guiyang City Statistical Yearbook, and the National Compilation of Information on Costs and Benefits of Agricultural Products for the years 2000–2020.

2.3. Methods

This study assessed the spatial and temporal changes in land use/land cover (LULC) in the study area using land use raster data of 30 m accuracy in conjunction with a land use type conversion matrix, land use dynamics, and LUI techniques. Next, this study used ArcGIS 10.6 to create a 3.5 km $\times$ 3.5 km fishing net and calculated the ESV for each land use type and the total ESV of the study area. The ESV was determined using Xie's method [19], in this research, in which the standard equivalence coefficient was adjusted by calculating the economic value produced by food crops per unit area of the study area. Subsequently, the adjusted standard equivalence coefficient was combined with the land use data to obtain the monetarily quantified ESV. The temporal and spatial variations of the ESV were characterized using the Theil index, gradient analysis, and hot spot analysis. Coupled coordination analysis, gradient analysis, and spatial autocorrelation analysis were used to explore the relationship between ESV and LUC. In order to delve further into the relationship between ESV and LUC, a polynomial fit was chosen to reveal the correlation between LUI and CCD (Figure 2).

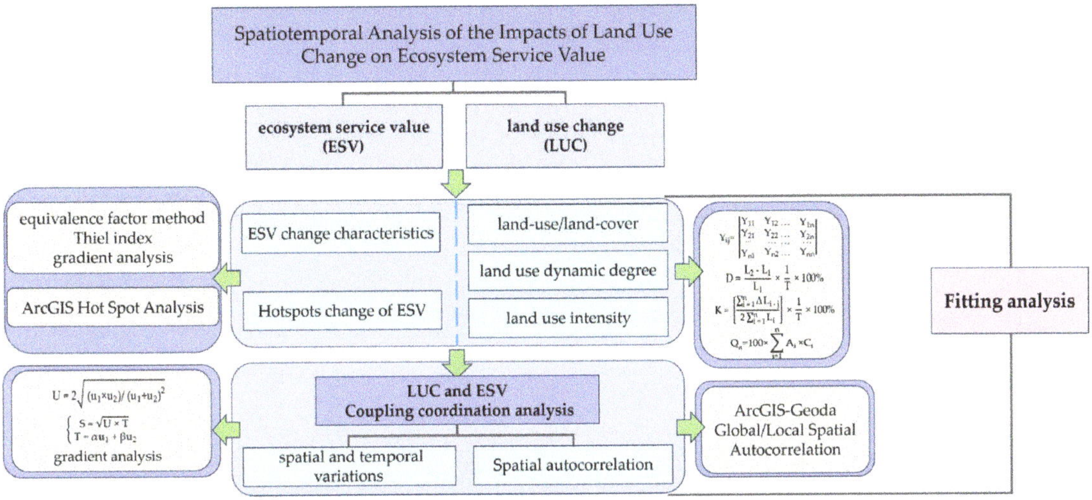

Figure 2. Research framework.

2.3.1. LUC Characteristics

(1) LULC

A land use-type transfer matrix was employed to clarify the land use characteristics and the transfer path between different land use categories:

$$Y_{ij} = \begin{vmatrix} Y_{11} & Y_{12} & \ldots & Y_{1n} \\ Y_{21} & Y_{22} & \ldots & Y_{2n} \\ \ldots & \ldots & \ldots & \ldots \\ Y_{n1} & Y_{n2} & \ldots & Y_{nn} \end{vmatrix} \quad (1)$$

where Y is the study area; i denotes the LULC in the initial stage of the study; j corresponds to the LULC in the terminal stage of the study; and n represents the number of land use types.

(2) Land use dynamic degree

Single and integrated dynamic land use models were proposed to better represent the land use coverage and interconversion intensity of each land use type [50]:

$$D = \frac{L_2 - L_1}{L_1} \times \frac{1}{T} \times 100\% \quad (2)$$

where D denotes the degree of single land use dynamics; the area of a specific land use type at the start and completion of the study are indicated by the variables L_1 and L_2, respectively; and T is the period of the study.

Integrated land use dynamics was utilized to characterize the degree of interconversion of each land use type:

$$K = \left[\frac{\sum_{i=1}^{n} \Delta L_{i-j}}{2\sum_{i=1}^{n} L_i} \right] \times \frac{1}{T} \times 100\% \quad (3)$$

where K represents the degree of integrated land use dynamics; L_i represents land use type i in the initial stage of the study; $\Delta L_{i\text{-}j}$ represents the absolute value of the area of type i land converted into another land type; and T is the period of the study.

(3) LUI

The LUI can be classified into four classes: unutilized land (class 1); water body, forest, and grassland (class 2); agricultural land (class 3); and construction land (class 4) [51,52]:

$$Q_a = 100 \times \sum_{i=1}^{n} A_i \times C_i \tag{4}$$

where Q_a denotes the combined LUI; A_i represents the level of LUI; and C_i corresponds to the proportion of land used for each land use type.

2.3.2. ESV Estimation

(1) Standard equivalent

The ESV equivalence factor can be calculated by taking 1/7 of the market value of the average crop yield in the study area [53]:

$$E_a = \frac{1}{7} \times \sum_{i=1}^{n} \frac{m_i p_i q_i}{M} \tag{5}$$

where E_a corresponds to the economic worth of the production service that a unit of a farming ecosystem may supply; i indicates the kind of crop; p_i represents the national average market price of the i crop; q_i denotes the output of i crop; m_i denotes the acreage of i crop; and M indicates the aggregate area of n crops.

$$ESV_\mathrm{i} = A_i \times \sum_{j=1}^{5} V_{ij} \tag{6}$$

$$ESV_n = \sum_{i=1}^{m} ESV_i \tag{7}$$

At every grid point, ESV_i represents the ESV for land use type i; A_i stands for the land use area; m denotes the number of land use types; V_{ij} represents the unit value of ESV category j of land use i; and ESV_n indicates the overall ESV of cell n.

(2) Theil index

The Theil index, known as a common economic index, can be used to illustrate the extent of the disparity between regions and to estimate the size of the inter-regional variations in the ESV [54]:

$$T_a = \sum_i \frac{ESV_i}{ESV} \ln \frac{ESV_i/ESV}{S_i/S} \tag{8}$$

where T_a stands for the inter-area Theil index; S_i represents the area of the i-th region; ESV indicates the overall ESV; and S denotes the entire area. The greater the variation in the ESV between regions, the higher the Theil coefficient.

(3) Gradient analysis

The circle method of gradient analysis enables the effect of circles on the spatial gradient of individual urban elements to be explored [55]. This research used ArcGIS 10.6 to construct buffers to form circles, and the element values were calculated separately for each circle:

$$D_i = \frac{\sum_{j=1}^{n} D_j}{\sum_{j=1}^{n} S_j} \tag{9}$$

where D_i indicates the average elemental value of the i-th circle; S_j denotes the area of cell j contained in the i-th circle; D_j represents the primary count of cell j; and n corresponds to the number of cells in each circle.

2.3.3. Coupled Coordination Relationship between LUI and ESV

The CCD between LUI and ESV was calculated by using the model for coupled coordination [56,57]:

$$U = 2\sqrt{(u_1 \times u_2)/(u_1 + u_2)^2} \tag{10}$$

$$\begin{cases} S = \sqrt{U \times T} \\ T = \alpha u_1 + \beta u_2 \end{cases} \tag{11}$$

where U describes the degree of coupling; U_1 and U_2 are the normalized values of LUI and ESV calculated by using the polar variance method of normalization, respectively; S stands for CCD; and T denotes the index of system comprehensive coordination; α represents the weight of LUI and β indicates the weight of ESV, and they are both set to 0.5 [58].

$$K'_{xi} = \frac{K_{xi} - \min(K_{xi})}{\max(K_{xi}) - \min(K_{xi})} \tag{12}$$

K_{xi} equals the initial value of the i-th indicator; K'_{xi} denotes the standardized indicator data.

According to current researches, the CCD can be classified as having a severe imbalance ($0 \leq U < 0.2$), moderate imbalance ($0.2 \leq U < 0.4$), essential coordination ($0.4 \leq U < 0.6$), reasonable coordination ($0.6 \leq U < 0.8$), and high coordination ($0.8 \leq U \leq 1$) [59].

2.3.4. The Polynomial Fit of the LUI to the CCD

This study explored how the degree of coupled coordination between LUI and ESV fluctuates with LUI by simulating the variation curves between LUI and CCD with the polynomial fit feature in Origin 2021 software. This study chose the second-order polynomial in this software, and the normalized LUI and CCD were the independent and dependent variables, respectively. The curve represented the coupling degree trends of the LUI and ESV as the LUI changed.

3. Results

3.1. *The LUC of Guiyang*

In Guiyang, construction land accounted for the majority of LULC between 2000 and 2020. Construction land generally increased as farmland, forest, grassland, and unused land diminished. The most significant changes occurred in the construction land and water body, which increased by 146.752% and 49.752%, respectively. Forest exhibited the least rate of change, dropping by 0.762% in 2020 compared to 2000. Throughout these 20 years, farmland, forest, and grassland were the three main land categories, accounting for the most significant percentage of the whole area (Figure 3). The regions where construction land expanded between 2000 and 2020 focused on Wudang, Baiyun, Guanshanhu, Xifeng, and Qingzhen; the most considerable percentage growth was in Guanshanhu, and the smallest was in Yunyan. The highest growth rate of farmland was in Xifeng (0.978%). Guanshanhu had the highest decline in farmland, at 44.529%. The forest areas in Baiyun showed the most excellent rate of improvement at 10.427%, and Nanming had the worst at −15.134%. The percentage increase in grassland areas was negative in all counties, with the lowest in Yunyan at −70.880%. Kaiyang underwent the most rapid expansion in the water area, at 1047.036%. Only Guanshanhu and Qingzhen had unused land by 2015; by 2020, only Qingzhen had unused land.

In 2000–2020, the most significant amount of land transferred in was construction land (377.0101 km^2), whereas the largest amount of land transferred out was farmland (418.775 km^2). The majority of forest had been turned into farmland, the majority of unused land into construction land, and the majority of grassland, water body, and construction land into forest. The conversion of the most incredible amount of grassland into farmland occurred between 2000 and 2005; the most incredible amount of grassland to forest conversion happened between 2005 and 2010; the greatest amount of farmland was converted into forest between 2010 and 2015; and the largest conversion of farmland into construction land occurred between 2015 and 2020 (Figure 4).

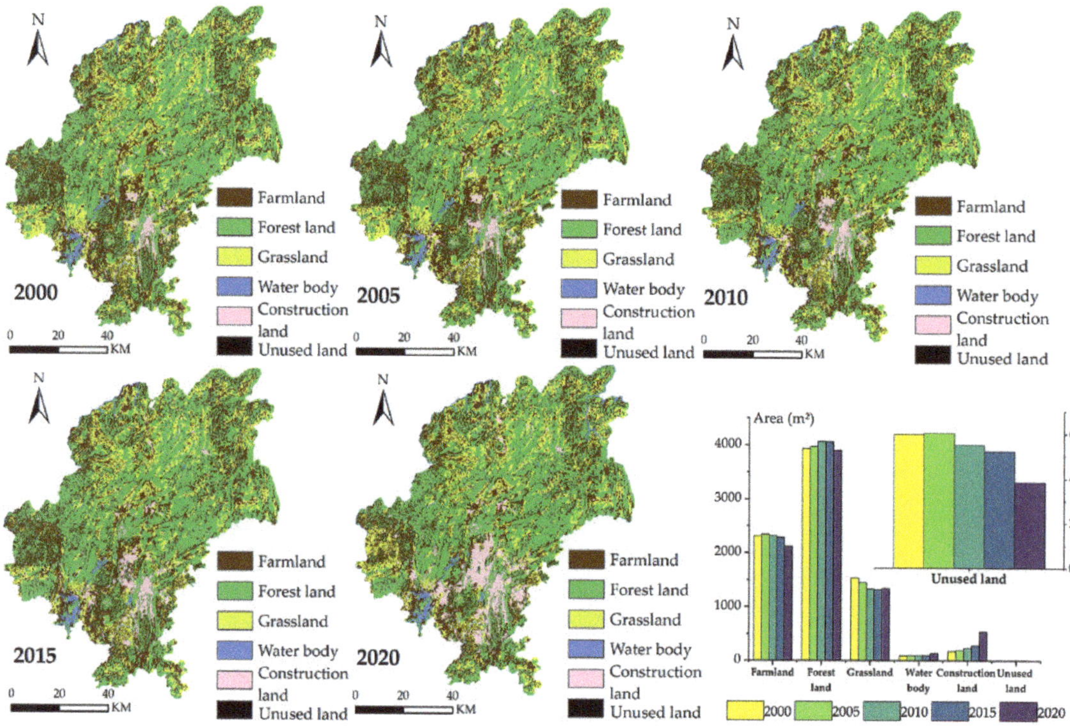

Figure 3. The characters of LULC in Guiyang.

Between 2000 and 2020, the integrated land use dynamic degree was 0.258%. Assessments of the alterations in land use have been conducted throughout time every five years. The percentage of integrated dynamics degree was 0.205%, 0.359%, 0.110%, and 0.816% in the years 2000–2005, 2005–2010, 2010–2015, and 2015–2020, respectively. The integrated dynamics of the various land use types exhibited a pattern that first increased, then decreased, then increased once again, and peaked at the fourth stage, according to the data. For each land use type, the single land use dynamic degree outcomes from high to low were in the following order: construction land > water body > forest > farmland > grassland > unused land (Figure 5). For both the construction land and the water body, the dynamic degree was 2.621% and 10.527%, respectively, and overall, only these two land types had positive dynamics from 2000 to 2020: −0.410% and −0.619% for the dynamics of farmland and grassland, respectively. Following the increase in farmland dynamics in 2000–2005, the 2005–2015 dynamics remained negative. The grassland dynamics were negative in 2000–2015; the 2015–2020 dynamics increased to 0.310%. Forest dynamics increased from 2000 to 2010 and continued to decline from 2010 to 2020. The unused land dynamics were only positive in 2000–2005; they were negative in 2005–2020 and substantially decreased in 2015–2020. In the period 2000–2020, LUI displayed a consistent increasing tendency. The LUI of Guiyang maintained an upward trend, with the largest increase, at 13.852%, in Yunyan. The LUI of Xiuwen, Xifeng, Kaiyang, and Wudang was at a lower level. The LUI of Baiyun, Guanshanhu, Huaxi, and Qingzhen was at a moderate level (Figure 5).

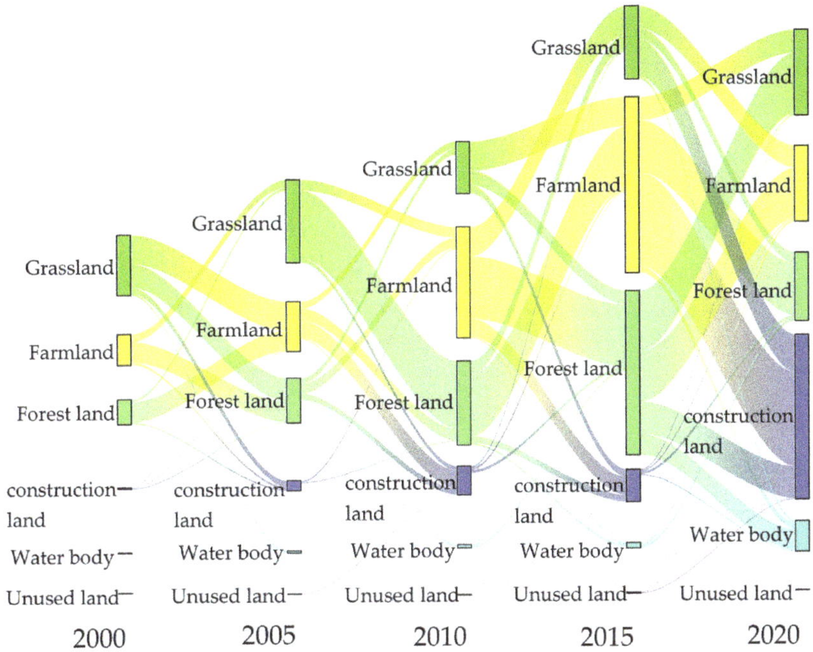

Figure 4. The changes in LULC in Guiyang.

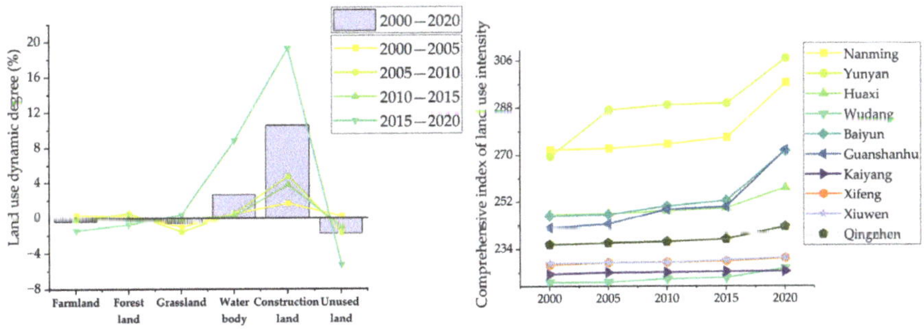

Figure 5. Single land use dynamic degree and comprehensive index of LUI.

3.2. Features of ESV Change

From 2000 to 2020, the ESV trended lower before rising again. From 2000 to 2015, it was declining, but the rate of decline slowed down over time (Figure 6a–e). The ESV over five periods was 2130.018, 2119.914, 2114.006, 2108.909, and 2149.387 million dollars, with a slight overall change of 0.909%. The northeast of Kaiyang, north of Xifeng, and south of Qingzhen were the primary locations with high ESVs; the low-ESV areas were mainly in Yunyan, Nanming, Baiyun, Guanshanhu, Huaxi, the west of Qingzhen, Xiuwen, the northwest of Xifeng, and the central part of Kaiyang. The Theil index was observed to increase, which indicates a gradual widening of the gap in ESV within the region. The reason for the increased inequality in ESV within regions could be the irrational development and utilization of land by humans, and the LULC study found that the gradual increase in land for construction came at the expense of the decrease in ecological lands, such as forest and grassland. The degree of economic development varied between regions, as did the degree of land exploitation and use. Figure 6a–e demonstrates that the ESV was usually lower in regions close to urban centers; the imbalance in inter-regional development affected the overall variability of the ESV in Guiyang city. The six main urban areas of Guiyang showed a downward tendency in ESV from 2000 to 2020, with the highest rate of decrease recorded in Yunyan, which reached −28.088%. Among the "one city and three counties", Kaiyang showed the most significant change, rising by 10.825% in the period 2015–2020; the value for Xifeng began to rise between 2015 and 2020, with growth rates of 0.411% and 0.127%, respectively. Between 2015 and 2020, there were increases of 2.095% in Qingzhen and 2.002% in Xiuwen. The water body experienced the most enormous growth in ESV, whereas the unused land saw the most significant decline. From 2000 to 2005, only grassland showed a decrease in ESV, which persisted until 2015. Farmland and unused land had a declining ESV starting in 2005 and continued to do so until 2020. The year 2010 was the starting point for the fall in forest land ESV. The ESV for water body grew between 2000 and 2020 (Table 1).

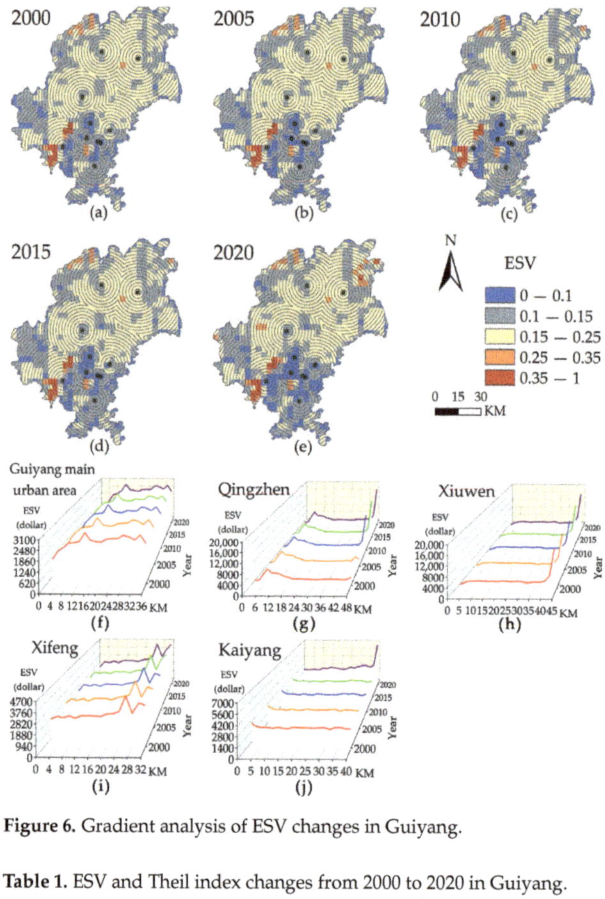

Figure 6. Gradient analysis of ESV changes in Guiyang.

Table 1. ESV and Theil index changes from 2000 to 2020 in Guiyang.

	Year	Farmland	Forest Land	Grassland	Water Body	Unused Land	Total
ESV (millions of dollars)	2000	166.964	1237.918	518.916	206.197	0.022	2130.018
	2005	169.306	1248.513	491.014	211.058	0.022	2119.914
	2010	167.364	1279.280	451.367	215.975	0.020	2114.006
	2015	165.337	1277.606	447.739	218.209	0.019	2108.909
	2020	153.285	1227.120	454.681	314.287	0.014	2149.387

Table 1. *Cont.*

	Year	Farmland	Forest Land	Grassland	Water Body	Unused Land	Total
Change rate (%)	2000–2005	1.403	0.856	−5.377	2.357	0.843	−0.474
	2005–2010	−1.147	2.464	−8.075	2.330	−8.596	−0.279
	2010–2015	−1.211	−0.131	−0.804	1.034	−5.241	−0.241
	2015–2020	−7.289	−3.952	1.550	44.030	−26.413	1.919
	2000–2020	−8.193	−0.872	−12.379	52.420	−35.726	0.909
Year		2000	2005	2010	2015	2020	
Theil index		0.0046	0.0052	0.0053	0.0056	0.0100	

The main six urban areas of Guiyang were considered as a whole, while Qingzhen, Xiuwen, Kaiyang, and Xifeng were viewed as a whole. Their commercial centers have been selected to draw buffer zones with an interval of 2 km (Figure 6f–j). Overall, the value trend in each area changing with the circle remained the same from 2000 to 2020. The main urban areas of Guiyang within the 2–10-km circles were considerably less valuable in 2020 than in 2000–2015 (Figure 6f). Kaiyang showed much-improved values after the 16-km circle compared to previous years, with the highest rise occurring inside the 40-km circle (Figure 6j). The 2-km circle values of Xifeng were considerably lower from 2010 onwards than in 2000–2010, and after the 28-km circle, they remained high in 2015–2020 before starting to fall in value inside the 30-km circle in 2010 (Figure 6i). The trend for Xiuwen in 2000–2020 was unchanged, and the values inside the 40-km circle in 2015 and 2020 were markedly higher than in the other years (Figure 6h). Qingzhen showed an upward trend in the 46–48-km circles in 2000, but the value in the 48-km circle fell again in 2005 and then rose sharply after 2010 until the beginning of the 46-km circle, growing each year (Figure 6g).

The south of Qingzhen, Guanshanhu, the north of Xiuwen, the north of Xifeng, and the east and west of Kaiyang were the primary locations of hot-spot changes in ESV in Guiyang. The cold-spot changes were primarily located around Yunyan, Nanming, and the south of Baiyun, as well as the Huaxi, Qingzhen, Xifeng, and Kaiyang border areas (Figure 7).

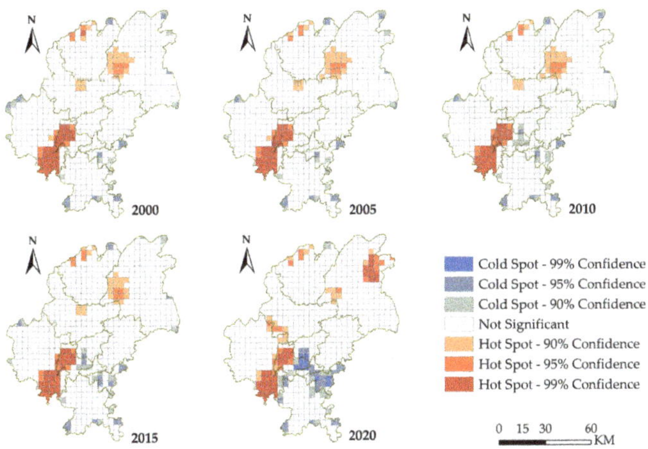

Figure 7. Hot spots for change in the ESV in Guiyang from 2000 to 2020.

3.3. *Evaluation of LUC and ESV Coupling Coordination*

The 2000–2020 LUI and ESV coupling coordination in Guiyang was mainly primary coordination and endangered dysfunction (Figure 8a–e). Primary coordination accounted for most of the coupling coordination in 2000 (Figure 8a), accounting for 21.989% of it, but was overtaken by endangered dysfunction in 2015 (Figure 8d), with endangered dysfunction reaching 21.020%. Severe dislocation and primary coordination regions gradually shifted to endangered dislocation. The areas of severe dislocation rose by 16.129% from 2005 to 2010 and progressively decreased from 2010 to 2020. The primary coordination areas shrank by 24.205% between 2000 and 2020. A growth of 29.206% in endangered dislocation regions occurred between 2000 and 2020, and the most dramatic change occurred between 2015 and 2020. Both intermediate coordination and severe dislocation were centered in Qingzhen city, but the proportion was tiny. The overall level of coupled coordination in Guiyang did not change much between 2000 and 2020, and neither did the tendencies within the districts. Qingzhen (Figure 8g), Xiuwen (Figure 8h), Kaiyang (Figure 8j), Xifeng (Figure 8i), and the main six urban areas of Guiyang (Figure 8f) were considered individual areas. Their economic hubs have been chosen to provide buffer zones spaced 2 km apart. The level of coupled coordination between the main urban areas of Guiyang and Qingzhen was characterized by a high level of coupled coordination in the areas surrounding the regional centers. Owing to their high degrees of urbanization and building, Qingzhen and the main metropolitan regions of Guiyang showed a high degree of interaction between LUI and ESV. Every Xiuwen circle had a CCD marked low in the center and high on both sides. Specifically, before the 8-km circle, the CCD of Xiuwen dropped with the distance from the regional center. Then, it fluctuated at a low level until a noticeable upswing occurred at the 32- and 42-km circles. Up to the 30-km circle, the CCD in Xifeng was very flat; nevertheless, the 30–32-km circle showed a definite downward trend. Kaiyang displayed a comparable pattern, with little change until the 28-km circle and a continuous decline in CCD levels after a brief rise in the 28–32-km circle.

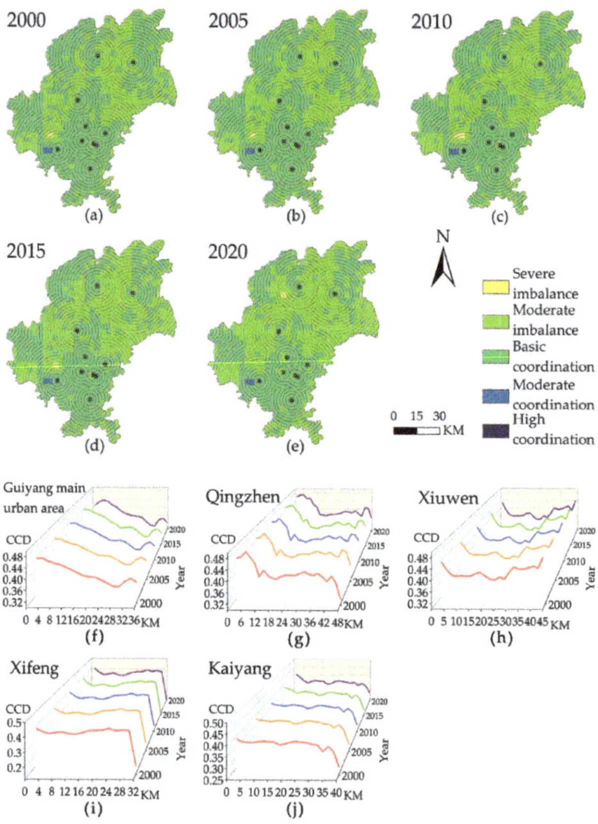

Figure 8. Gradient analysis of CCD changes in Guiyang.

A global autocorrelation analysis of CCD indices for LUI and ESV in Guiyang was conducted using ArcGIS10.6 software. For each of the five time periods from 2000 to 2020, the global Moran's I value of the Guiyang CCD was > 0.26 with $p < 1$. Significant regional variances and a somewhat positive spatial correlation were observed in the degree of coupling coordination, with a maximum value of 0.282 in 2015 and a minimum value of 0.261 in 2020. Low–low and high–high aggregation types were found to be the primary characteristics of Guiyang, according to the results of the localized spatial autocorrelation. Huaxi, Yunyan, Nanming, Guanshanhu, Baiyun, southern Wudang, southern and northern Qingzhen, the northeastern edge of Xifeng, and the central part of Kaiyang were the prominent locations of high–high aggregation; east Qingzhen, northeastern Wudang, and the western and eastern borders of Kaiyang were all areas of low-low aggregation (Figure 9).

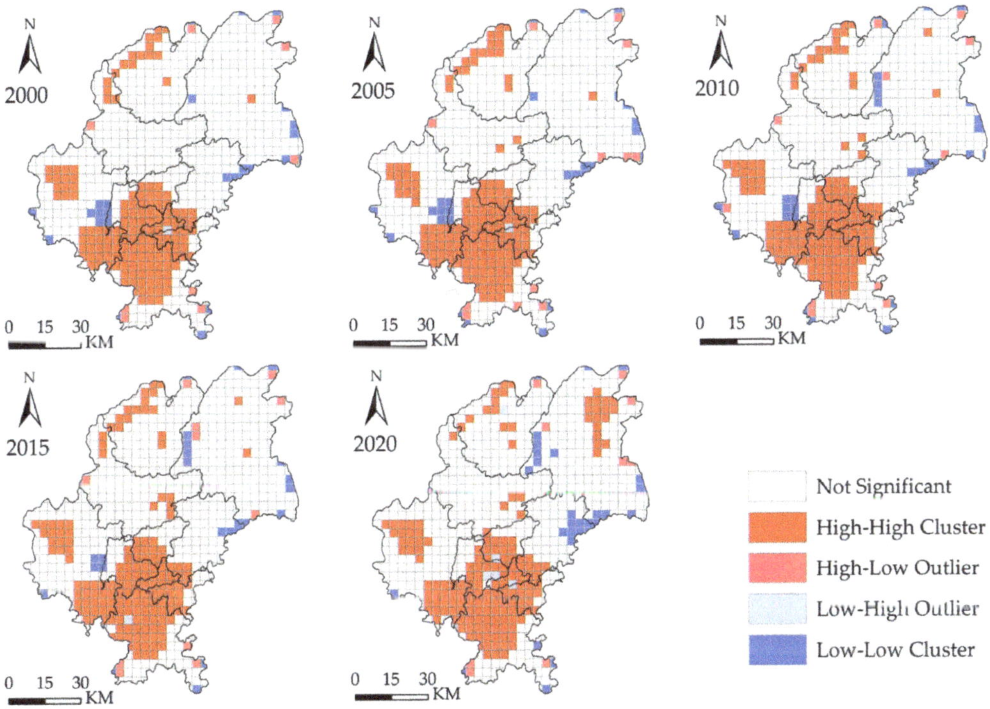

Figure 9. Local indicators of the spatial association of CCD in Guiyang from 2000 to 2020.

3.4. Polynomial Fitting Analyses of the LUI and CCD

The polynomial-fitted R^2 values were all >0.6, and the *p*-value was <0.000, suggesting that the equation has some explanatory effect on the change process in LUI and its coupling coordination with ESV (Table 2). Specifically, an inflection point appeared in all five periods of the fitting curve from 2000 to 2020, and the peak values of the CCD index appeared when the LUI index was 0.5, 0.497, 0.526, 0.496, and 0.499, respectively, beginning to change from an upward trend to a downward trend. The fitted curve (Figure 10) had an inverted U-shape, where the level of coupling between LUI and ESV peaked and then declined as LUI grew. This reveals that from the onset of LUI growth, the relationship between LUI and ESV progressively became stronger. However, as LUI continued to increase, the coupling index progressively declined following its highest value, which suggests that at this point, the interaction between LUI and ESV was less intense.

Table 2. Expression and inflection points between LUI and CCD.

		2000	2005	2010	2015	2020
		Expression: y = Intercept + B1 × x + B2 × x^2				
R^2		0.619	0.643	0.706	0.675	0.825
P		0.000	0.000	0.000	0.000	0.000
Intercept		0.25953 ± 0.00449	0.26097 ± 0.00437	0.26385 ± 0.00364	0.25066 ± 0.00397	0.26452 ± 0.00233
B1		0.90071 ± 0.03268	0.90544 ± 0.03124	0.87213 ± 0.02591	0.95957 ± 0.0292	0.87428 ± 0.01744
B2		−0.90321 ± 0.05142	−0.90964 ± 0.04779	−0.82857 ± 0.03721	−0.96629 ± 0.04218	−0.87525 ± 0.02233
Inflection	LUI	0.500	0.497	0.526	0.496	0.499
	CCD	0.483	0.486	0.493	0.489	0.484

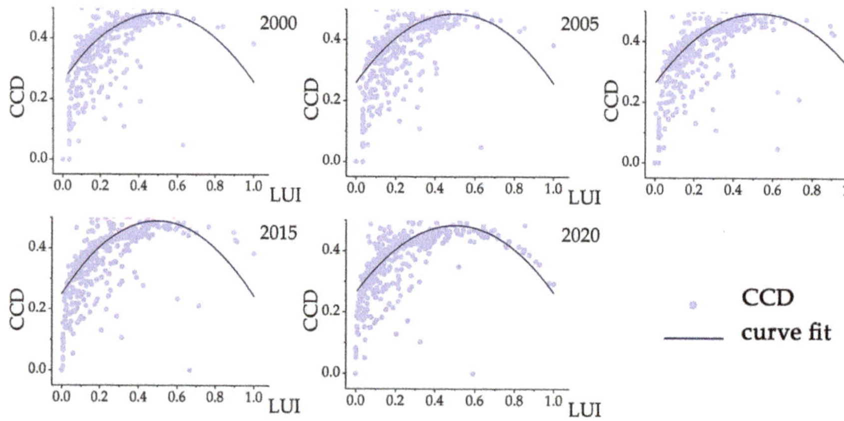

Figure 10. Fitting relationships between LUI and CCD from 2000 to 2020.

4. Discussion

4.1. Dynamics of Land Use, ESV, and Their Coupling Relationship in Guiyang

The patterns of urban ecosystems have been dramatically impacted by modifications to the urban land use structure brought about by the current imperative for fast urban expansion [60]. This analysis shows that between 2000 and 2020, the amount of land used for construction and the growth of water body in Guiyang dominated the variation in land use, while the amount of farmland, forest land, grassland, and unused land generally reduced. Farmland, forest, and grassland were the primary input categories used for expanding construction land. Farmland and forest land were the main drivers of the growth of the water body category. The growth of construction land in Guiyang resulted in the loss of farmland, forest land, and grassland, which reduced the ESV from 2130.018 to 2108.909 million dollars, or 0.991%, between 2000 and 2015. However, because of the growth of the watershed from 2015 to 2020, the ESV increased by 1.919% to 2149.387 million dollars. The loss of agriculture, forest, and grassland along with urbanization were all contributing factors to the overall drop in ESV; nevertheless, the preservation and restoration of water body drove an increase in the ESV. This finding was consistent with previous studies conducted in Southwest China and indicated that the extension of the water body area was the primary cause of the increase in ESV [61]. The water body area in the Yunnan–Guizhou Plateau expanded between 2001 and 2020 [62]. Between 2000 and 2020, Guizhou Province did not experience a notable uptick in precipitation; the development of water conservation projects was the primary cause of the increase in water area [63]. The Qianzhong Water Conservancy Hub project has effectively increased the amount of surface water storage in Guiyang. Compared with other land use types, the water body category had a more significant coefficient of value per unit area; therefore, even though water body was not very large overall, it substantially influenced changes in the ESV.

The LUI can have a direct or indirect impact on biodiversity. Increases in LUI may result in biodiversity loss, land degradation, higher carbon emissions, and other environmental effects [64–66]. The urban center areas of Guiyang, Yunyan, and Nanming had a much higher LUI than other places because of the density of their construction land and a small area of ecological land. Xifeng, Xiuwen, Kaiyang, and Wudang had less LUI and ecological space as they were less urbanized overall. As a mountainous city, geographic factors and government policies influenced and restricted urban expansion and LUC in Guiyang [67]. In Guiyang, since 2000, economic development has been increasing, and its ESV was low in the central city. The regions of Kaiyang, Qingzhen, Xifeng, Xiuwen, Huaxi, and Wudang have been included in most high ESV areas. These regions produced more ESV because they had a lower proportion of construction land and were more prosperous in ecological resources than Guanshanhu, Baiyun, Yunyan, and Nanming. The ESV of the Guiyang districts showed different fluctuations with distance from the regional center, and the fluctuation trends did not change much over time, which indicated stability in the inter-regional LUC. The ESV was often lower near the regional center, showing a negative impact of urbanization on the ecosystem. The ESV was greater and even exhibited a considerable increase in the regions far from the regional center, which suggested that these locations had superior ecological environments. Overall, LUI and ESV climbed by 2.93% and 0.909%, respectively, between 2000 and 2020, with primary harmonization and endangered dislocation accounting for most of their CCD-level types. Unreasonable land usage was one major factor that contributed to the low level of coupled coordination between LUI and ESV. One can determine how benign their interaction was by examining the CCD between LUI and ESV. It is essential to actively develop the economy while concentrating on preserving and rehabilitating the natural environment.

4.2. Polynomial-Fitted Relationship between LUI and CCD

It was evident from the inverted U-shaped curve of the Guiyang LUI and LUI–ESV CCD that the two phases of their relationship could be separated. During the initial phase, the degree of coupling coordination rose in tandem with the LUI. Land use still negatively affected the permitted range of ecosystems, and the LUI was not surprising. The close interaction between land use and ecosystem services persisted, and there was still potential to enhance the degree of coordination between the two. The second stage could be conceptualized as a large-scale extension of urbanization, and it was characterized by a rise in construction land, a corresponding increase in the pace of urbanization, an invasion of ecological space, and the degradation of the natural environment. The CCD between ecosystems and land use declined as a result of all these issues. At this point, the amount of LUI–ESV coupling coordination peaked and then began to display a declining trend as the LUI reached a particular level and continued to grow. When the LUI reached a high level, the LUI–ESV coupling coordination reduced in degree. This could be explained by the fact that the LUI gradually grew with the development of urbanization and that the harm produced by humans to the ecological environment affected the services and functions of the ecosystem, which resulted in a continual drop in the degree of LUI–ESV coupling coordination.

In summary, the polynomial-fitted relationship established the mutual influence, dependency, and limitations between land use and LUI–ESV CCD. Accordingly, the government should implement ecological protection measures at various stages of urbanization to prevent the negative effects of irrational land use on ecosystems. Future planning of Guiyang for ecological environments and urban development needs to consider the evolutionary trend in this relation. Plans for the development of urban areas should create ecological networks and delineate the limits of ecological control zones. Encouraging communities to conserve and restore natural resources through appropriate financial incentives is one way to support the preservation of existing ecosystems.

4.3. Limitations and Future Research

The way that humans interact with the environment produces a pattern of land use, which is essential for ecosystem services [68]. Ecosystem sustainability and the advancement of human society are closely linked. This study estimated the ESV of Guiyang from 2000 to 2020 using the equivalence coefficient approach as a means for quantifying the benefits produced by the ecosystem. However, data accuracy, such as land use and food production, affected the value of the monetized ESV. In future studies, the characteristics of karst regions should be considered, and the relevant literature should be referred to construct a revised model in order to provide a more accurate estimate of the ESV [69,70]. In addition, this study explored the relationship between land use and ESV, but it has not yet analyzed the driving factors that affect the relationship between land use and ESV. Techniques such as Bayesian spatiotemporal hierarchy models [71], econometric models [72], and geographic probes [73] can be utilized to gain additional insights into the variables that affect ecosystem services and land use. Meanwhile, the characterization of karst ecosystem services will be further considered in future studies.

5. Conclusions

This study analyzed the LUC and ESV in Guiyang from 2000 to 2020 from both temporal and spatial viewpoints with the 30 m × 30 m land use raster data and the fishing net of 3.5 km × 3.5 km. The CCD model was used to quantify the benign CCD and the synergistic connection between LUI and ESV. The findings in this study demonstrated that the main cause of changes in the land use categories in Guiyang was the growth of the water body and construction land categories, with farmland, grassland, and forest making up the bulk of the categories of land output. The larger watershed was the main factor that drove the increase in the Guiyang ESV, which first trended downward before reversing and trending upward. The coupling coordination level of LUI and ESV was mainly dominated by primary coordination and near-dissonance, which indicated that the two were primarily constrained by each other at a low level. The fitting examination of the coupling coordination levels of LUI and LUI–ESV revealed an inverted, U-shaped curve relationship with a polynomial fit. Throughout the early growth stage of LUI, there was a rising trend in both the degree of benign coupling and the interaction between land use and ecosystems. The land use-disturbed ecosystem functioned, to some extent, as a result of the ongoing increase in LUI, which lowered the degree of benign coupling between the two. The importance of water body for ESV growth cannot be ignored during urbanization, and there is a need to focus on the protection of water body during future urbanization. According to the level of urbanization and development and considering the needs of social development and ecological sustainability, the relevant departments should formulate appropriate development plans. Governments must promptly control human interference with ecosystem services until land use intensity has caused severe damage to ecosystem services. The results of this study were crucial to understanding how ecosystem services and land use are related, as well as how these services change as a result of urbanization. This study offers a theoretical foundation for the administration and design of ecosystems during a subsequent urbanization phase.

Author Contributions: Conceptualization, L.Y., Q.L. and H.J.; methodology, Q.L.; software, Q.L. and Q.H.; verification, L.Y., H.J. and Q.H.; formal analysis, Q.L.; survey, Q.L.; resources, L.Y.; data management, Q.H.; writing—original draft Preparation, Q.L.; writing—review and editing, L.Y.; visualization, H.J.; supervision, L.Y.; project management, L.Y., H.J. and Q.H.; capital acquisition, L.Y. All authors have read and agreed to the published version of the manuscript.

Funding: This work was supported by a project from the National Natural Science Foundation of China, "the Rural spatial restructuring in poverty-stricken mountainous areas of Guizhou based on spatial equity: A case study of the Dianqiangui Rocky Desertification Area", grant number 41861038; the Youth Fund Project of Humanities and Social Sciences Research of the Ministry of Education, "Research on farmers livelihood differentiation and farmland marginalization in karst areas of Guizhou province", grant number 18YJC630216; the major project fund for social science and humanities of Guizhou University, grant number GDZT2013-05; the National Natural Science Foundation of China, "Research on the Evidence Chain Construction from the Analysis of the Investigation Documents", grant number 62166006; the National Key Research and Development Program, "Research and Development of Emergency Response and Collaborative Command System with Holographic Perception of Traffic Network Disaster", grant number 2020YFC1512002; and the National Natural Science Foundation of China Major Program, grant numbers 42192580 and 42192583.

Institutional Review Board Statement: Not applicable.

Informed Consent Statement: Not applicable.

Data Availability Statement: Data is contained within the article.

Conflicts of Interest: The authors declare no conflicts of interest.

References

1. Holling, C.S. Resilience and stability of ecological systems. *Annu. Rev. Ecol. Syst.* **1973**, *4*, 1–23. [CrossRef]
2. Costanza, R.; d'Arge, R.; de Groot, R.; Farber, S.; Grasso, M.; Hannon, B.; Limburg, K.; Naeem, S.; O'Neill, R.V.; Paruelo, J.; et al. The value of the world's ecosystem services and natural capital. *Nature* **1997**, *387*, 253–260. [CrossRef]
3. Stoeckl, N.; Hicks, C.C.; Mills, M.; Fabricius, K.; Esparon, M.; Kroon, F.; Kaur, K.; Costanza, R. The economic value of ecosystem services in the Great Barrier Reef: Our state of knowledge. *Ann. N. Y. Acad. Sci.* **2011**, *1219*, 113–133. [CrossRef] [PubMed]
4. Duraiappah, A.K. Ecosystem Services and Human Well-being: Do Global Findings Make Any Sense? *BioScience* **2011**, *61*, 7–8. [CrossRef]
5. Millennium Ecosystem Assessment. *Ecosystems and Human Well-Being*; Island Press: Washington, DC, USA, 2005; Volume 5.
6. Kumar, P. *The Economics of Ecosystems and Biodiversity: Ecological and Economic Foundations*; Routledge: London, UK, 2012.
7. De Groot, R.; Brander, L.; Van Der Ploeg, S.; Costanza, R.; Bernard, F.; Braat, L.; Christie, M.; Crossman, N.; Ghermandi, A.; Hein, L. Global estimates of the value of ecosystems and their services in monetary units. *Ecosyst. Serv.* **2012**, *1*, 50–61. [CrossRef]
8. Felipe-Lucia, M.R.; Soliveres, S.; Penone, C.; Fischer, M.; Ammer, C.; Boch, S.; Boeddinghaus, R.S.; Bonkowski, M.; Buscot, F.; Fiore-Donno, A.M. Land-use intensity alters networks between biodiversity, ecosystem functions, and services. *Proc. Natl. Acad. Sci. USA* **2020**, *117*, 28140–28149. [CrossRef] [PubMed]
9. Hardaker, A.; Pagella, T.; Rayment, M. Integrated assessment, valuation and mapping of ecosystem services and dis-services from upland land use in Wales. *Ecosyst. Serv.* **2020**, *43*, 101098. [CrossRef]
10. Tusznio, J.; Pietrzyk-Kaszyńska, A.; Rechciński, M.; Olszańska, A.; Grodzińska-Jurczak, M. Application of the ecosystem services concept at the local level–Challenges, opportunities, and limitations. *Ecosyst. Serv.* **2020**, *42*, 101077. [CrossRef]
11. Costanza, R.; De Groot, R.; Braat, L.; Kubiszewski, I.; Fioramonti, L.; Sutton, P.; Farber, S.; Grasso, M. Twenty years of ecosystem services: How far have we come and how far do we still need to go? *Ecosyst. Serv.* **2017**, *28*, 1–16. [CrossRef]
12. Tobias, D.; Mendelsohn, R. Valuing ecotourism in a tropical rain-forest reserve. *Ambio* **1991**, *20*, 91–93. Available online: https://www.jstor.org/stable/4313783 (accessed on 19 June 2023).
13. Wainger, L.A.; King, D.M.; Mack, R.N.; Price, E.W.; Maslin, T. Can the concept of ecosystem services be practically applied to improve natural resource management decisions? *Ecol. Econ.* **2010**, *69*, 978–987. [CrossRef]
14. Xiao, Y.; Huang, M.; Xie, G.; Zhen, L. Evaluating the impacts of land use change on ecosystem service values under multiple scenarios in the Hunshandake region of China. *Sci. Total Environ.* **2022**, *850*, 158067. [CrossRef] [PubMed]
15. Fioramonti, D.L. *How Numbers Rule the World: The Use and Abuse of Statistics in Global Politics*; Zed Books Ltd.: London, UK, 2014.
16. Jiang, W.; Wu, T.; Fu, B. The value of ecosystem services in China: A systematic review for twenty years. *Ecosyst. Serv.* **2021**, *52*, 101365. [CrossRef]
17. Liu, W.; Zhan, J.; Zhao, F.; Yan, H.; Zhang, F.; Wei, X. Impacts of urbanization-induced land-use changes on ecosystem services: A case study of the Pearl River Delta Metropolitan Region, China. *Ecol. Indic.* **2019**, *98*, 228–238. [CrossRef]
18. Schägner, J.P.; Brander, L.; Maes, J.; Hartje, V. Mapping ecosystem services' values: Current practice and future prospects. *Ecosyst. Serv.* **2013**, *4*, 33–46. [CrossRef]
19. Xie, G.; Zhang, C.; Zhen, L.; Zhang, L. Dynamic changes in the value of China's ecosystem services. *Ecosyst. Serv.* **2017**, *26*, 146–154. [CrossRef]
20. Zhang, B.; Li, W.; Xie, G. Ecosystem services research in China: Progress and perspective. *Ecol. Econ.* **2010**, *69*, 1389–1395. [CrossRef]
21. Liu, Y.; Hou, X.; Li, X.; Song, B.; Wang, C. Assessing and predicting changes in ecosystem service values based on land use/cover change in the Bohai Rim coastal zone. *Ecol. Indic.* **2020**, *111*, 106004. [CrossRef]
22. Costanza, R.; De Groot, R.; Sutton, P.; Van der Ploeg, S.; Anderson, S.J.; Kubiszewski, I.; Farber, S.; Turner, R.K. Changes in the global value of ecosystem services. *Glob. Environ. Change* **2014**, *26*, 152–158. [CrossRef]
23. Xie, G.; Zhen, L.; Lu, C.-X.; Xiao, Y.; Chen, C. Expert knowledge based valuation method of ecosystem services in China. *J. Nat. Resour.* **2008**, *23*, 911–919.
24. Xie, G.-D.; Lu, C.X.; Leng, Y.-F.; Zheng, D.; Li, S. Ecological assets valuation of the Tibetan Plateau. *J. Nat. Resour.* **2003**, *18*, 189–196.

25. Clements, H.S.; De Vos, A.; Bezerra, J.C.; Coetzer, K.; Maciejewski, K.; Mograbi, P.J.; Shackleton, C. The relevance of ecosystem services to land reform policies: Insights from South Africa. *Land Use Policy* **2021**, *100*, 104939. [CrossRef]
26. Gomes, E.; Inácio, M.; Bogdzevič, K.; Kalinauskas, M.; Karnauskaitė, D.; Pereira, P. Future land-use changes and its impacts on terrestrial ecosystem services: A review. *Sci. Total Environ.* **2021**, *781*, 146716. [CrossRef]
27. Anley, M.A.; Minale, A.S.; Haregeweyn, N.; Gashaw, T. Assessing the impacts of land use/cover changes on ecosystem service values in Rib watershed, Upper Blue Nile Basin, Ethiopia. *Trees For. People* **2022**, *7*, 100212. [CrossRef]
28. Yang, Y.; Wang, K.; Liu, D.; Zhao, X.; Fan, J. Effects of land-use conversions on the ecosystem services in the agro-pastoral ecotone of northern China. *J. Clean. Prod.* **2020**, *249*, 119360. [CrossRef]
29. Lawler, J.J.; Lewis, D.J.; Nelson, E.; Plantinga, A.J.; Polasky, S.; Withey, J.C.; Helmers, D.P.; Martinuzzi, S.; Pennington, D.; Radeloff, V.C. Projected land-use change impacts on ecosystem services in the United States. *Proc. Natl. Acad. Sci. USA* **2014**, *111*, 7492–7497. [CrossRef]
30. Qiu, H.; Hu, B.; Zhang, Z. Impacts of land use change on ecosystem service value based on SDGs report--Taking Guangxi as an example. *Ecol. Indic.* **2021**, *133*, 108366. [CrossRef]
31. Deng, X.; Gibson, J. Sustainable land use management for improving land eco-efficiency: A case study of Hebei, China. *Ann. Oper. Res.* **2020**, *290*, 265–277. [CrossRef]
32. Bryan, B.A.; Gao, L.; Ye, Y.; Sun, X.; Connor, J.D.; Crossman, N.D.; Stafford-Smith, M.; Wu, J.; He, C.; Yu, D. China's response to a national land-system sustainability emergency. *Nature* **2018**, *559*, 193–204. [CrossRef] [PubMed]
33. Turner, K.G.; Anderson, S.; Gonzales-Chang, M.; Costanza, R.; Courville, S.; Dalgaard, T.; Dominati, E.; Kubiszewski, I.; Ogilvy, S.; Porfirio, L. A review of methods, data, and models to assess changes in the value of ecosystem services from land degradation and restoration. *Ecol. Model.* **2016**, *319*, 190–207. [CrossRef]
34. Baral, H.; Keenan, R.J.; Fox, J.C.; Stork, N.E.; Kasel, S. Spatial assessment of ecosystem goods and services in complex production landscapes: A case study from south-eastern Australia. *Ecol. Complex.* **2013**, *13*, 35–45. [CrossRef]
35. Fu, B.; Zhang, L.; Xu, Z.; Zhao, Y.; Wei, Y.; Skinner, D. Ecosystem services in changing land use. *J. Soils Sediments* **2015**, *15*, 833–843. [CrossRef]
36. Fang, Z.; Ding, T.; Chen, J.; Xue, S.; Zhou, Q.; Wang, Y.; Wang, Y.; Huang, Z.; Yang, S. Impacts of land use/land cover changes on ecosystem services in ecologically fragile regions. *Sci. Total Environ.* **2022**, *831*, 154967. [CrossRef]
37. Wang, X.; Dong, X.; Liu, H.; Wei, H.; Fan, W.; Lu, N.; Xu, Z.; Ren, J.; Xing, K. Linking land use change, ecosystem services and human well-being: A case study of the Manas River Basin of Xinjiang, China. *Ecosyst. Serv.* **2017**, *27*, 113–123. [CrossRef]
38. Ye, Y.; Bryan, B.A.; Connor, J.D.; Chen, L.; Qin, Z.; He, M. Changes in land-use and ecosystem services in the Guangzhou-Foshan Metropolitan Area, China from 1990 to 2010: Implications for sustainability under rapid urbanization. *Ecol. Indic.* **2018**, *93*, 930–941. [CrossRef]
39. Wang, X.; Wu, J.; Liu, Y.; Hai, X.; Shanguan, Z.; Deng, L. Driving factors of ecosystem services and their spatiotemporal change assessment based on land use types in the Loess Plateau. *J. Environ. Manag.* **2022**, *311*, 114835. [CrossRef]
40. Li, Y.; Liu, W.; Feng, Q.; Zhu, M.; Yang, L.; Zhang, J.; Yin, X. The role of land use change in affecting ecosystem services and the ecological security pattern of the Hexi Regions, Northwest China. *Sci. Total Environ.* **2023**, *855*, 158940. [CrossRef] [PubMed]
41. Liu, Y.; Yuan, X.; Li, J.; Qian, K.; Yan, W.; Yang, X.; Ma, X. Trade-offs and synergistic relationships of ecosystem services under land use change in Xinjiang from 1990 to 2020: A Bayesian network analysis. *Sci. Total Environ.* **2023**, *858*, 160015. [CrossRef]
42. Zheng, H.; Peng, J.; Qiu, S.; Xu, Z.; Zhou, F.; Xia, P.; Adalibieke, W. Distinguishing the impacts of land use change in intensity and type on ecosystem services trade-offs. *J. Environ. Manag.* **2022**, *316*, 115206. [CrossRef]
43. Wen, Z.; Zheng, H.; Smith, J.R.; Zhao, H.; Liu, L.; Ouyang, Z. Functional diversity overrides community-weighted mean traits in linking land-use intensity to hydrological ecosystem services. *Sci. Total Environ.* **2019**, *682*, 583–590. [CrossRef]
44. Xu, Y.; Tang, H.; Wang, B.; Chen, J. Effects of land-use intensity on ecosystem services and human well-being: A case study in Huailai County, China. *Environ. Earth Sci.* **2016**, *75*, 1–11. [CrossRef]
45. Chillo, V.; Vázquez, D.P.; Amoroso, M.M.; Bennett, E.M. Land-use intensity indirectly affects ecosystem services mainly through plant functional identity in a temperate forest. *Funct. Ecol.* **2018**, *32*, 1390–1399. [CrossRef]
46. Liu, S.; Wang, Z.; Wu, W.; Yu, L. Effects of landscape pattern change on ecosystem services and its interactions in karst cities: A case study of Guiyang City in China. *Ecol. Indic.* **2022**, *145*, 109646. [CrossRef]
47. Chen, J.; Yu, J.; Bai, X.; Zeng, Y.; Wang, J. Fragility of karst ecosystem and environment: Long-term evidence from lake sediments. *Agric. Ecosyst. Environ.* **2020**, *294*, 106862. [CrossRef]
48. Liu, Y.; Huang, X.; Yang, H.; Zhong, T. Environmental effects of land-use/cover change caused by urbanization and policies in Southwest China Karst area–A case study of Guiyang. *Habitat Int.* **2014**, *44*, 339–348. [CrossRef]
49. He, N.; Zhou, Y.; Wang, L.; Li, Q.; Zuo, Q.; Liu, J. Spatiotemporal differentiation and the coupling analysis of ecosystem service value with land use change in Hubei Province, China. *Ecol. Indic.* **2022**, *145*, 109693. [CrossRef]
50. Pan, N.; Guan, Q.; Wang, Q.; Sun, Y.; Li, H.; Ma, Y. Spatial differentiation and driving mechanisms in ecosystem service value of arid region: A case study in the middle and lower reaches of Shule River Basin, NW China. *J. Clean. Prod.* **2021**, *319*, 128718. [CrossRef]
51. Tian, Y.; Xu, D.; Song, J.; Guo, J.; You, X.; Jiang, Y. Impacts of land use changes on ecosystem services at different elevations in an ecological function area, northern China. *Ecol. Indic.* **2022**, *140*, 109003. [CrossRef]
52. Zhuang, D.; Liu, J. Study on the model of regional differentiation of land use degree in China. *J. Nat. Resour.* **1997**, *12*, 105–111.

53. Gao, X.; Shen, J.; He, W.; Zhao, X.; Li, Z.; Hu, W.; Wang, J.; Ren, Y.; Zhang, X. Spatial-temporal analysis of ecosystem services value and research on ecological compensation in Taihu Lake Basin of Jiangsu Province in China from 2005 to 2018. *J. Clean. Prod.* **2021**, *317*, 128241. [CrossRef]
54. Chen, J.; Li, T. Changes of spatial variations in ecosystem service value in China. *Beijing Da Xue Xue Bao* **2019**, *55*, 951–960.
55. Jiao, L. Urban land density function: A new method to characterize urban expansion. *Landsc. Urban Plan.* **2015**, *139*, 26–39. [CrossRef]
56. Ding, T.; Chen, J.; Fang, Z.; Chen, J. Assessment of coordinative relationship between comprehensive ecosystem service and urbanization: A case study of Yangtze River Delta urban Agglomerations, China. *Ecol. Indic.* **2021**, *133*, 108454. [CrossRef]
57. Xiao, R.; Lin, M.; Fei, X.; Li, Y.; Zhang, Z.; Meng, Q. Exploring the interactive coercing relationship between urbanization and ecosystem service value in the Shanghai–Hangzhou Bay Metropolitan Region. *J. Clean. Prod.* **2020**, *253*, 119803. [CrossRef]
58. Huang, J.; Na, Y.; Guo, Y. Spatiotemporal characteristics and driving mechanism of the coupling coordination degree of urbanization and ecological environment in Kazakhstan. *J. Geogr. Sci.* **2020**, *30*, 1802–1824. [CrossRef]
59. Zhang, H.; Wang, Y.; Wang, C.; Yang, J.; Yang, S. Coupling analysis of environment and economy based on the changes of ecosystem service value. *Ecol. Indic.* **2022**, *144*, 109524. [CrossRef]
60. Zhang, X.; Ren, W.; Peng, H. Urban land use change simulation and spatial responses of ecosystem service value under multiple scenarios: A case study of Wuhan, China. *Ecol. Indic.* **2022**, *144*, 109526. [CrossRef]
61. Wang, Y.; Zhang, Z.; Chen, X. Spatiotemporal change in ecosystem service value in response to land use change in Guizhou Province, southwest China. *Ecol. Indic.* **2022**, *144*, 109514. [CrossRef]
62. Wang, Z.-J.; Liu, S.-J.; Li, J.-H.; Pan, C.; Wu, J.-L.; Ran, J.; Su, Y. Remarkable improvement of ecosystem service values promoted by land use/land cover changes on the Yungui Plateau of China during 2001–2020. *Ecol. Indic.* **2022**, *142*, 109303. [CrossRef]
63. Wang, Y.; Zhang, Z.; Chen, X. The dominant driving force of Forest change in the Yangtze River basin, China: Climate variation or anthropogenic activities? *Forests* **2022**, *13*, 82. [CrossRef]
64. Lange, M.; Feilhauer, H.; Kühn, I.; Doktor, D. Mapping land-use intensity of grasslands in Germany with machine learning and Sentinel-2 time series. *Remote Sens. Environ.* **2022**, *277*, 112888. [CrossRef]
65. Le Provost, G.; Thiele, J.; Westphal, C.; Penone, C.; Allan, E.; Neyret, M.; Van Der Plas, F.; Ayasse, M.; Bardgett, R.D.; Birkhofer, K. Contrasting responses of above-and belowground diversity to multiple components of land-use intensity. *Nat. Commun.* **2021**, *12*, 3918. [CrossRef]
66. Semenchuk, P.; Plutzar, C.; Kastner, T.; Matej, S.; Bidoglio, G.; Erb, K.-H.; Essl, F.; Haberl, H.; Wessely, J.; Krausmann, F. Relative effects of land conversion and land-use intensity on terrestrial vertebrate diversity. *Nat. Commun.* **2022**, *13*, 615. [CrossRef] [PubMed]
67. Hu, C.; Wu, W.; Zhou, X.; Wang, Z. Spatiotemporal changes in landscape patterns in karst mountainous regions based on the optimal landscape scale: A case study of Guiyang City in Guizhou Province, China. *Ecol. Indic.* **2023**, *150*, 110211. [CrossRef]
68. Guo, P.; Zhang, F.; Wang, H. The response of ecosystem service value to land use change in the middle and lower Yellow River: A case study of the Henan section. *Ecol. Indic.* **2022**, *140*, 109019. [CrossRef]
69. Li, M.; Liu, S.; Wang, F.; Liu, H.; Liu, Y.; Wang, Q. Cost-benefit analysis of ecological restoration based on land use scenario simulation and ecosystem service on the Qinghai-Tibet Plateau. *Glob. Ecol. Conserv.* **2022**, *34*, e02006. [CrossRef]
70. Su, K.; Wei, D.-Z.; Lin, W.-X. Evaluation of ecosystem services value and its implications for policy making in China–A case study of Fujian province. *Ecol. Indic.* **2020**, *108*, 105752. [CrossRef]
71. Yang, R.; Ren, F.; Xu, W.; Ma, X.; Zhang, H.; He, W. China's ecosystem service value in 1992–2018: Pattern and anthropogenic driving factors detection using Bayesian spatiotemporal hierarchy model. *J. Environ. Manag.* **2022**, *302*, 114089. [CrossRef] [PubMed]
72. Luo, Q.; Zhou, J.; Li, Z.; Yu, B. Spatial differences of ecosystem services and their driving factors: A comparation analysis among three urban agglomerations in China's Yangtze River Economic Belt. *Sci. Total Environ.* **2020**, *725*, 138452. [CrossRef] [PubMed]
73. Zhang, L.; Hu, B.; Zhang, Z.; Liang, G. Research on the spatiotemporal evolution and mechanism of ecosystem service value in the mountain-river-sea transition zone based on "production-living-ecological space"—Taking the Karst-Beibu Gulf in Southwest Guangxi, China as an example. *Ecol. Indic.* **2023**, *148*, 109889. [CrossRef]

land

Article

Spatial Optimization and Temporal Changes in the Ecological Network: A Case Study of Wanning City, China

Shisi Zou [1], Rong Fan [2] and Jian Gong [1,*]

[1] Department of Land Resource Management, School of Public Administration, China University of Geosciences, Wuhan 430074, China; zoushisi@cug.edu.cn
[2] Introduction for School of Physical Education, China University of Geosciences, Wuhan 430074, China; fanrong@cug.edu.cn
* Correspondence: gongjian@cug.edu.cn

Abstract: Ecological networks serve as vital tools for safeguarding biodiversity and ensuring regional ecological stability. This study, conducted in Wanning City, employs minimum-area threshold analysis to pinpoint crucial ecological sources while extracting potential ecological corridors using the minimum cumulative resistance model. Our investigation delves into the ecological network's elements and structural transformations within Wanning City, spanning the period from 2000 to 2020, and assesses the priorities for ecological network preservation. The findings of our research reveal noteworthy spatial disparities in the distribution of ecological sources across Wanning City. Furthermore, the ecological corridors display sparse patterns in the north and denser patterns in the south. Over the two decades from 2000 to 2020, Wanning's ecological resources exhibited a discernible trend of contraction and fragmentation, accompanied by an uneven spatial distribution. The average path length of the ecological corridors has increased, indicative of reduced biological flow efficiency. Correspondingly, the structural accessibility of the ecological network has decreased, signifying a decline in landscape connectivity. Based on our analysis, we propose an ecological protection and restoration framework denoted as "One Belt, Four Sources, Eight districts, multiple corridors, and multiple points". Therefore, with the Shangxi–Jianling, Liulianling, Nanlin, and Jiexin nature reserves as the core area, and Houan Town, Damao Town, Changfeng Town, and Liji Town as the key restoration areas, we have proposed an ecological protection and restoration pattern.

Keywords: ecological network; ecological source; ecological corridor; network structure; spatial restoration; Wanning City

1. Introduction

Urbanization and industrialization constitute primary drivers of land use and land cover changes, particularly as rural-to-urban migration escalates. The global urbanization rate has risen from 46.69% to 56.2%, while China's urbanization rate has surged from 36.22% to 63.89% between 2000 and 2020. Projections indicate that China's urbanization rate may reach 75% by 2050 [1]. This rapid urbanization has spurred unprecedented expansion of cities, accompanied by intensified human activities that exert substantial pressure on natural ecosystems. Consequently, global landscape patterns have undergone substantial alterations, leading to a surge in ecological and environmental issues, such as the urban heat island effect, environmental pollution, habitat loss, diminished biodiversity, and declining ecological functionality [2–4]. As a result, mitigating the adverse impact of urbanization on the environment and ecology, while also addressing the requirements of urban development, enhancing the quality of human habitats, and promoting sustainable development of regional ecological environments, has gained widespread attention [5–7].

As a pivotal national strategy, the exploration of ecological security patterns plays a critical role in resolving conflicts between land development and ecological conservation [8].

Ecological networks present a natural approach to address the challenges of habitat loss and declining biodiversity by offering a spatial framework for the conservation and sustainable management of natural ecosystems. They are emerging as a vital means to quantify regional ecological security [9,10]. These networks represent integral regional habitats where landscape units are organically linked to create a comprehensive network comprising regional environmental and landscape elements. An ecological network serves as the fundamental spatial framework required for sustaining ecosystem functions and essential ecological components, effectively coupling landscape structure, ecological processes, and functions [11–13].

The use of landscape ecology and graph theory in studying ecological networks has gained popularity. Leveraging landscape ecology, this research approach allows for the spatial measurement of landscapes based on existing data, such as land use changes, thereby overcoming limitations arising from incomplete information on aspects like predation, reproduction, and species migration. Currently, there is growing interest in the application of landscape ecology and graph theory to investigate ecological networks. The research approach, which integrates landscape ecology and ecological network analysis, enables the spatial assessment of landscapes using available data, such as land use change data. This method addresses challenges associated with the limited availability of information on species predation, reproduction, and migration and facilitates the long-term monitoring of habitat changes in terms of both quantity and quality. This technique enables the multi-temporal observation of habitat quantities and qualities [14–17]. Graph theory simplifies complex landscape systems into network diagrams, with ecological sources as nodes and ecological corridors as edges. This method directly reflects the structural topology and complexity of the ecological network. Most previous ecological network studies have primarily concentrated on constructing and optimizing regional ecological networks, evaluating ecological network structures, and assessing landscape connectivity based on ecological networks [15,18–20]. Nevertheless, certain aspects, such as identifying ecological sources and evaluating ecological network structures, have room for improvement. For the identification of ecological source areas, most methods either directly select nature reserves or ecological protection areas as ecological sources or base their choices on morphological spatial pattern analysis, leading to no scientific threshold for selecting ecological sources. As for ecological network structure evaluation, researchers often use landscape pattern indicators to assess the spatial distribution of ecological patches and corridors or to optimize ecological network structures and robustness. However, comprehensive analyses of the overall structure and characteristics remain limited [2]. Existing evaluations of ecological network structure predominantly focus on the current state of ecological networks, with relatively few studies considering structural changes over extended timeframes.

Regarding the evaluation of ecological network structure, researchers often use landscape pattern indicators to evaluate the spatial allocation of ecological patches and corridors or evaluate the robustness and optimize the structure of ecological networks. However, detailed analyses of the overall structure and characteristics remain scarce [2]. Currently, the evaluation of the ecological network structure usually focuses on the current ecological network structure, and, to date, relatively few studies have considered the structural succession of a series of ecological networks over longer timescales. Hence, it is imperative to conduct a thorough examination of the holistic framework and attributes of the ecological network at the municipal and county levels.

Located in the southeast of Hainan Island, Wanning City plays a pivotal role in the "Dongyi Group", with "Qionghai–Wanning" forming the core of Hainan Province's territorial spatial planning. It represents a critical nexus for Hainan Province's future urban development. In recent years, the city's land use and landscape patterns have undergone significant transformations due to development, construction, and human activities, impacting regional ecosystem functions and stability [21]. In light of future regional coordinated and integrated development, numerous policies and plans are poised to have a profound influence on Wanning City's land use layout. This study aims to

(1) analyze the spatiotemporal evolution of landscape patterns and assess changes in the ecological network structure in Wanning City and (2) present recommendations, thereby providing a scientific foundation for ecological preservation and restoration in the territorial spatial planning of Wanning City.

2. Materials and Methods

2.1. Study Area

Wanning City is situated in the southeastern part of Hainan Island, bordering the South China Sea to the east, Qiongzhong Li and Miao Autonomous County to the west, Lingshui Li Autonomous County to the south, and Qionghai City to the north (18°35′–19°06′ N, 110°00′–110°37′ E). The terrain varies from high in the west to low in the east. Wanning City enjoys a tropical oceanic monsoon climate with ample rainfall, mild temperatures, and minimal temperature fluctuations. Covering a total land area of 1904.17 km^2, the city comprises 12 townships, including Wancheng, Changfeng, Houan, Damao, and one Xinglong overseas Chinese farm [21]. As of 2020, the city's population was approximately 632,700. Blessed with favorable natural geographic conditions, Wanning City boasts rich natural resources and remarkable ecological advantages. It hosts eight nature reserves, a national forest park, and a natural green oxygen zone, which feature numerous mountainous areas, including Dongshanling, Liulianling, and Jianling. However, in recent years, the combined impact of development, construction, and human activities has resulted in significant changes to cultivated, constructed, and forested lands. The landscape has exhibited a fragmentation trend, leading to issues such as habitat degradation and habitat reduction.

2.2. Data

The ecological resistance surface model of Wanning City was developed by utilizing four types of spatial data: land use/land cover (LULC), road network, elevation (DEM), and habitat suitability assessment. Additionally, a regional ecological network was created. (1) LULC data from 2000 and 2020 were sourced from the GlobeLand30 dataset (http://www.globallandcover.com, accessed on 11 November 2023). This dataset, available free of charge, provides data at a 30 m resolution, encompassing ten land types, including cultivated land, forests, grasslands, and artificial surfaces. (2) Route network data were acquired from Beijing University's 2000 geographic data platform (https://geodata.pku.edu.cn, accessed on 11 November 2023) and the 2020 road network data from OpenStreetMap (https://www.openstreetmap.org, accessed on 11 November 2023). Road networks were categorized into railways and national roads based on traffic conditions. (3) DEM data were derived from the geospatial data cloud (http://www.giscloud.cn, accessed on 11 November 2023). The elevation data were utilized to calculate the slope, and, in combination with the slope, the elevation mobile search window was used to calculate the morphology index (TPI). TPI classified the landscape into valley, low slopes, gentle slopes, steep slopes, and ridges, creating six classes (http://www.jennessent.com/arcview/tpi.htm, accessed on 20 July 2023). All spatial data were converted into a raster format (WGS-84 projection) with a spatial resolution of 30 m.

2.3. Methods

This study focuses on Wanning City, situated in Hainan Province, to identify significant ecological source areas through threshold analysis and analyze the structural changes in the ecological network from the perspective of "ecological source area–ecological corridor–network structure". Ecological networks were constructed for the years 2000 and 2020, identifying ecological pinch points and obstacles affecting landscape connectivity using landscape graph theory and circuit theory. Circuit theory is the study of the relationship between the current and various resistors within a circuit board. This method or model of circuit theory yields three significant outcomes, ecological corridor (or ecological circulation channel), ecological pinch point, and ecological barrier point, which are collectively referred to as ecological nodes. The study proposes an ecological network space optimization scheme.

2.3.1. Assessment of Habitat Suitability and Identification of Ecological Source Areas

Previous methods typically identified ecological source areas through two approaches. First, nature reserves or areas designated as ecological protection zones were directly chosen as ecological source areas. Second, ecological patches were identified based on references and selected if they exceeded a certain minimum area threshold [14,18]. However, these methods often relied on subjective factors and neglected the scale effect of the landscape. In this study, a comprehensive index system was created to evaluate habitat suitability (ranging from 0 to 1), incorporating factors such as the LULC type, road networks, landform, proximity to water sources, and distance from main traffic arteries (Table 1). Based on this index, a threshold area was set at intervals of 2 ha, ranging from 2 to 40 ha, to analyze changes in the number of ecological patches, total area, and habitat suitability within the threshold. The analysis revealed a rapid decline in both the number and total area of ecological patches as the threshold increased (Figure 1). At a threshold of 32 ha, the decline in the number and total area of ecological patches began to stabilize, while the mean habitat suitability value increased slightly with the rising threshold, consistently reaching approximately 0.9. Therefore, 32 ha was selected as the minimum threshold area for ecological source areas in Wanning City.

Table 1. Habitat suitability and ecological resistance factors and their scores in Wanning City.

Factor	Resistance Factor	Coefficient	Habitat Suitability
Land use	Farmland	50	0.5
	Woodland	1	1
	Grassland	20	0.9
	Wetland	20	0.8
	Waters	100	0
	Construction land	100	0
	Bare land	70	0.2
Road	Railway	90	0
	Expressway	60	0
Distance from a water source	<100 m	1	0.9
	100–200 m	10	0.8
	200–500 m	20	0.7
	>500 m	40	0.6
Distance from major traffic arteries	<100 m	90	0.1
	100–200 m	70	0.3
	200–500 m	50	0.5
	500–1000 m	40	0.7
	>1000 m	10	0.9
Geomorphic morphological index	Valley	1	0.9
	Low slope	10	0.8
	Gentle slope	20	0.6
	Steep slope	30	0.5
	Extremely steep slope	50	0.3

2.3.2. Resistance Surface Construction

Ecological resistance signifies the level at which specific landscape features hinder or facilitate species movement between habitat patches. This factor primarily relies on the land cover type and the extent of human disturbance. The construction of the ecological resistance surface model incorporated various factors such as LULC type, roads, topography, distance from water sources, and proximity to major traffic arteries (refer to Table 1) [16,22,23]. The ecological resistance value reflects the presumed relationship between ecological variables and the difficulty of animal movement across pixels and serves as the basis for extracting ecological corridors. The ecological resistance coefficient ranged between 1 and 100, where a coefficient of 1 denotes an ideal environment with minimal movement costs. The coefficient increases as the cost of movement rises, as observed in railways, which act as significant constraints to the migratory movement of terrestrial animals. Consequently, a drag coefficient of 90 was adopted.

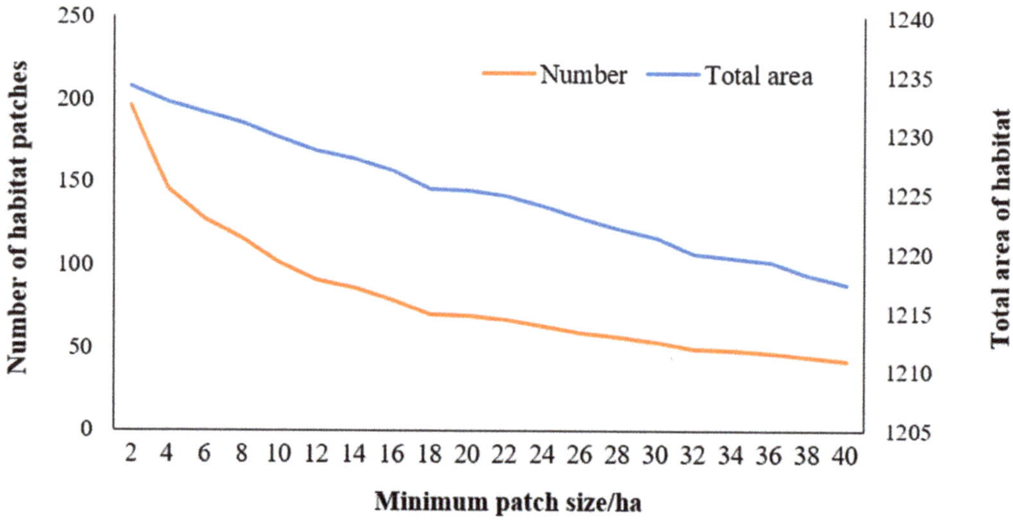

Figure 1. Threshold analysis for minimum ecological source areas.

2.3.3. Ecological Corridor Identification

Utilizing the minimum cumulative resistance model, this study simulated the biological migration path by computing the minimum cumulative resistance between the source and the other patches (minimum cost) to identify potential ecological corridors [24]. The formula for the minimum cumulative resistance model is as follows:

$$MCR = f_{min} \sum_{j=n}^{i=m} (D_{ij} \times R_i), \tag{1}$$

where MCR is the minimum cumulative resistance value and f_{min} represents the relationship between landscape elements and landscape units. D_{ij} represents the Euclidean distance between the species from source j and unit i in the landscape unit. R_i represents the cost of landscape unit i for the species diffusion process [25], which is related to the setting of resistance factors and their coefficients.

2.3.4. Ecological Network Structure

Five landscape indicators, namely, the number of ecological source areas (PN), total area (TA), largest patch index (LPI), area-weighted mean proximity index (AWMPI), and degree of landscape division (DIVISION), were selected to assess the changes in landscape patterns within the habitat patches (refer to Table 2). Sustaining adequate habitat quantity and quality forms the foundation for supporting population reproduction and ecological flow, thereby playing a key role in maintaining biodiversity. The PN and TA indices, which are directly related to habitat loss and fragmentation, were chosen to depict the abundance of ecological source areas. The LPI serves as a measure of the dominance of a particular landscape type, indirectly reflecting the direction and intensity of human activity. The AWMPI gauges the concentration of regional landscape patterns concerning landscape fragmentation. As the AWMPI decreases with increased patch dispersion, the DIVISION focuses on measuring the degree of dispersion of individual distributions of different patches in a specific landscape type. A smaller mean proximity index (MPI) value indicates a higher degree of dispersion among patches of the same type, reflecting an increased level of landscape fragmentation.

The presence of potential ecological corridors connecting distinct habitats may serve as channels for animal migration and diffusion, playing a significant role in enhancing regional landscape connectivity and constituting the fundamental framework for ecological networks. The number of ecological corridors (L) and average path length (MAL) were employed to evaluate changes in the characteristics of the ecological corridors (refer to Table 2). A longer MAL signifies a lower flow efficiency of organisms.

Three indices, namely network closure (α index), line point rate (β index), and network connectivity (γ index), were selected to describe the structure of the ecological networks [26]. The α index serves as a measure of closed loops in a network, occurring when there are more than one connection paths between two nodes. It provides an alternative migration path for organisms needing to evade disturbances or predators. The β and γ indices were used to assess the average number of links and the degree of connectivity of the nodes. Higher values of these three indices signify a greater connectivity of the ecological network, indicating a more comprehensive network structure. The overall connectivity index (OCI) and probability of connectivity (PC) [27] were selected to measure the landscape connectivity of ecological networks.

Table 2. Ecological network elements and structural indices.

Factor		Index	Formula (References [23,27,28])
Ecological source		P	—
		Ta	$\sum a_{ij}$
		LPI	$\frac{max(a_{ij})}{A_L} \times 100$
		AWMPI	$\frac{\sum_{j=1}^{n}\sum_{s=1}^{n}\frac{a_{ijs}}{{h_{ijs}}^2}\left(\frac{a_{ij}}{TA}\right)}{PN}$
		DIVISION	$1-\sum\left(\frac{a_j}{TA}\right)^2$
Ecological corridor		L	—
		MAL	—
Network performance analysis	Network structure	α	$\frac{L-N+1}{2N-5}$
		β	$\frac{L}{N}$
		γ	$\frac{L}{3(N-2)}$
	Network connectivity	IIC	$\frac{\sum_{i=1}^{n}\sum_{j=1}^{n}a_i a_j/(1+nl_{ij})}{A_L^2}$
		PC	$\frac{\sum_{i=1}^{n}\sum_{j=1}^{n}a_i a_j p_{ij}^*}{A_L^2}$

2.3.5. Ecological Network Space Optimization and Ecological Protection Priority Evaluation

An essential approach in ecological network research involves identifying ecological pinch points and obstacle points, followed by identifying areas for ecological network protection and restoration. The ecological pinch point is a significant regional landscape area that requires protection due to its high population density and environmental susceptibility. Habitat fragmentation and degradation in this area may lead to the discontinuity and disappearance of ecological corridors, thereby reducing connectivity between ecological sources and affecting the movement, predation, and migration of species. The ecological barrier refers to the area where species movement is impeded by ecological sources. During simulation, the potential restoration value of the regional landscape after the barrier point is removed and evaluated. Effective ecological restoration in this area can significantly improve the connectivity of the regional landscape. The identified ecological pinch points serve as the key nodes connecting the regional ecological source areas, playing the role of an ecological network hub for critical protection. Ecological restoration is performed with the restoration of ecological obstacles as the central task to optimize the connectivity of ecological networks, promote the flow of factors, and enhance the stability of regional ecosystem services and functions.

The significance of spatial elements in ecological networks was assessed using landscape diagram theory, leading to the proposal of a multilevel protection scheme for ecological networks. Landscape connectivity serves as a vital indicator of the degree of connection between regional landscape patches. The PC index is commonly employed to evaluate the overall connectivity of regional landscapes, considering habitat attributes and the diffusion ability and probability of species. The dPC indices, based on PC development, are frequently used to assess the relative importance of individual landscape patches:

$$dPC(\%) = \frac{PC_{all} - PC^{k}_{remove}}{PC_{all}} \times 100, \quad (2)$$

where the more significant the dPC value, the higher the importance of patch k (ecological source k). PC_{all} is the PC values of all the patches in the original landscape of the study area, that is, the PC values between all ecological sources. PC^{k}_{remove} represents the value of PC, in the study area after removing patch k. In this study, the importance of landscape patches was calculated using Conefor software 2.6. Linkage Mapper evaluated the importance of ecological corridors based on current centrality, categorizing the importance of ecological sources and corridors into four levels: very important, relatively important, important, and generally important.

3. Results

3.1. Spatial Distribution of the Ecological Network

3.1.1. Spatial Distribution of Habitat Suitability and the Ecological Resistance Surface

Figure 2a,c illustrate the spatial distribution of habitat suitability values in Wanning City, showcasing higher values in the western region and lower values in the eastern region. In 2000, Sangengluo, Nanqiao, Beidai, and Longjiu in Wanning City exhibited the highest habitat suitability, while Damao, Houan, Hele, and the western area of Wancheng displayed the lowest habitat suitability. By 2020, the western townships, specifically Sangengluo, Nanqiao, and Longwu, continued to exhibit relatively high habitat suitability, while the eastern townships, such as Canning, Changfeng, Damao, Houan, and Hele, experienced a significant decline in habitat suitability, becoming concentrated around the central urban area. When analyzing the spatial distribution of the ecological resistance surface model in Wanning City (Figure 1), we can note that the ecological resistance value exhibited a lower magnitude in the western region and a higher magnitude in the eastern region. This trend displayed a gradual decrease emanating from the central urban area and the main transportation route. From 2000 to 2020, the ecological resistance values of Wancheng, Changfeng, Damao, and Liji towns in the eastern region significantly increased, while the western region experienced varying degrees of increase in the ecological resistance values of Nanqiao, Sangenluo, and Beifang.

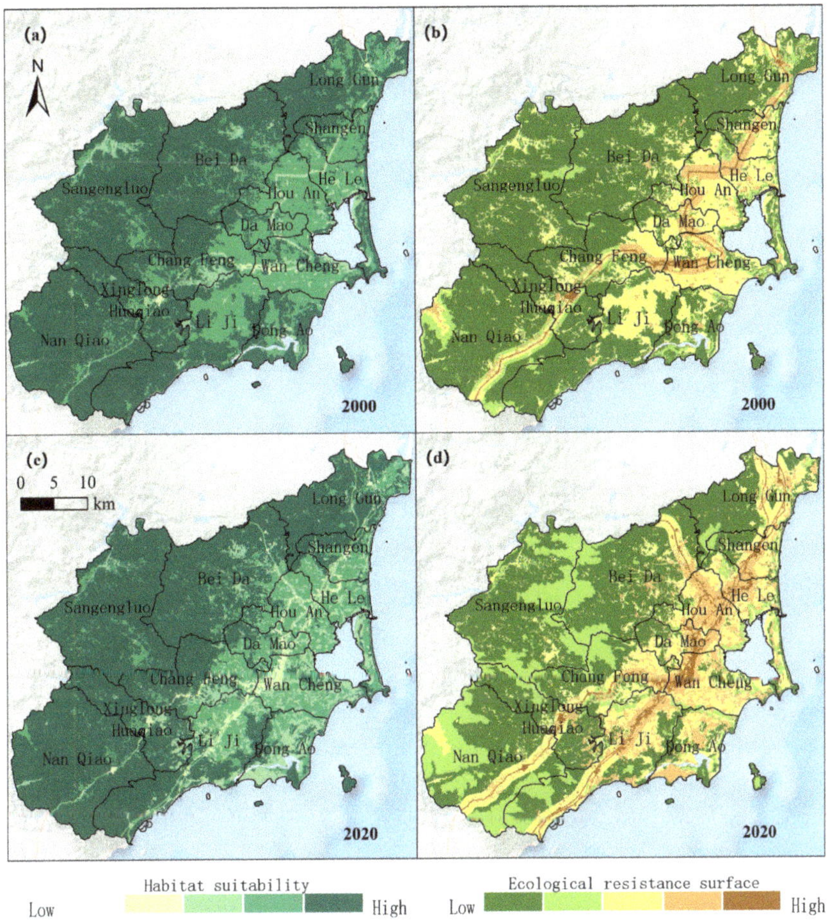

Figure 2. Distribution map of habitat suitability and the ecological resistance surface in Wanning City from 2000 to 2020. ((**a**,**c**): habitat suitability for 2000 and 2020; (**b**,**d**): ecological resistance surface for 2000 and 2020).

3.1.2. Analysis of Spatial Distribution of the Ecological Network

The ecological source areas of Wanning City were predominantly concentrated in the west and south, with relatively fewer ecological source areas in the middle and east (Figure 3). The eastern portion of the city is characterized by a mountainous terrain featuring expansive forests and grassy areas, whereas the central and western areas comprise hilly or flat landscapes with developed and cultivated land. Consequently, the number, size, and distribution of ecological source areas exhibit significant variations. Regarding the spatial distribution of ecological corridors, Wanning's ecological corridors are primarily situated in the central and eastern parts of the city, with sparse distribution in the north and dense distribution in the south. In 2000, Wanning's ecological corridor spanned Longgu, Shangen, Houan, Hele, Damao, Wancheng, and Liji. By 2020, ecological corridors in Wanning City were concentrated in Damao, Changfeng, and Wancheng, with reductions in other regions. Over the period from 2000 to 2020, several small ecological sources in Wancheng, Liji, and Dongao disappeared, leading to a reduction in the corridors.

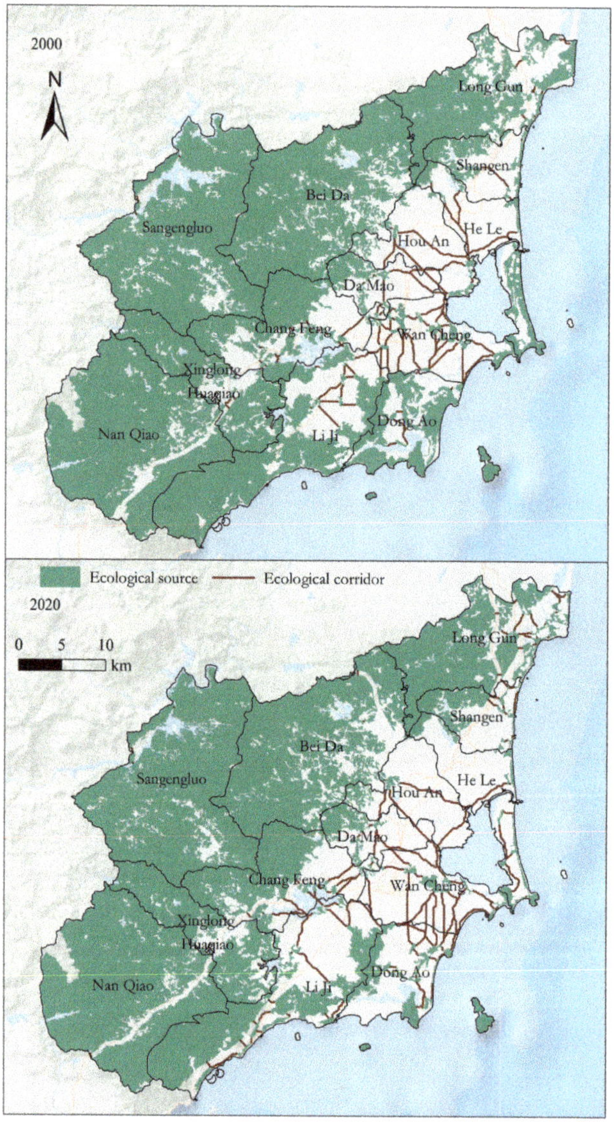

Figure 3. Spatial distribution of ecological networks in Wanning from 2000 to 2020.

3.2. Analysis of the Change in Ecological Network Structure

3.2.1. Ecological Source Area

Between 2000 and 2020, the number of patches in Wanning City's ecological source area increased from 49 to 61, while the total area decreased by 61.62 km^2 (Table 3). This indicates that the ecological source area in the study area became more fragmented, a trend further substantiated by the change in the AWMPI. The AWMPI for the ecological source area of Wanning City was highest in 2000, suggesting that the patches in the ecological source area were closer and more spatially continuous. However, the decrease in the AWMPI in 2020 indicates a reduction in connectivity between patches, contributing to increased landscape fragmentation. Over the same period, the LPI decreased from 46.92% to 40.03%, indicating a reduction in the dominance of habitat patches in Wanning City, indirectly reflecting increased human activity disturbance within the habitat. An analysis of the DIVISION index reveals that the degree of landscape segmentation in the ecological source area of Wanning City was severe between 2000 and 2020, resulting in increased dispersion of landscape patches.

Table 3. Structural changes in the ecological network in Wanning City from 2000 to 2020.

Factor		Index	2000	2020
Ecological source		Ecological source	49	61
		T.A./km^2	1219.97	1158.35
		LPI	46.92	40.03
		AWMPI	4254	2793
		DIVISION	0.76	0.82
Ecological corridor		Ecological corridor	94	114
		MAL/km	1.91	2.30
Network performance analysis	Network performance analysis	α	0.495	0.462
		β	1.918	1.869
		γ	0.667	0.644
	Network connectivity	IIC	0.289	0.244
		PC	0.372	0.335

3.2.2. Ecological Corridor

The number of ecological corridors in Wanning City increased from 94 in 2000 to 114 in 2020 (Table 3). Simultaneously, the MAL in 2020 was higher compared to that in 2000, growing from 1.91 km in 2000 to 2.30 km in 2020. This suggests reduced organism flow efficiency in the study area due to several factors. Firstly, the shrinking and fragmentation of ecological sources, especially the decreasing trend in the number and area of ecological sources in Wancheng, Liji, and Dongao, have led to the severe fragmentation of ecological sources in some areas. Consequently, habitats that could have been directly connected now rely on further habitats for connectivity. As a result, the number and length of corridors have increased. Secondly, the growing urbanization in Wanning City has heightened ecological resilience in the central urban area, resulting in a reduction in the distance between ecological sources.

3.2.3. Network Topology

The α, β, and γ indices in 2000 were significantly higher than those in 2020 (Table 3), indicating that the ecological network structure of Wanning City in 2000 was superior to that in 2020. In 2000, the α index of the ecological network in Wanning City was 0.495, while in 2020, it decreased to 0.462, signifying a reduction in closed loops and a shift toward a more linear network structure in 2020. Changes in the β index reveal that the average number of connections at each network node in 2020 was lower than in 2000, reflecting reduced network accessibility. The γ index of the ecological network in Wanning City decreased from 0.667 in 2000 to 0.664 in 2020, indicating reduced connectivity and network effectiveness. Changes in landscape connectivity indices (PC and IIC) of the ecological network suggest diminished landscape connectivity and habitat accessibility for species in the study area. The loss of ecological source areas in the eastern part of Wanning City directly contributed to a reduction in closed loops and the structural accessibility of the ecological network. The reduction and fragmentation of the habitat patch area led to decreased landscape connectivity. Longer corridor distances between ecological sources, increased ecological resistance due to urbanization and human activities, and a subsequent rise in resistance cost for potential ecological corridors all impacted network connectivity.

3.3. Ecological Network Space Optimization and Restoration Countermeasures

Consistent with established ecological principles, we identified ecological barriers to locate areas of degradation and damage within the broader landscape. The current intensity, symbolized by color, gradually increased from blue to red (Figure 4a). The red areas represent regions with the most significant current intensity and the highest ecological restoration potential. These areas are primarily situated in Damao, Houan, Changfeng, Liji, and Wancheng. Ecological pinch points were determined using the current theory, and we established a 5 km buffer as the foundational component of the ecological corridor. The distribution of current density within the ecological corridor in Wanning City ranged from yellow to red (Figure 4b), indicating a progressive increase in intensity. The red areas signify locations with the most substantial current density, signifying ecological pinch points within Wanning City. Ecological pinch points, with ecological importance, were predominantly concentrated in the southwest and central areas of Wanning City, particularly in Liji, Dongao, Wancheng, and Damao.

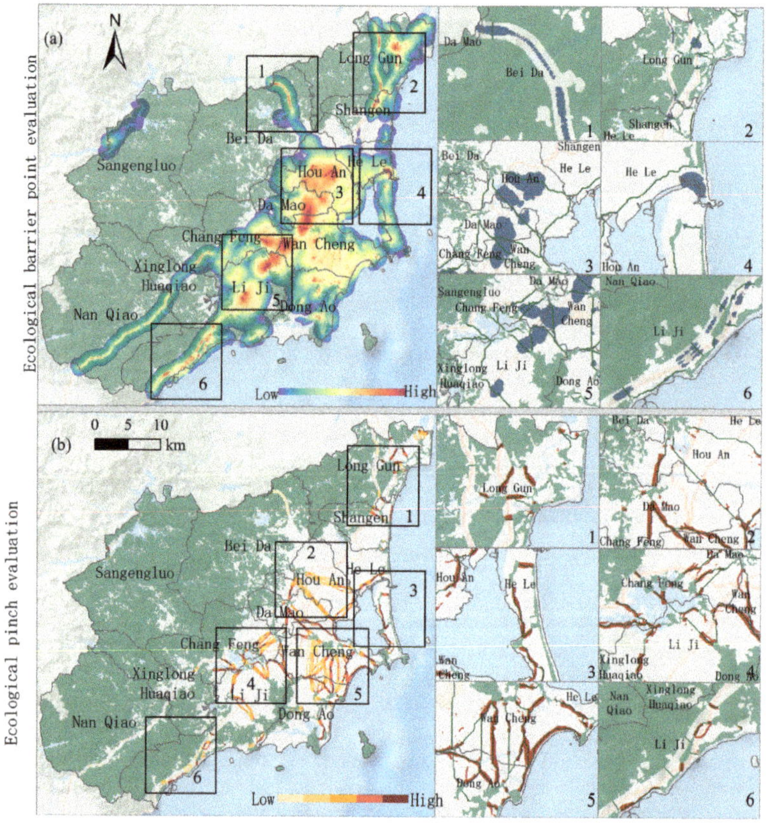

Figure 4. Optimization pattern of ecological networks in Wanning City. ((**a**). ecological barrier point evaluation; (**b**). ecological pinch evaluation).

Based on the identification of ecological pinch points and obstacles, coupled with the prioritization of ecological network elements, we introduce an ecological network protection and restoration plan known as "One Belt, Four Sources, Eight Districts, Multiple Corridors, and Multiple Points" (Figure 5a). "One Belt" alludes to the coastal ecological landscape belt, safeguarding coastal resources, defense forests, and sandy beaches. "Four sources" pertains to the four primary ecological source areas encompassing Shangxi–Jianling, Liulianling, Nanlin, and Jiaxin nature reserves, national forest parks, mountains, and inland seas. These areas hold great significance for soil and water conservation, biodiversity preservation, and soil and water conservation within Wanning City. The "eight districts" encompass ecological restoration zones concentrated in the towns of Houan, Damao, Changfeng, and Liji. "Multi-corridor and multi-point" refer to the establishment of numerous ecological corridors and crucial ecological nodes, relying on reservoirs, wetlands, islands, and mountains to preserve regional landscape connectivity.

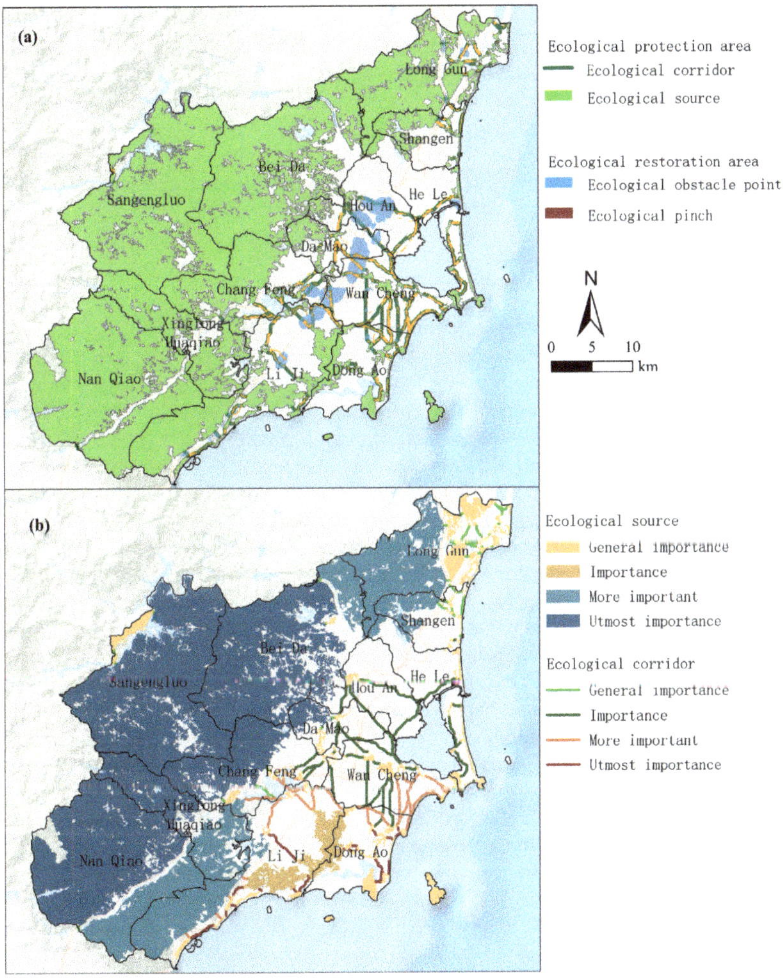

Figure 5. Distribution of ecological spaces for protection and restoration in Wanning City. ((**a**). ecological protection and restoration area (**b**) ecological source and corridor).

Considering the importance value of ecological source patch dPC and the current centrality value of the ecological corridor, we categorized the significance of ecological network elements into four levels: significant, relatively important, meaningful, and generally important. The most vital ecological sources were situated in Sangengluo and Nanqiao (Figure 5b). The crucial ecological sources were located in the Longgu and Xinglong Overseas Chinese Farm. The highly important ecological sources were primarily found in Liji, while the critical ecological sources were widely distributed in Longgu, Damao, Houan, Wancheng, and other central and eastern regions. The most crucial ecological corridors were found in Liji and Dongao, the highly important ecological corridors were mainly distributed in Wancheng and Changfeng, the highly important ecological corridors were concentrated in Damao and other central and eastern areas, and the highly important ecological corridors were mainly situated in the north of Wanning City.

4. Discussion and Conclusions

Ecological networks play a crucial role in maintaining regional ecological security patterns and are essential for the sustainable development of regional ecological environments. The quality of their structure is of utmost importance in this regard. Understanding the structure of these networks is essential for safeguarding biodiversity and sustaining an ecological equilibrium. This study constructed and analyzed the ecological network of Wanning City for the years 2000 and 2020, with a focus on examining changes in the region's ecological network in terms of the spatial arrangement of landscape elements, the structure of the landscape, and connectivity. We have devised an ecological network optimization strategy for Wanning City, presenting a regional ecological preservation and restoration blueprint. The main conclusions drawn from this study are as follows.

(1) The study reveals a distinct pattern of habitat suitability across Wanning City, with higher suitability in the west and lower suitability in the east. Notably, between 2000 and 2020, the habitat suitability of regions such as Wancheng, Changfeng, Damao, Houan, and Le, to the east of Wanning City, experienced significant decreases. The areas exhibiting lower habitat suitability expanded outward from the center at Wancheng. Concurrently, the ecological resistance in Wanning City displayed a similar spatial trend, with lower values in the west and higher values in the east. Notably, Wancheng, Changfeng, Damao, and Liji emerged as high-resistance areas. Over the same period, an increase in ecological resistance was observed in the eastern region, accompanied by varying degrees of increases in Nanqiao, Sangengluo, and Beiduo in the western region, signifying heightened resistance to species movement.

(2) The ecological source areas in Wanning City displayed notable regional variations, with higher concentrations found in the western and southern sectors, while the central and eastern regions exhibited fewer ecological source areas. The ecological corridors were predominantly located in the central and eastern parts of the city, with a less dense presence in the northern area. This divergence can be attributed to the prevalence of mountains and extensive forest and grassland in the eastern region, in contrast to the dominance of hills and plains, characterized by urbanization and human activities, in the central and western areas. This disparity highlights the substantial spatial heterogeneity within the ecological network.

(3) The in-depth scrutiny of the ecological network structure reveals several critical findings. Between 2000 and 2020, ecological source areas within Wanning City decreased, indicating a trend toward fragmentation, reduced structural accessibility, and diminished landscape connectivity within the ecological network. The number of patches in the ecological source area increased while the total area diminished, signifying the dwindling dominance of habitat patches. This trend corresponds with changes in the AWMPI, LPI, and DIVISION landscape pattern indices. Furthermore, during the same period, the number of ecological corridors increased, but the average path length also increased. This signifies a decrease in the efficiency of biological flow within ecological corridors, primarily due to two factors. Firstly, habitat fragmentation caused by the shrinking of the ecological source area necessitates connecting more distant habitats, and secondly, increased urbanization and human activities have elevated ecological resistance, lengthening the shortest path between ecological sources. Furthermore, the analysis of ecological network performance reveals that the ecological network structure was better suited in 2000 than in 2020, with significantly higher values for the α, β, and γ indices, as well as for the PC and IIC indices, representing the network's structure. The results indicate a reduction in the number of loops and structural accessibility of the ecological network in Wanning City, leading to a transition from a ring-like structure in 2000 to a more radial one in 2020, ultimately reducing the landscape connectivity and habitat accessibility of the network.

(4) To protect crucial ecological spaces and restore degraded and damaged areas, we propose an ecological network space protection and restoration plan, termed "one belt, four sources, eight districts, multiple corridors, and multiple points". The restoration of territorial space will be focused on areas in Houan, Damao, Changfeng, and Liji. Emphasis will be placed on protecting and restoring ecological lands, such as forestland and grassland, in the north-central and eastern regions of the city to mitigate fragmentation. The goal is to enhance the connectivity of the landscape ecological network by increasing the number and size of ecological patches in regions like Wancheng, Damao, Dongao, and other central and eastern areas, while also restoring connectivity between ecological sources. Additionally, we aim to establish multiple ecological corridors and important ecological nodes based on reservoirs, wetlands, islands, and mountains, connected by the landscape of the Weiyu region. To ensure the protection and control of ecological resources, ecological protection spaces are classified into four levels: significant, relatively important, necessary, and generally essential, based on the importance of ecological source patches and the current centrality value of ecological corridors.

In conclusion, this research provides valuable insights for scientifically assessing critical areas for territorial ecological restoration, identifying essential regions for ecological preservation, and optimizing the allocation of natural resources.

Author Contributions: S.Z. and J.G. designed the research; R.F. processed the data; S.Z. analyzed the data; S.Z. and J.G. wrote the paper. All authors have read and agreed to the published version of the manuscript.

Funding: This research was funded by the National Social Science Foundation of China and Fundamental Research Funds for the Central Universities, China University of Geosciences (Wuhan).

Data Availability Statement: The authors have not obtained permission to publish the data. Therefore, the data can be obtained from the corresponding author upon reasonable request.

Conflicts of Interest: The authors declare no conflicts of interest. The funding sponsors had no role in the design of the study; in the collection, analysis, or interpretation of data; in the writing of the manuscript; or in the decision to publish results.

References

1. Gu, C.; Guan, W.; Liu, H. Chinese urbanization 2050: S.D. modeling and process simulation. *Sci. China Earth Sci.* **2017**, *60*, 1067–1082. [CrossRef]
2. Liu, X.; Wei, M.; Zeng, J.; Zhang, S. Ecological network analysis and construction: A case study of the urban agglomeration of the Min River Delta, China. *Resour. Sci.* **2021**, *43*, 357–367. [CrossRef]
3. Wu, D.; Li, H.; Ai, N.; Tao, H.; Gu, J. Predicting spatiotemporal changes in land use and habitat quality based on CA-Markov: A case study in central Ningxia, China. *Chin. J. Eco-Agric.* **2020**, *28*, 1969–1978. [CrossRef]
4. Gao, J.; Gong, J.; Li, J. Effects of source and sink landscape pattern on land surface temperature: An urban heat island study in Wuhan City. *Prog. Geog.* **2019**, *38*, 1770–1782. [CrossRef]
5. Wang, Z.; Ya, S.; Pu, H.; Mofakkarul, I.; Ou, L. Simulation of spatiotemporal variation of land use in mountainous-urban fringes based on improved CA-Markov model. *Trans. Chin. Soc. Agric. Eng.* **2020**, *36*, 239–248. [CrossRef]
6. Bielecka, E. GIS spatial analysis modeling for land use change. A bibliometric analysis of the intellectual base and trends. *Geosciences* **2020**, *10*, 421. [CrossRef]
7. Zhao, C.; Jensen, J.L.R.; Weaver, R. Global and local modeling of land use change in the border cities of Laredo, Texas, USA and Nuevo Laredo, Tamaulipas, Mexico: A comparative analysis. *Land* **2020**, *9*, 347. [CrossRef]
8. Huang, M.; Yue, W.; Feng, S.; Cai, J. Analysis of spatial heterogeneity of ecological security based on MCR model and ecological pattern optimization in the Yuexi county of the Dabie Mountain Area. *J. Nat. Resour.* **2019**, *34*, 771–784. [CrossRef]
9. Peng, J.; Zhao, H.; Liu, Y.; Wu, J. Research progress and prospects on regional ecological security pattern construction. *Geogr. Res.* **2017**, *36*, 407–419.
10. Zhou, C.; Wu, Y. A planning support tool for layout integral optimization of urban blue–green infrastructure. *Sustainability* **2020**, *12*, 1613. [CrossRef]
11. Saputra, M.H.; Lee, H.S. Prediction of land use and land cover changes for North Sumatra, Indonesia, using an artificial-neural-network-based cellular automaton. *Sustainability* **2019**, *11*, 3024. [CrossRef]

12. Walker, N.J.; Schaffer-Smith, D.; Swenson, J.J.; Urban, D. Improved connectivity analysis using multiple low-cost paths to evaluate habitat for the endangered San Martin Titi monkey (*Plecturocebus oenanthe*) in North-Central Peru. *Landsc. Ecol.* **2019**, *34*, 1859–1875. [CrossRef]
13. Heinonen, T. Developing landscape connectivity in commercial boreal forests using minimum spanning tree and spatial optimization. *Can. J. For. Res.* **2019**, *49*, 1198–1206. [CrossRef]
14. Cao, Z.; Sun, Y.; Xie, G.; Qiu, P. Study on the evolution of the ecological network in the Haikou coastal zone. *Acta Ecol. Sin* **2020**, *40*, 1044–1054.
15. Yang, J.; Zeng, C.; Cheng, Y. Spatial influence of ecological networks on land use intensity. *Sci. Total Environ.* **2020**, *717*, 137151. [CrossRef]
16. He, J.; Pan, Y.; Liu, D. Analysis of the wetland ecological pattern in Wuhan City from the perspective of ecological network. *Acta Ecol. Sin* **2020**, *40*, 3590–3601.
17. Zhao, W.; Han, Z.; Yan, X.; Zhong, J. Ecological security pattern construction based on the multi-scenario trade-off of ecosystem services: A case study of Wafangdian, Dalian. *J. Nat. Resour.* **2020**, *35*, 546–562.
18. Fang, Y.; Wang, J.; Huang, L.; Zhai, T. Determining and identifying critical areas of ecosystem preservation and restoration for territorial spatial planning based on ecological security patterns: A case study of Yantai city. *J. Nat. Resour.* **2020**, *35*, 190–203.
19. Shi, F.; Liu, S.; An, Y.; Sun, Y.; Dong, S.; Wu, X. Changes of landscape fragmentation and connectivity with urbanization: A case study of Kunming City. *Acta Ecrol. Olog. Sin* **2020**, *40*, 3303–3314.
20. Shi, F.; Liu, S.; An, Y.; Sun, Y. Biodiversity conservation of mountains-rivers-forests-farmlands-lakes-grasslands using an ecological network: A case study on the Zuoyoujiang river basin in Guangxi Province, China. *Sheng Tai Xue Bao* **2019**, *39*, 8930–8938. [CrossRef]
21. Wei, Z.; Luo, G.; Lu, Y. Optimization of landscape pattern in Wanning City based on ecological perspective. *Mod. Hortic.* **2018**, *X*, 108–111.
22. Liu, X.; Li, J.; Zhou, Y.; Chen, Z.; Ding, Y. Analysis of landscape ecological pattern evolution and ecological network structure optimization for Shanghai. *Resour. Environ. Yangtze Basin* **2019**, *28*, 2340–2352.
23. Liu, X.P.; Zhang, Z.; Li, L.Y.; Li, M.X. Comprehensive evaluation of the evolution of ecological network structure in Tianjin, China, from a multi-dimensional perspective. *Ying Yong Sheng Tai Xue Bao.* **2021**, *32*, 1554–1562. [CrossRef] [PubMed]
24. Li, X. *The Delineation of the Ecological Red Line Considers the Pattern and Function of the Ecological Network*; Wuhan University: Wuhan, China, 2017.
25. Li, W.; Ma, L.; Zang, Z.; Gao, J.; Li, J. Construction of ecological security patterns based on the ecological red line in Erhai Lake Basin of southwestern China. *J. Beijing For. Univ.* **2018**, *40*, 85–95. [CrossRef]
26. Wu, L. *Study on the Optimal Layout of the Ecological Network of the Oasis in Hexi Corridor*; Gansu Agricultural University: Lanzhou, China, 2016.
27. Saura, S.; Torné, J. CONEFOR SENSINODE 2.2: A software package for quantifying the importance of habitat patches for landscape connectivity. *Environ. Model Softw.* **2009**, *24*, 135–139. [CrossRef]
28. Mcgarigal, K.; Marks, B.J. *Fragstats—Spatial Pattern Analysis Program for Quantifying Landscape Structure*; US Department of Agriculture, Forest Service, Pacific Northwest Research Station: Washington, DC, USA, 1995; p. 351.

land

MDPI

Article

Spatial–Temporal Changes and Driving Mechanisms of Ecological Environmental Quality in the Qinghai–Tibet Plateau, China

Zhan Shen and Jian Gong *

Department of Land Resource Management, School of Public Administration, China University of Geosciences (Wuhan), Wuhan 430074, China; szshenzhan@cug.edu.cn
* Correspondence: gongjian@cug.edu.cn

Abstract: This study examines the evolution of eco-environmental quality and its driving forces in the Qinghai-Tibet Plateau, with a particular focus on the Qinghai Lake region (QLR). By employing principal component analysis (PCA) on nearly 20 years of remote sensing data, we reveal the dynamic characteristics of ecological quality in this sensitive area. The results indicate that the ecological quality of the QLR has exhibited significant fluctuations over the past two decades, influenced by multiple factors such as climate change, human activities, and policy adjustments. Specifically, the fluctuations in ecological quality are closely associated with key ecological indicators, including the Normalized Difference Vegetation Index (NDVI), Land Surface Temperature (LST), Wetness Index (WET), and Normalized Differential Bare Soil Index (NDBSI). Vegetation cover and moderate humidity have substantial positive effects on ecological quality, while high temperatures and dry soil conditions exert negative impacts.

Keywords: ecological quality; spatiotemporal evolution; influencing factor; random forest; Qinghai-Tibet Plateau

1. Introduction

The ecological environment serves as a crucial foundation for human survival and development [1,2]. However, due to economic growth and social progress, human activities have increasingly placed pressure on the environment. Unsustainable human practices have threatened regional ecological balance, and the conflict between ecological protection and economic development has become more pronounced. Therefore, conducting long-term assessments of ecological quality and identifying influencing factors are of great practical significance for promoting harmonious development between the natural environment, economy, and society [3]. The assessment of ecological environmental quality and its changes is a critical foundation for ecosystem management and serves as a significant indicator for measuring the effectiveness of regional ecological civilizational development. Although relevant evaluation criteria have been established and extensive evaluation efforts have been undertaken, the adequacy of accounting for natural ecosystem differences in these evaluations remains questionable [4,5]. Whether the evaluation results accurately reflect the true ecological quality of a region's environment remains a topic of discussion. Xu Hanqiu [6–8] proposed the Remote Sensing-based Ecological Index (RSEI), which is derived from the analysis of remote sensing image data. The RSEI evaluates the quality of regional ecological environments, and its research outcomes, based on image data, can accurately capture spatial pattern characteristics. Over the past decade, it has been widely utilized and refined. The ecosystem types included in the study are relatively straightforward, allowing for model reconstruction that aligns more closely with local conditions when enhancing and applying the index. However, among the four component indicators of the RSEI, the degree of correlation and the trend of change between vegetation coverage and the RSEI demonstrate a high level of consistency, indicating that the index is fundamentally rooted in the quality of vegetation cover [9–11]. Thus, when characterizing a single ecosystem type, it can be regarded as homogeneous, fulfilling the requirements for assessing ecological environmental quality. However, in regions with multiple ecosystem types, addressing the limitations of the Remote Sensing Ecological Index (RSEI) remains a central focus of ongoing research [11,12].

As a vital component of the Qinghai–Tibet Plateau, the evolution of ecological environmental quality in the QLR significantly influences the ecological security of the entire plateau. This region is abundant in natural resources, serves as a habitat for numerous rare plant and animal species, and is a critical area for water conservation [13,14]. However, in recent years, the ecological quality of the QLR has significantly changed due to the combined effects of climate change, overgrazing, land development, and other factors. Some areas have even experienced severe ecological degradation. Therefore, evaluating the long-term evolution of ecological quality in this region and exploring its underlying driving forces is crucial for developing effective and scientifically sound ecological protection and restoration strategies.

The objective of this research is to comprehensively and systematically evaluate the ecological quality of the QLR over the past 20 years by utilizing remote sensing technology, geographic information systems (GIS), and statistical analysis methods. The aim is to reveal dynamic evolution patterns and conduct an in-depth analysis of the key factors influencing changes in ecological quality. Firstly, the Remote Sensing Ecological Index (RSEI) will serve as the primary metric for assessing ecological quality. The RSEI is a comprehensive assessment index based on remote sensing technology that integrates multiple ecological indicators, including the Normalized Difference Vegetation Index (NDVI), Land Surface Temperature (LST), Wetness Index (WET), and Normalized Difference Bare Soil Index (NDBSI), to provide a holistic reflection of the overall status of regional ecological quality. By calculating the RSEI values for different years, we can intuitively observe the trends and spatial distribution characteristics of ecological environmental quality in the QLR. Specifically, we will (1) identify the characteristics of changes in ecological environmental quality across various temporal and spatial scales; (2) determine the main driving factors influencing these changes and their interactions; (3) evaluate the effectiveness of ecological protection measures; and (4) propose scientifically sound and reasonable recommendations for ecological protection and restoration strategies. These research outcomes are not only significant for the ecological environment protection of the QLR but also provide a valuable reference for assessing and safeguarding the ecological environmental quality of the Qinghai–Tibet Plateau and other similar regions worldwide.

2. Materials and Methods

2.1. Research Area

Located on the northeastern edge of the Qinghai–Tibet Plateau, the QLR is a closed inland basin surrounded by mountains. It was formed by the fault collapse between Datong Mountain, Riyue Mountain, and Nanshan Mountain, which are part of the Qilian Mountains. This region is situated in the central and northern parts of Qinghai Province, bordering Gansu Province to the north, Haidong City to the east, and the Haixi Mongolian and Tibetan Autonomous Prefecture to the west. It spans latitudes between 36°00′ and 38°15′ north and longitudes between 99°50′ and 102°40′ east. The specific geographical location is illustrated in Figure 1. The region primarily falls under the administrative jurisdiction of "three prefectures and four counties" in Qinghai Province, which includes Haiyan and Gangcha County of the Haibei Tibetan Autonomous Prefecture, Tianjun County of the Haixi Mongolian Tibetan Autonomous Prefecture, and Gonghe County of the Hainan Tibetan Autonomous Prefecture. Covering a total area of 55,700 square kilometers, it accounts for 7.99% of the province's total land area. Qinghai Lake, located between 36°15′ and 38°20′ north latitude and 97°50′ and 101°20′ east longitude, is the largest saline lake in China. Over the past decade, it has maintained an average elevation of 3193.15 m. The expansive grasslands and numerous snow-capped peaks surrounding the lake contribute to its status as an internationally significant wetland. The Qinghai Lake Basin acts as a natural barrier, helping to control the spread of desertification to the west and ensuring ecological security in the northeastern region of the Qinghai–Tibet Plateau. Qinghai Lake stretches 106 km in length from east to west, 63 km in width from north to south, and has a circumference of 360 km. The lake is elongated in shape, narrow from north to south, and slightly elliptical. The water in the lake is cold, brackish, and bitter, with a specific gravity lower than that of seawater but slightly higher than that of freshwater. Its salt content is 1.25%, and the pH level is 9.2.

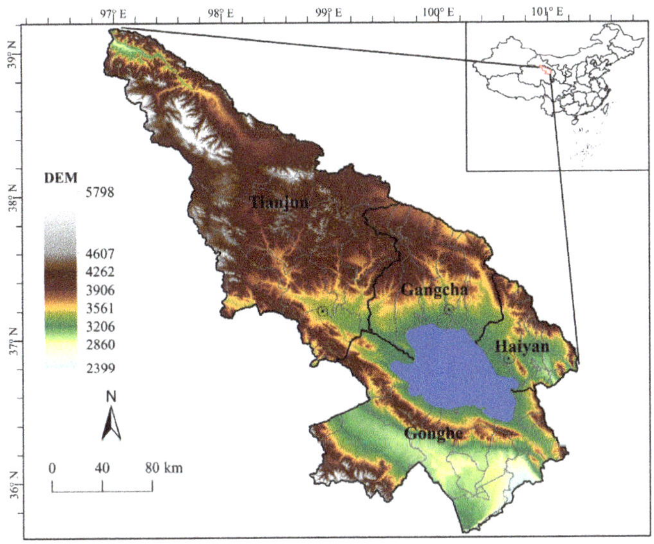

Figure 1. Study area location map.

The QLR experiences a typical plateau continental climate characterized by drought, limited rainfall, dryness, strong winds, prolonged cold periods, and intense solar radiation. The climate varies significantly across the seasons, featuring high winds and sandstorms in spring, cool but brief summers, and long, cold winters. The area receives ample solar radiation and abundant light resources. Rainfall is scarce, with precipitation and heat occurring simultaneously during certain periods, resulting in distinct dry and wet seasons. Situated between the Loess Plateau in the northeast of Qinghai Province, the region surrounding Qinghai Lake borders the Hehuang Valley to the east, the Qilian Mountains to the north, and the Gonghe Basin to the south. The elevation of the entire region ranges from 3294 m to 5174 m, with an average elevation of 3900 m. The terrain is relatively flat with gentle slopes, and the area is higher in the northwest and lower in the southeast. Mountains occupy about 68.6% of the total area, while valleys and lake plains make up a smaller proportion, accounting for 31.4% of the region's total area. The mountains are steep, the valleys are densely covered, and ice erosion terrain is widespread. The landforms are complex and diverse, consisting mainly of alluvial plains, lacustrine plains, undulating mountain ranges, ice platforms, and modern glaciers. The vertical distribution of soil in the QLR is pronounced. In Jinshawan, the Gahai area, and near Bird Island, floating aeolian sand and fixed aeolian sand are prevalent. The lake plain is predominantly chestnut soil, while low mountains and hills are mainly covered with mountain chestnut soil, mountain steppe meadow soil, and mountain meadow soil. The region exhibits a diverse range of vegetation types, showcasing the coexistence of warm and alpine vegetation. The primary vegetation types include alpine meadow, alpine steppe, sandy vegetation, salt meadow, cold desert steppe, and swamp steppe.

2.2. Data Source

This study mainly uses MODIS remote sensing image, administrative boundary data, DEM data, etc. The data sources are shown in Table 1.

Table 1. Data sources.

DATA	Resolution	Year	Source
DEM	30 m	2000	SRTM (Shuttle Radar Topography Mission, SRTM)
Administrative Boundary Data	vector	2020	China National Center for Basic Geographic Information (http://www.ngcc.cn/ (accessed on 9 December 2024))
Slope	30 m	2020	SRTM (Shuttle Radar Topography Mission, SRTM)
MOD09A1 8-Day	500 m	2000–2020	GEE remote sensing cloud platform
MOD11A2 LST 8-Day	1 km	2000–2020	GEE remote sensing cloud platform
MOD13A1 16-Day	500 m	2000–2020	GEE remote sensing cloud platform

2.3. Research Methods

2.3.1. Research Framework

RSEI is a remote sensing-based ecological environmental quality evaluation system, emphasizing four key indices, the WET, NDVI, LST, and NDBSI, which are utilized to assess regional ecological quality. By leveraging the robust cloud data storage and high-performance online analysis capabilities of the GEE spatiotemporal remote sensing cloud service platform, the overall framework consists of three primary steps. First, remote sensing image data, population data, and economic data are collected and pre-processed. Next, the water area is identified using the WET, NDVI, LST, and NDBSI indices for the study area in the QLR. The ecological quality is then derived using the PCA method. Finally, the ecological indicators in the QLR are analyzed to examine the temporal and spatial differences and dynamic changes in the RSEI, as well as to investigate the driving forces behind the RSEI.

2.3.2. RSEI Model

The NDVI is widely used to monitor vegetation growth, as it directly reflects the quality of the regional ecological environment. LST is closely associated with vegetation growth, crop yield, the surface water cycle, urbanization, and various natural phenomena and processes, as well as human activities. LST serves as a heat index that indicates the condition of the surface ecological environment. The Kauth–Thomas transformation method, a linear transformation based on multispectral imaging, is employed to derive moisture components that provide information on soil and the WET. The dryness index, which indicates soil dryness, can have detrimental effects on the ecological environment. Given that urban construction land predominates in our study area, the dryness index was represented by combining the Bare Soil Index (SI) and the Construction Index (IBI) into the NDBSI.

The NDVI is a measure used to assess vegetation health. The WET quantifies moisture levels in the soil. LST indicates the temperature of the Earth's surface, while the NDBSI helps differentiate between built-up areas and soil.

2.3.3. Random Forest Algorithm

In the study of ecological quality driving factors, the random forest algorithm has proven to be an effective tool for identifying key influencing factors, owing to its robust classification and regression capabilities [6,13–16]. Six driving factors were employed to analyze the ecological quality of the QLR: the WET, NDVI, LST, NDBSI, slope, and elevation. The random forest algorithm trains the data by constructing multiple decision trees, enhancing the prediction accuracy and robustness of the model through majority voting or averaging outputs. During the construction of each decision tree, the algorithm randomly selects specific features from the six driving factors for segmentation. This method helps prevent overfitting and improves the model's generalization ability. The core strength of the algorithm lies in its capacity to automatically assess the importance of each feature for classification or regression objectives, thereby identifying the critical factors that influence ecological quality. When analyzing the drivers of ecological quality, the random forest algorithm initially performs multiple samplings of the training dataset and constructs several decision tree models. Subsequently, the importance of each feature is evaluated by counting the frequency of its occurrence as a split node across all trees and measuring the corresponding reduction in Gini impurity. Features with high importance are typically closely linked to changes in ecological quality. For instance, the NDVI reflects the status of vegetation cover, and its significant importance indicates that vegetation growth has a substantial positive effect on ecological quality. By employing this method, the random forest algorithm uncovers the underlying mechanisms driving changes in ecological quality in the QLR, providing a scientific foundation for the formulation of ecological protection measures.

3. Results

3.1. PCA: Principal Component Analysis

Figure 2 illustrates the trend in the contribution of PC1 to ecological quality in the QLR over the past 20 years. This principal component encompasses four ecological indicators, dryness, greenness, heat, and humidity, and is derived through PCA to reflect the overall ecological quality of the region. As shown in Figure 2, the ecological quality of the QLR has experienced significant fluctuations over the past two decades. Notably, around 2003, 2007, and 2012, the contribution of PC1 reached distinct peaks, indicating relatively high ecological quality during these years. These peaks may be attributed to the combined effects of climate conditions, environmental protection policies, and human activities during these periods. Conversely, in years such as 2006, 2011, and 2016, the contribution of the PC1 was relatively low, reflecting a decline in ecological quality. This decline can mainly be attributed to climate change resulting from drought or excessive rainfall, as well as human activities such as overgrazing and land development, which have negative impacts on the ecological environment. It is noteworthy that the overall trend of the line chart suggests that the ecological quality of QLR appears to exhibit cyclical fluctuations. These fluctuations are closely associated with the natural climate change cycle in the region, the implementation and adjustment of ecological protection policies, and the impact of human economic activities.

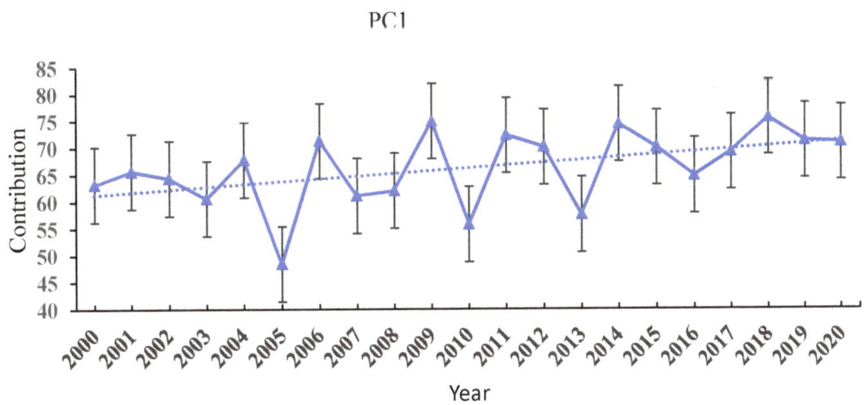

Figure 2. PC1 principal component contribution degree.

In summary, the ecological quality of QLR has demonstrated significant fluctuation characteristics over the past 20 years. To maintain and enhance the ecological quality of the region, it is imperative to develop targeted ecological protection and restoration measures that consider multiple factors such as climate change, policy adjustments, and human activities. Additionally, ecological monitoring and assessment efforts should be intensified to promptly detect issues and implement effective intervention measures.

Figure 3 illustrates that the ecological quality of the QLR is influenced by four key ecological factors: the NDVI, LST, WET, and NDBSI. The degree of contribution of the PCA eigenvalues visually demonstrates this relationship. Our analysis of Figure 3 reveals that the NDVI and WET, as positive factors, exhibit a significant positive correlation with ecological quality (RSEI). The NDVI, or the Normalized Vegetation Index, is a crucial indicator for measuring vegetation growth and coverage. In the QLR, the high contribution of the NDVI indicates that vegetation cover plays a vital role in enhancing ecological quality. Lush vegetation not only improves ecosystem stability but also promotes biodiversity, thereby establishing a solid foundation for regional ecological balance. Therefore, protecting and increasing vegetation cover is an effective strategy to improve the ecological quality of the area. Additionally, the WET positively impacts ecological quality. Moderate humidity helps maintain soil water balance, promotes plant root development and nutrient absorption, and thereby enhances the self-recovery ability of the ecosystem. The high contribution degree of humidity in the QLR underscores the importance of water management in ecological protection. On the other hand, heat (LST) and dryness (NDBSI), as negative factors, are negatively correlated with ecological quality (RSEI). LST reflects surface temperatures, and high temperatures not only limit plant growth but can also exacerbate soil erosion and water scarcity problems, posing challenges to ecosystems. As a measure of dryness, a high value of the NDBSI indicates dry soil, which is detrimental to vegetation growth and ecosystem health. Therefore, controlling surface temperature and maintaining soil moisture are crucial measures to mitigate these negative effects. In summary, the ecological quality of the QLR is the result of the combined influence of multiple factors. Through scientific management and rational intervention, such as strengthening vegetation protection, optimizing water resources allocation, and controlling surface temperature, the ecological quality of the region can be effectively improved, and sustainable development can be achieved.

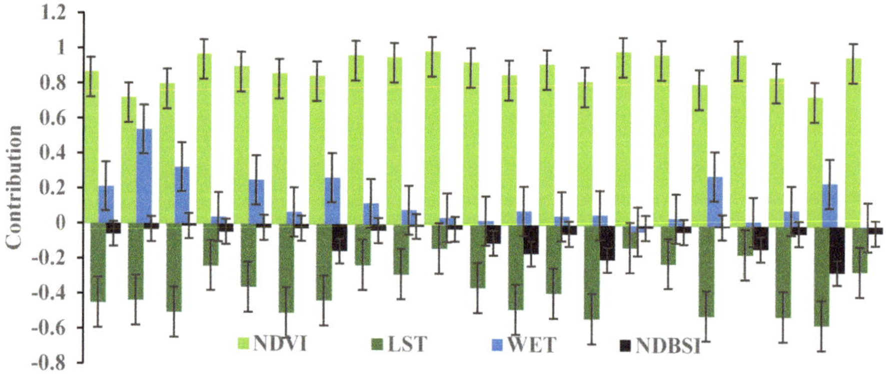

Figure 3. Characteristic value contribution degree of four ecological elements.

3.2. *Temporal and Spatial Distribution of Ecological Quality in the Qinghai Lake Region*

Figures S1 and S2 illustrate the spatial distribution of the RSEI ecological quality in the QLR over the past 20 years, highlighting the temporal and spatial characteristics of ecological quality in the region. As an ecological quality assessment tool based on remote sensing technology, the RSEI effectively demonstrates dynamic changes in ecological quality by integrating various ecological indicators. The transition from red to blue signifies a progressive improvement in ecological quality, ranging from Poor (0–0.2) to Excellent (0.8–1). Red areas are predominantly found in regions with high human activity, such as urban centers and agricultural lands, where ecological quality is at its lowest due to vegetation destruction and exposed soil. Fair (0.2–0.4) areas are more common in semi-urbanized regions, such as villages and pastures, which display slightly better ecological quality but continue to be affected by overgrazing and agricultural practices. As the colors transition to green and blue, ecological quality gradually improves. Moderate (0.4–0.6) areas primarily consist of natural grasslands and woodlands, where vegetation cover is high and ecosystems are relatively stable, although degradation risks persist. Good (0.6–0.8) areas are found in nature reserves and around lakes, characterized by healthy ecosystems and high ecological quality. The Excellent (0.8–1) areas, represented in blue, are natural treasures such as alpine meadows and deserts, boasting extremely high vegetation cover and extremely healthy ecosystems.

From a time series perspective, the ecological quality of the QLR has exhibited a fluctuating trend over the years, rather than remaining unchanged. During some years, such as 2000, 2005, and 2010, areas with poor ecological quality were concentrated, which was attributed to climate change and intensified human activities. However, in years like 2015 and 2020, areas with better ecological quality increased significantly, thanks to the effective implementation of ecological protection measures and improved climatic conditions.

In summary, the ecological quality of the QLR has faced both challenges and opportunities over the past 20 years. In regions with low ecological quality, it is essential to strengthen ecological protection and reduce human activities' interference with the natural environment. At the same time, areas with good ecological quality should also be continuously monitored and managed to ensure the healthy and sustainable development of their ecosystems. Through scientific planning and effective management, we are confident that the ecological quality of the QLR will further improve in the future.

As evident from Table 2, the area with "Poor" ecological quality (indicating low ecological quality) constituted a relatively large proportion in 2000, amounting to 628.5 km^2. However, this proportion exhibited a significant downward trend in subsequent years. By 2020, the proportion of the "Poor" grade had decreased to 6.80%, indicating an overall substantial improvement in the ecological quality of the QLR. Simultaneously, the area shares of "Fair" (general ecological quality) and "Moderate" (medium ecological quality) also underwent varying degrees of change. Specifically, the share of the "Fair" rating increased initially but began to decline in recent years, while the share of the "Moderate" rating started to decrease in 2015 after a period of increase. The changing trends of these two levels suggest that the ecological quality of the QLR gradually stabilized and improved after a period of fluctuation. It is noteworthy that the area proportions of "Good" (indicating good ecological quality) and "Excellent" (indicating excellent ecological quality) have shown a significant upward trend in recent years. In particular, the "Good" rating increased from 0.43% in 2000 to 2.93% in 2020, while the "Excellent" rating increased from 0.33% to 3.78%. The improvement in these two levels fully demonstrates the remarkable achievements in the ecological protection and restoration of the QLR.

Table 2. Area and proportion of ecological quality level in Qinghai Lake region (unit: km^2/%).

Ecological Quality	2003		2008		2013		2018		2023	
Poor	628.5	1.00%	6273.75	9.95%	18,677	29.63%	33,176	52.63%	4286.25	6.80%
Fair	269.75	0.43%	5975.5	9.48%	19,261.75	30.56%	31,681	50.26%	5844.5	9.27%
Moderate	382.75	0.61%	4374.5	6.94%	33,511.25	53.16%	22,921.75	36.36%	1851.25	2.94%
Good	270	0.43%	9299	14.75%	33,963.25	53.88%	17,659.25	28.01%	1847.75	2.93%
Excellent	210.25	0.33%	7490.75	11.88%	27,265.5	43.25%	25,692.5	40.76%	2380.25	3.78%

Overall, the changes in the area and proportion of ecological quality grades in the QLR, as presented in the table, fully reflect the efforts and accomplishments made in ecological protection and restoration in this region. Although the area with low ecological quality was relatively large in the early stage, the ecological quality of the QLR has significantly improved through years of management and protection, with the proportion of good and excellent ecological quality areas increasing annually. This change is not only conducive to ecological environment protection and sustainable development in the QLR but also plays a positive role in promoting ecological balance and environmental protection in the surrounding region and even the entire country.

Figure 4 is a bar chart that directly illustrates the change in the proportion of ecological quality grades, ranging from Poor (0–0.2) to Excellent (0.8–1), in the QLR from 2000 to 2020. In the early stage, the ecological quality of the QLR was generally low, with Poor and Fair levels (0.2–0.4) accounting for a relatively high proportion. This reflected the fragility of the ecosystem and the severe environmental pressure it faced. However, over time, this situation has undergone a radical transformation. The proportion of Moderate (0.4–0.6), Good (0.6–0.8), and Excellent grades has increased annually, particularly the Good grades, which surpassed 30% around 2015, and the Excellent grades, which also reached about 10%. This indicates that the ecological quality of the QLR has achieved a qualitative leap. Behind this positive change lies the great importance attached to environmental protection by the government and all sectors of society, who have made unremitting efforts. The implementation of a series of ecological protection projects, such as the protection of lakes and surrounding wetlands, the promotion of ecological restoration projects, and a reduction in human activities in the ecological environment, have jointly facilitated the restoration and improvement of the ecosystem. Additionally, climate change has also brought positive effects to the QLR, including increased precipitation, rising lake water levels, and improved vegetation coverage, which have provided favorable conditions for ecosystem restoration. Nevertheless, despite the overall improvement in the ecological quality of the QLR, potential problems and challenges still require vigilance. While the proportion of Poor and Fair grades has fallen sharply, it still fluctuated in some years, indicating that the ecosystem remains unstable and vulnerable to external factors. Furthermore, the relatively high Moderate level suggests that continuous efforts are still needed for ecosystem restoration. Although the ratings for Good and Excellent have improved, there is still potential for further enhancement of the ecosystem. The broken line chart in Figure 5 illustrates the trend of ecological quality changes in four major counties surrounding the QLR: Gangcha County, Tianjun County, Haiyan County, and Gonghe County. The ecological quality in these four counties has shown a year-by-year improvement, aligning with the overall trend observed in the QLR. Notably, the advancements in ecological quality in Gangcha County and Haiyan County are particularly significant, while progress in Tianjun County and Gonghe County has been relatively slower. This disparity indicates that, when promoting regional ecological protection, it is essential to implement differentiated protection measures tailored to local conditions.

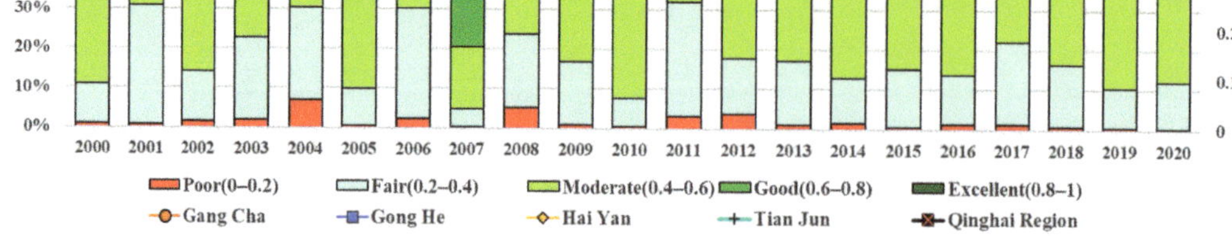

Figure 4. Trend map of ecological quality ratio in Qinghai Lake region.

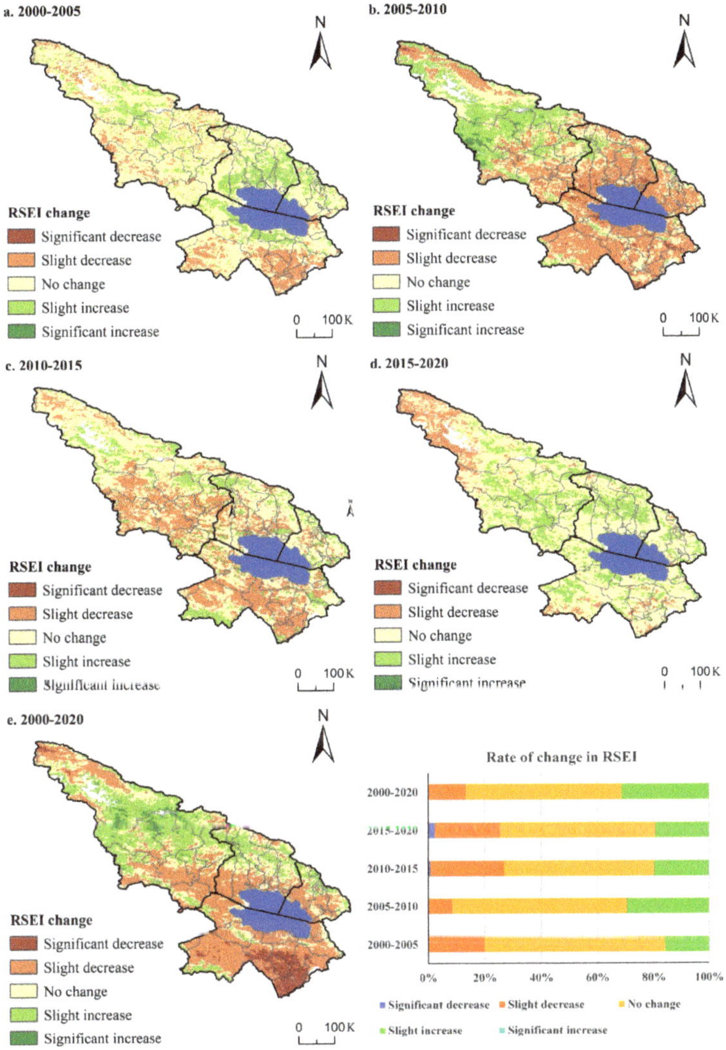

Figure 5. Spatial distribution map of ecological quality change in Qinghai Lake region.

3.3. Dynamic Monitoring of Ecological Quality in Qinghai Lake Region

To further analyze the spatial differences in RSEI changes, we categorized the RSEI changes into five groups based on the changes in the RSEI index at 5-year intervals.

Through Figure 5, we observed the spatial variation characteristics of the ecological quality (RSEI) in the QLR from 2000 to 2020 at six-year intervals. This analysis not only reveals the dynamic evolution of ecological quality but also provides insights into the interactions between human activities and the natural environment.

From a temporal perspective, the ecological quality of the area around the QLR has undergone a transformation from a slight decline to gradual improvement over the past 20 years. During the period from 2000–2005 to 2005–2010, the region experienced a slight decline in ecological quality, evident by the widespread distribution of orange (slight decline) and yellow (no change) areas in the figure. Notably, the northwest and south-east QLR exhibited significant ecological deterioration (red area), potentially linked to increased human activity intensity and natural environment vulnerability during this period. However, over time, the ecological quality of the QLR began to show positive signs of change. Between 2005–2010 and 2010–2015, while areas of slight decline and no change still dominated, areas of significant deterioration decreased, and areas of significant improvement (dark green) gradually increased in the northeast and southwest parts of the lake, indicating gradual ecological recovery in these areas. This change may be attributed to the heightened awareness of ecological environmental protection by the local government and all sectors of society, as well as the effective implementation of various ecological protection measures. In the period from 2010–2015 to 2015–2020, the improvement trend in the ecological quality of the QLR became more pronounced. The area of significant improvement increased significantly, particularly in the northeast and southwest of the lake, becoming bright spots for ecological quality enhancement. Although areas with slight decline and no change still accounted for a certain proportion, areas with significant decline had decreased significantly, indicating that the overall trend of ecological quality deterioration had been effectively contained. From a spatial distribution perspective, the change in ecological quality around the QLR exhibits notable regional characteristics. The northwest and southeast of the lake, as ecologically sensitive areas, have experienced significant deterioration and subsequent efforts to improve their ecological quality. Conversely, the northeast and southwest are relatively stable, with a remarkable trend of improvement in recent years, which may be closely related to the topography, climate conditions, and the effectiveness of ecological protection measures in these regions.

In summary, the ecological quality of the QLR has undergone a transition from a slight decline to gradual improvement over the past 20 years. This change not only reflects the resilience of the region's ecological environment but also demonstrates the potential for humans to enhance the natural environment through scientific management and effective intervention. To further promote the improvement of ecological quality around the QLR, we should continue to strengthen the implementation of ecological protection measures, optimize resource allocation, promote the coordinated development of ecology and economy, and collectively preserve this precious green home.

3.4. Driving Analysis of Ecological Quality Change in Qinghai Lake Region

Table 3 illustrates the extent of influence exerted by the variables WET, LST, NDVI, NDBSI, SLOPE, and DEM on the variation in the RSEI. By employing the random forest algorithm to analyze the changes in ecological quality (RSEI) in the QLR, we have uncovered the driving mechanism of several key variables on its evolution. Firstly, the NDVI serves as the primary indicator of vegetation cover, exerting a direct and significant influence on the ecological quality of the QLR. The abundance of vegetation not only enhances soil retention and reduces soil erosion but also enriches biodiversity and improves the overall stability of the ecosystem. Secondly, the WET, another crucial factor, follows closely in its contribution. Moderate humidity is an essential element for maintaining ecosystem health, directly related to plant growth and biodiversity preservation. The significant influence of humidity change on ecological quality suggests that the rational management and utilization of water resources should be prioritized in ecological protection. The NDBSI reflects the degree of surface exposure. Although its contribution is less than the previous two factors, an increase in bare soil area often indicates a decrease in vegetation cover, which negatively impacts ecological quality. This suggests that attention should be paid to the potential destruction of vegetation by human activities and climate change, and effective measures should be taken to reduce the area of bare soil. The SLOPE (slope) and DEM (digital elevation model) reveal the drivers of ecological quality change from a topographical perspective. Soil and water loss are prone to occurring in areas with large slopes, which affects vegetation growth. However, due to the cold climate, vegetation growth is limited in high-altitude areas, resulting in relatively low ecological quality. These findings highlight the important role of topographic factors in ecological conservation and the need to develop conservation measures tailored to local conditions. Finally, LST has the smallest contribution to the change in ecological quality, but high temperatures can still stress vegetation and affect ecological quality. Therefore, in the context of climate warming, monitoring and regulating surface temperature to reduce its negative impact on vegetation is also a crucial aspect of ecological protection.

Table 3. Analysis of driving factors of ecological quality change in Qinghai Lake region.

Factors	LST	WET	NDVI	NDBSI	SLOPE	DEM
RMSE_noise	0.09	0.12	0.09	0.18	0.17	0.15
Contribution	0.41	0.25	0.12	0.06	0.24	0.23

In summary, the driving mechanism of ecological quality evolution in the QLR is complex and diverse, with vegetation cover, humidity, bare soil degree, slope, elevation, and surface temperature jointly influencing the ecosystem. To improve and enhance ecological quality, it is necessary to comprehensively consider all factors and implement scientific and effective protection measures to promote the sustainable development of the ecological environment in the QLR.

4. Discussion

(1) Rigorous Implementation of Land Use Planning and Continuous Optimization of Land Use Structure.

In alignment with the socio-economic development trends surrounding the QLR, as well as the regional disparities in land resources and environmental conditions, a scientific general plan for land use should be developed to strictly regulate land use practices. Taking into account land use control regulations and the fragile ecological environment, the QLR is categorized into ecological protection zones, basic farmland protection zones, basic grassland protection zones, and urban development zones. Ecological protection zones prioritize the conservation of woodland, water, and wetland ecosystems, which are essential for climate regulation, gas regulation, water conservation, and biodiversity protection in the QLR [1,2,17]. By implementing differentiated land use policies, the shoals of Qinghai Lake and significant river basins can be designated as ecological environmental protection zones, effectively enhancing the ecosystem service value of the lake's surrounding area. Basic farmland and grassland protection zones aim to promote local agriculture and animal husbandry development, improve the living conditions of farmers and herdsmen, and ensure soil formation and protection, food production, and raw material supply. Urban development zones should reasonably determine the scale of urban development and allocate urban development space within permitted and conditionally permitted construction areas, without encroaching on basic farmland and grassland. The scale, structure, and layout of land use significantly impact ecological and environmental issues such as the land ecological landscape, ecosystem types, vegetation, and biodiversity. Therefore, it is essential to rationally adjust the land use structure [2,18–20], balance the relationship between agricultural land and construction land, ensure the preservation of cultivated and forest land areas, maintain grassland and water area sizes, protect wetland areas, scientifically control construction land area, and continuously enhance ecosystem service value.

(2) Restricted Development of Inland Beaches and Effective Protection of Critical Wetland Resources.

The QLR, located at the intersection of China's eastern monsoon region, western arid region, and the Qinghai–Tibet Plateau, is recognized as a national nature reserve and is included in international wetland inventories due to its unique geographical position and fragile ecological environment. Wetland ecosystems perform vital functions such as gas regulation, climate control, water conservation, waste treatment, biodiversity protection, and cultural recreation, while also serving as habitats for numerous plant and animal species. However, over the past two decades, the wetland area surrounding the QLR has decreased by 3570 square kilometers, resulting in a declining trend in ecosystem service value in this region. Therefore, it is imperative to develop appropriate wetland protection measures, strictly limit the development of beaches along the QLR, prioritize the protection of wetlands in significant river basins surrounding the lake, and continuously enhance their recreational and cultural value [21].

(3) Implementation of Engineering and Biological Measures to Mitigate Soil Erosion and Desertification.

In areas prone to severe soil erosion, engineering measures should be employed to straighten and dredge large gullies, sandbanks, rocks, and mounds blocking water flow, while regulating gullies and streams to flow into the main river. In regions with severe grassland desertification, the restoration and expansion of forest and grass vegetation should be prioritized, establishing windbreak and sand-fixing forest belts, implementing fencing and grazing prohibitions, etc. Through managing existing vegetation and reducing human disturbance, the natural renewal and restoration of degraded vegetation can be achieved, promoting the development of desert grass and natural plants, and minimizing shifting sand activities in the region. For areas with severe grassland degradation, improving natural grasslands and constructing artificial grasslands should be primary strategies. While protecting existing forest and grass vegetation, comprehensive management measures should be adopted for damaged ground, and forest land after logging should be promptly renewed through rotational sealing and grazing [22–24]. The "double package of grass and livestock" responsibility system should be implemented in grassland areas to balance livestock and grassland resources, maintaining "buffer grassland" or "mobile grassland" to slow grassland degradation. Addressing the primary ecological issues around the lake area requires adapting to local conditions, strengthening the construction of agricultural land ecological environments, developing ecological agriculture and animal husbandry, and achieving the sustainable utilization of agricultural land resources.

(4) Establishment of a Reasonable Ecological Compensation Mechanism to Encourage Farmer and Herdsman Participation in Ecological Protection.

Currently, Qinghai Province only provides ecological compensation for ecological migrants in Sanjiangyuan, excluding other areas and ecological protection behaviors, which hinders the achievement of an effective protection mechanism. Ecological compensation, as an institutional arrangement to adjust interest relationships between subjects and protect the ecological environment, serves as an incentive measure for environmental protection, with the participation and support of herdsmen being crucial for realizing this mechanism. Therefore, based on the value of ecosystem services in townships surrounding the QLR, combined with the significance of ecological protection and the willingness of farmers and herdsmen to contribute, a reasonable ecological compensation mechanism should be established. This mechanism should provide fair and reasonable ecological compensation and incentives for the decline in well-being caused by abandoning economic development and changing lifestyles, encouraging farmers and herdsmen to actively engage in ecological protection efforts [23].

5. Conclusions

After assessing and analyzing the evolution of the ecological environmental quality of the QLR over the past 20 years, the following conclusions have been drawn:

(1) The ecological quality of the region surrounding the QLR has exhibited notable fluctuations in the past two decades. These fluctuations are closely tied to multiple factors, including climate change, human activities, and policy adjustments. Specifically, key ecological factors such as vegetation cover, surface temperature, humidity, and dryness have significant impacts on ecological quality.

(2) Despite fluctuations, the overall ecological quality of the region around the QLR has shown an improving trend over the past 20 years. The RSEI ecological quality spatial distribution map indicates that the proportion of high-quality ecological areas is increasing annually, reflecting the effective implementation of ecological protection measures and the improvement of climate conditions.

(3) Analysis using the random forest algorithm reveals that vegetation cover and humidity are the primary factors contributing to the improvement of ecological quality, while bare soil area, slope, and elevation are the key factors leading to a reduction in ecological quality. These findings provide a scientific foundation for formulating targeted ecological protection and restoration measures.

(4) The changes in ecological quality around the QLR exhibit notable regional characteristics. The trends in ecological quality variation differ across regions, which may be related to factors such as terrain, climate, and the intensity of human activities. Therefore, it is essential to consider regional differences and adopt protection measures tailored to local conditions when formulating ecological protection strategies.

Despite the overall improvement in the ecological quality of the area around QLR, it still faces numerous challenges, including the uncertainty of climate change and the ongoing impact of human activities. Consequently, it is crucial to strengthen ecological monitoring and assessment efforts in the future, promptly identify issues, and implement effective intervention measures. Simultaneously, it is necessary to continue promoting the implementation and adjustment of ecological protection policies to ensure the sustainable development of the ecological environment. In summary, this study has revealed the evolution patterns and driving mechanisms of ecological environmental quality in the region around the QLR, providing an important reference for formulating scientific and reasonable ecological protection and restoration measures. In the future, we should continue to strengthen research and exploration in related fields, contributing our wisdom and efforts to the protection of the ecological environment of the Qinghai–Tibet Plateau.

Supplementary Materials: The following supporting information can be downloaded at: https://www.mdpi.com/article/10.3390/land13122203/s1, Figure S1. Spatial distribution map of ecological quality in Qinghai Lake region. Figure S2. Spatial distribution map of ecological quality in Qinghai Lake region.

Author Contributions: Conceptualization, Z.S. and J.G.; methodology, Z.S.; software, Z.S.; validation, Z.S. and J.G.; formal analysis, Z.S. and J.G.; data curation, Z.S. and J.G.; writing—original draft preparation, Z.S. and J.G.; writing—review and editing, Z.S.; visualization, Z.S.; supervision, J.G.; project administration, J.G.; funding acquisition, J.G. All authors have read and agreed to the published version of the manuscript.

Funding: This research was funded by Key Project from the National Natural Science Foundation of China (Grant No. 42071254).

Data Availability Statement: The authors can provide the data upon reasonable request. The data are not publicly available due to privacy restrictions.

Acknowledgments: The authors are particularly grateful to all researchers for providing data support for this study.

Conflicts of Interest: The authors declare no conflicts of interest.

References

1. Estoque, R.C.; Murayama, Y.; Myint, S.W. Effects of landscape composition and pattern on land surface temperature: An urban heat island study in the megacities of Southeast Asia. *Sci. Total Environ.* **2017**, *577*, 349–359. [CrossRef] [PubMed]
2. Firozjaei, M.K.; Fathololoumi, S.; Weng, Q.; Kiavarz, M.; Alavipanah, S.K. Remotely sensed urban surface ecological index (RSUSEI): An analytical framework for assessing the surface ecological status in urban environments. *Remote Sens.* **2020**, *12*, 2029. [CrossRef]
3. Arab, S.T.; Noguchi, R.; Matsushita, S.; Ahamed, T. Prediction of grape yields from time-series vegetation indices using satellite remote sensing and a machine-learning approach. *Remote Sens. Appl. Soc. Environ.* **2021**, *22*, 100485. [CrossRef]
4. Chen, W.; Wang, J.; Ding, J.; Ge, X.; Han, L.; Qin, S. Detecting Long-Term Series Eco-Environmental Quality Changes and Driving Factors Using the Remote Sensing Ecological Index with Salinity Adaptability (RSEISI): A Case Study in the Tarim River Basin, China. *Land* **2023**, *12*, 1309. [CrossRef]
5. Gong, C.; Lyu, F.; Wang, Y. Spatiotemporal change and drivers of ecosystem quality in the Loess Plateau based on RSEI: A case study of Shanxi, China. *Ecol. Indic.* **2023**, *155*, 111060. [CrossRef]

6. Hu, X.; Xu, H. A new remote sensing index for assessing the spatial heterogeneity in urban ecological quality: A case from Fuzhou City, China. *Ecol. Indic.* **2018**, *89*, 11–21. [CrossRef]
7. Xu, H.; Wang, M.; Shi, T.; Guan, H.; Fang, C.; Lin, Z. Prediction of ecological effects of potential population and impervious surface increases using a remote sensing based ecological index (RSEI). *Ecol. Indic.* **2018**, *93*, 730–740. [CrossRef]
8. Xu, H.; Wang, Y.; Guan, H.; Shi, T.; Hu, X. Detecting ecological changes with a remote sensing based ecological index (RSEI) produced time series and change vector analysis. *Remote Sens.* **2019**, *11*, 2345. [CrossRef]
9. Hasanah, A.; Wu, J. Exploring dynamics relationship between carbon emissions and eco-environmental quality in Samarinda Metropolitan Area: A spatiotemporal approach. *Sci. Total Environ.* **2024**, *927*, 172188. [CrossRef]
10. Li, W.; Samat, A.; Abuduwaili, J.; Wang, W. Spatiotemporal Pattern, Evolutionary Trend, and Driving Forces Analysis of Ecological Quality in the Irtysh River Basin (2000–2020). *Land* **2024**, *13*, 222. [CrossRef]
11. Liao, W.; Jiang, W. Evaluation of the Spatiotemporal Variations in the Eco-environmental Quality in China Based on the Remote Sensing Ecological Index. *Remote Sens.* **2020**, *12*, 2462. [CrossRef]
12. Liu, Y.; Xu, W.; Hong, Z.; Wang, L.; Ou, G.; Lu, N.; Dai, Q. Integrating three-dimensional greenness into RSEI improved the scientificity of ecological environment quality assessment for forest. *Ecol. Indic.* **2023**, *156*, 111092. [CrossRef]
13. Li, Y.; Tian, H.; Zhang, J.; Lu, S.; Xie, Z.; Shen, W.; Zheng, Z.; Li, M.; Rong, P.; Qin, Y. Detection of spatiotemporal changes in ecological quality in the Chinese mainland: Trends and attributes. *Sci. Total Environ.* **2023**, *884*, 163791. [CrossRef] [PubMed]
14. Tang, Q.; Hua, L.; Tang, J.; Jiang, L.; Wang, Q.; Cao, Y.; Wang, T.; Cai, C. Advancing ecological quality assessment in China: Introducing the ARSEI and identifying key regional drivers. *Ecol. Indic.* **2024**, *163*, 112109. [CrossRef]
15. Kamusoko, C.; Gamba, J. Simulating urban growth using a Random Forest-Cellular Automata (RF-CA) model. *ISPRS Int. J. Geo-Inf.* **2015**, *4*, 447–470. [CrossRef]
16. Zhang, H.; Liu, Y.; Li, X.; Feng, R.; Gong, Y.; Jiang, Y.; Guan, X.; Li, S. Combing remote sensing information entropy and machine learning for ecological environment assessment of Hefei-Nanjing-Hangzhou region, China. *J. Environ. Manag.* **2023**, *325*, 116533. [CrossRef]
17. Cheng, K.; He, K.; Fu, Q.; Tagawa, K.; Guo, X. Assessing the coordination of regional water and soil resources and ecological-environment system based on speed characteristics. *J. Clean. Prod.* **2022**, *339*, 130718. [CrossRef]
18. Bonilla-Bedoya, S.; Mora, A.; Vaca, A.; Estrella, A.; Herrera, M.Á. Modelling the relationship between urban expansion processes and urban forest characteristics: An application to the Metropolitan District of Quito. *Comput. Environ. Urban Syst.* **2020**, *79*, 101420. [CrossRef]
19. Ding, C.; Zhao, X. Land market, land development and urban spatial structure in Beijing. *Land Use Policy* **2014**, *40*, 83–90. [CrossRef]
20. Grafius, D.R.; Corstanje, R.; Harris, J.A. Linking ecosystem services, urban form and green space configuration using multivariate landscape metric analysis. *Landsc. Ecol.* **2018**, *33*, 557–573. [CrossRef]
21. Keeler, B.L.; Hamel, P.; McPhearson, T.; Hamann, M.H.; Donahue, M.L.; Meza Prado, K.A.; Arkema, K.K.; Bratman, G.N.; Brauman, K.A.; Finlay, J.C.; et al. Social-ecological and technological factors moderate the value of urban nature. *Nat. Sustain.* **2019**, *2*, 29–38. [CrossRef]
22. Johnston, C.A.; McIntyre, N.E. Effects of cropland encroachment on prairie pothole wetlands: Numbers, density, size, shape, and structural connectivity. *Landsc. Ecol.* **2019**, *34*, 827–841. [CrossRef]
23. Tang, L.; Ke, X.; Zhou, T.; Zheng, W.; Wang, L. Impacts of cropland expansion on carbon storage: A case study in Hubei, China. *J. Environ. Manag.* **2020**, *265*, 110515. [CrossRef] [PubMed]
24. Wang, J.; Liu, D.; Ma, J.; Cheng, Y.; Wang, L. Development of a large-scale remote sensing ecological index in arid areas and its application in the Aral Sea Basin. *J. Arid. Land* **2021**, *13*, 40–55. [CrossRef]

MDPI AG
Grosspeteranlage 5
4052 Basel
Switzerland
Tel.: +41 61 683 77 34

Land Editorial Office
E-mail: land@mdpi.com
www.mdpi.com/journal/land

www.ingramcontent.com/pod-product-compliance
Lightning Source LLC
Chambersburg PA
CBHW061832210926
51850CB00006B/870
9783725848294